Mapping species distributions: spatial inference and prediction

Maps of species distributions or habitat suitability are required for many aspects of environmental research, resource management, and conservation planning. These include biodiversity assessment, reserve design, habitat management, and restoration, species and habitat conservation plans and predicting the effects of environmental change on species and ecosystems. The proliferation of methods and uncertainty regarding their effectiveness can be daunting to researchers, resource managers, and conservation planners alike. Franklin summarizes the methods used in species distribution modeling (also called niche modeling) and presents a framework for spatial prediction of species distributions based on the attributes (space, time, scale) of the data and questions being asked. The framework links theoretical ecological models of species distributions to spatial data on species and environment, and statistical models used for spatial prediction. Provides practical guidelines to students, researchers, and practitioners in a broad range of environmental sciences including ecology, geography, conservation biology, and natural resources management.

JANET FRANKLIN has been a Professor of Biology and Adjunct Professor of Geography at San Diego State University, where she was on the faculty from 1988–2009. In 2009 she joined the faculty of Arizona State University as a Professor in the Schools of Geographical Sciences and Life Sciences. She received the Bachelor's degree on Environmental Biology (1979), the Master of Arts (1983), and the Ph.D. (1988) in Geography, all from the University of California at Santa Barbara. Her research interests include biogeography, landscape ecology, plant ecology, biophysical remote sensing, digital terrain analysis, and geographic information science. She has conducted research on plant community composition, structure, dynamics and spatio-temporal patterns in Mediterranean–climate ecosystems, deserts, tropical dry forests and rain forests.

She was the Editor of *The Professional Geographer* (1997–2000), Board Member of *Landscape Ecology* (2000–2005), and Associate Editor of the *Journal of Vegetation Science* (1999–2006). She is currently a Board Member of *Ecology*, and *Diversity and Distributions*. She has published more than 80 refereed book chapters and papers in journals *Ecological Applications*, *Ecological Modelling*, *Journal of Vegetation Science*, *Ecology*, *Diversity and Distributions*, *Journal of Tropical Ecology* and *Conservation Biology*. She has received research support from NSF, NASA, USGS, Forest Service, California State Parks, National Geographic Society, and others.

ECOLOGY, BIODIVERSITY AND CONSERVATION

The world's biological diversity faces unprecedented threats. The urgent challenge facing the concerned biologist is to understand ecological processes well enough to maintain their functioning in the face of the pressures resulting from human population growth. Those concerned with the conservation of biodiversity and with restoration also need to be acquainted with the political, social, historical, economic, and legal frameworks within which ecological and conservation practice must be developed. The new *Ecology, Biodiversity, and Conservation* series will present balanced, comprehensive, up-to-date, and critical reviews of selected topics within the sciences of ecology and conservation biology, both botanical and zoological, and both "pure" and "applied." It is aimed at advanced final-year undergraduates, graduate students, researchers, and university teachers, as well as ecologists and conservationists in industry, government and the voluntary sectors. The series encompasses a wide range of approaches and scales (spatial, temporal, and taxonomic), including quantitative, theoretical, population, community, ecosystem, landscape, historical, experimental, behavioural, and evolutionary studies. The emphasis is on science related to the real world of plants and animals rather than on purely theoretical abstractions and mathematical models. Books in this series will, wherever possible, consider issues from a broad perspective. Some books will challenge existing paradigms and present new ecological concepts, empirical or theoretical models, and testable hypotheses. Other books will explore new approaches and present syntheses on topics of ecological importance.

The Ecology of Phytoplankton
C. S. Reynolds

Invertebrate Conservation and Agricultural Ecosystems
T. R. New

Risks and Decisions for Conservation and Environmental Management
Mark Burgman

Nonequilibrium Ecology
Klaus Rohde

Mapping species distributions

Spatial inference and prediction

JANET FRANKLIN

*School of Geographical Sciences and School of Life Sciences, Arizona
State University*
Formerly at the *Departments of Biology and Geography, San Diego
State University*

With contributions by
JENNIFER A. MILLER

*Department of Geography and the Environment, University of Texas,
Austin, Texas, USA*

CAMBRIDGE
UNIVERSITY PRESS

CAMBRIDGE
UNIVERSITY PRESS

University Printing House, Cambridge CB2 8BS, United Kingdom

Cambridge University Press is part of the University of Cambridge.

It furthers the University's mission by disseminating knowledge in the pursuit of education, learning and research at the highest international levels of excellence.

www.cambridge.org
Information on this title: www.cambridge.org/9780521700023

© Cambridge University Press 2010

First published 2009
7th printing 2014

A catalogue record for this publication is available from the British Library

ISBN 978-0-521-87635-3 Hardback
ISBN 978-0-521-70002-3 Paperback

To Serge, Savannah, and Connor

Contents

Preface

Maps of actual or potential species distributions or habitat suitability are required for many aspects of environmental research, resource management, and conservation planning. These applications include biodiversity assessment, biological reserve design, habitat management and restoration, species and habitat conservation plans, population viability analysis, environmental risk assessment, invasive species management, community and ecosystem modeling, and predicting the effects of global environmental change on species and ecosystems. In recent years a burgeoning number of statistical and related methods have been used with mapped biological and environmental data in order to model, or, in some way, spatially interpolate species distributions, and other biospatial variables of interest, over large spatial extents. This practice is known as species distribution modeling (SDM). It has also been referred to as environmental, bioclimatic, or species niche modeling, and habitat suitability modeling, but, in this book, the term SDM will be preferred.

The proliferation of modeling methods applied to SDM, and conflicting results regarding their efficacy and relative merits, is daunting to researchers and resource analysts alike. The lack of integration of modeling and Geographic Information System (GIS) tools can impede the effective implementation of SDM. This book summarizes the key components of, and various approaches to, this problem that have been applied worldwide. This comprehensive summary provides guidance to novice species distribution modelers and also a review of current practices for more advanced practitioners. The book is organized according to a framework for modeling species distributions that has three parts: the ecological, data, and statistical models. The ecological model includes ecological theory used to link environmental predictors to species distributions according to a response function. The data model includes the decisions made regarding how data for modeling are collected and measured. The statistical model includes the choice of modeling methods and decisions required during model fitting and evaluation.

The elements of SDM are: a conceptual model of the abiotic and biotic factors controlling species distributions in space and time; data on species occurrences in geographical space; digital maps of environmental variables representing those factors thought to control species distributions; a quantitative or rule-base model linking species occurrence to the environmental predictors; a geographic information system (GIS) for applying the model rules to the environmental variable maps in order to produce a map of predicted species occurrence; and, data and methods for evaluating the error or uncertainty in the predictions.

This book discusses each of these elements. It then concludes with a framework for mapping species distributions from biological survey data, statistical models and digital maps of the environment. That framework is based on the attributes (space, time, scale) of the data and questions being asked. The framework links ecological theories of species distributions to the spatial data and statistical models used in empirical studies. This provides practical guidelines for model formulation, calibration, evaluation, and application.

Acknowledgments

I thank Alan Crowden for encouraging me to write this book, providing mentorship, and offering excellent guidance and suggestions for improving the book. I am extremely grateful to Alexandra Syphard for reading and commenting on the entire manuscript and for developing several of the figures. I also greatly appreciate the considerable efforts of several others who reviewed all or significant portions of the book at various stages, and offered tremendously useful suggestions and corrections, including Mike Austin, Jane Elith, Jennifer Miller, Helen Regan and Michael Usher. In addition to them, I also thank Simon Ferrier for allowing me to draw on his expertise and graciously responding to my many questions.

The people who have inspired and nurtured my passion for this topic are too numerous to mention without fear of leaving someone out. They include many of the authors whose work I cite as well as many of my former students and mentors. I will single out a few who set me on this path and those who offered wisdom, timely insights, good questions, and stimulating discussions along the way (in alphabetical order): Richard Aspinall, Mike Austin, Mark Burgman, Frank Davis, Tom Edwards, Jane Elith, Simon Ferrier, Robert Fisher, Mike Goodchild, Antoine Guisan, John Leathwick, Brian Lees, Brendan Mackey, Ross Meentemeyer, Joel Michaelsen, Jennifer Miller, Gretchen Moisen, Aaron Moody, Helen Regan, John Rotenberry, Andrew Skidmore, Peter Scull, Alan Strahler, Alexandra Syphard, Kim Van Niel, John Wilson, Brendan Wintle, and Nick Zimmermann.

I invited Jennifer Miller to coauthor Chapter 6, contributing the material on spatial autocorrelation and statistical species distribution models. I appreciate the breadth and depth she has added to this topic and I believe that the readers will benefit from her expertise.

I would like to thank all of the participants of the Riederalp 2008 workshop on species distribution modeling for sharing their ideas and friendship. I also thank the students who participated in my species distribution modeling seminars at San Diego State University in 2003, 2005 and 2007. A number of students showed great enthusiasm for species distribution modeling and through their own research have taught me new things about it, including Paul McCullough, Dawn Lawson, Matt Guilliams, and Katherine Wejnert.

I thank my Editor at Cambridge University Press, Dominic Lewis, and Editorial Assistants, Alison Evans and Rachel Eley, for their support and assistance.

My long-time friend and colleague, David Steadman, has provided constant, unwavering support and encouragement for this book project, and even feigned interest in the topic, for which I am profoundly grateful. Finally, I sincerely thank my family for their endless patience, love and support, and for being proud of what I do for a living.

The writing of this book was supported, in part, by National Science Foundation grant BCS-0452389. Any opinions, findings, and conclusions or recommendations expressed in this material are those of the author and do not necessarily reflect the views of the National Science Foundation (NSF).

Part I

History and ecological basis of species distribution modeling

Recent decades have seen an explosion of interest in species distribution modeling. This has resulted from a confluence of the growing need for information on the geographical distribution of biodiversity and new and improved techniques and data suitable for addressing this information need – remote sensing, global positioning system technology, geographic information systems, and statistical learning methods. Developments in this area are occurring so rapidly that it was difficult to know when to stop writing this book. It is very challenging to write about a moving target. For the same reason, however, it was the right time to summarize the foundations of, and recent developments in this enterprise called species distribution modeling (SDM). This book provides an introduction to SDM for beginners in the field and for those wishing to use such models in environmental assessment and biodiversity conservation, while providing a significant reference on current practice for researchers.

In this Part, Chapter 1 establishes some basic terminology used to describe species' distribution modeling, describes a framework for implementing SDM that will be used as an organizing principle for the book, and reviews the problems and applications that have motivated a growing interest in SDM. Much of this book describes a modeling approach that links species location information with environmental data (Part II) in order to quantify the distributions of species on environmental gradients and map those distributions onto geographical space using a model (Part III). Before going into detail about this quantitative approach to describing and modeling species–environment relationships, Chapter 2 reviews other kinds of data and methods that can and have been used to more directly determine species distributions. A growing number of species atlas projects provide wall-to-wall, direct observational data of species distributions at certain scales. Cartographic interpolation and geostatistical techniques can be applied to species location data without reference to environmental data. Chapter 2 discusses the strengths and limitations of using species distribution data derived from these inter-polation approaches. Chapter 3 presents the ecological concepts that underlie our understanding of how environmental factors limit species distributions and determine species diversity. Based on this conceptual understanding, suitable data (Part II) and models (Part III) can be chosen and implemented to describe species–environment relationships, and, importantly, the assumptions and limitations of the resulting models and their predictions can be explicitly understood (Part IV).

1 · *Species distribution modeling*

Prediction is very difficult, especially about the future.
Niels Bohr

1.1 Introduction

What are predictive maps of species distributions? Why make them?
How? Environmental scientists increasingly need to use local measurements to assess change at landscape, regional and global scales, and statistical or simulation models are often used to extrapolate environmental
data in space (Miller *et al.*, 2004; Peters *et al.*, 2004). *Species distribution
modeling* (SDM) is just one example of this, but an increasingly important
one – SDM extrapolates species distribution data in space and time, usually based on a statistical model. Developing a species distribution model
begins with observations of species occurrences, and with environmental variables thought to influence habitat suitability and therefore species
distribution. The model can be a quantitative or rule-based model and, if
the fit is good between the species distribution and the predictors that are
examined, this can provide insight into species environmental tolerances
or habitat preferences. It also provides the opportunity to make a spatial prediction. *Predictive mapping*, or geographical extrapolation using the
model, results in a spatially explicit "wall-to-wall" prediction of species
distribution or habitat suitability (Fig. 1.1). Maps of environmental predictors, or their surrogates, must be available in order for predictive
mapping to be implemented (Franklin, 1995).

The purpose of this book is to describe the process of species distribution modeling, review the significant advancements that have been
made recently in this field, and try to provide guidelines and a framework
for choosing data, methods and study design that are appropriate to a
given SDM application. This book provides an introduction to SDM for
beginners in the field who wish to use such models in environmental
analyses while also providing a comprehensive review of current practice

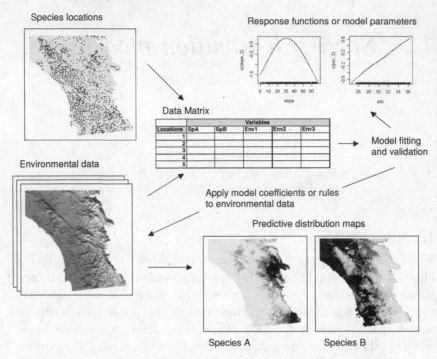

Species locations

Response functions or model parameters

Data Matrix

Model fitting
and validation

Environmental data

Apply model coefficients or rules
to environmental data

Predictive distribution maps

Species A   Species B

Fig. 1.1. The steps in species distribution modeling and predictive mapping. Spatially explicit (georeferenced) species occurrence data are linked with digital maps of environmental predictors and values of environmental predictors for species locations are extracted. A statistical, machine learning, or other type of model describing the relationship between species occurrence and environmental data is developed. The parameters or response functions are evaluated and these coefficients or decision rules are applied to environmental maps yielding spatial predictions of species distribution or habitat suitability.

for more advanced practitioners. There are a wide variety of applications of SDM ranging from basic to applied science. It is because of this that there are many choices that have to be made about what data and model to use. Given the range of subsequent decisions that will be made on the basis of SDMs, it is important to understand how they are constructed, using what types of applicable data, the assumptions behind them and their limitations. In the remainder of this chapter I will define some key terms, highlight some recent publications that illustrate the growing interest in SDM, outline the process of species distribution modeling along with the organization of the remainder of this book, and review some of the main uses of SDM.

1.2 What is in a name?

It seems a straightforward question, but what do species distribution models *describe*, and, literally, what do they *predict*? What variable is depicted in the resulting map? (This is discussed further in Chapter 4.) The terminology that has developed in this field was somewhat confusing when I reviewed it in 1995, with individual studies using different terms, and has grown even more complex. *Species distribution models*, of the kind discussed in this book, have been said to describe both (a) the species niche, and (b) the suitability of habitat to support a species.

1.2.1 Niche models

"Species niche model," "ecological niche model" or even "niche-theory model" are terms that have been used to describe SDM. These models have variously been described as estimating the fundamental (potential) niche, realized (actual) niche, the multivariate species niche (Rotenberry *et al.*, 2006), or, when conditioned only on climate variables, the "climatic niche" (Chapter 3 discusses this further).

The original climate niche modeling system (BIOCLIM) actually referred to the "climate profile" of a species or other entity (Busby, 1991). The term climate (or bioclimatic) envelope modeling (Heikkinen *et al.*, 2006) has been used even when envelope methods are not used (see Chapter 8) and when environmental predictors include variables other than climatic, especially when the models are applied to climate change questions. The maps resulting from the application of climatic niche models are often referred to as predictions of (species) geographical *range* (Graham *et al.*, 2004). Some authors even distinguish between ecological niche models, which they define as models of potential distribution, and species distribution models of actual distributions (Peterson *et al.*, 2008).

I prefer the term "species distribution model" to any of the variations on "niche model" (in spite of their widespread use) because SDM more accurately describes the modeling process and resulting model. Although niche theory strongly underpins SDM (Chapter 3), SDMs describe empirical correlations between species distributions and environmental variables. While these models should always be evaluated for "ecological realism," that is, consistency with ecological knowledge of limiting factors and species response curves (Chapter 3), the data rarely allow the true species niche to be fully specified or confirmed. Further, these models are frequently used to predict the geographical distribution of a species, rather than to study the characteristics of its distribution in environmental (niche) space.

1.2.2 Habitat suitability models

SDMs are also referred to as habitat suitability models (e.g., Hirzel *et al.*, 2006; Ray & Burgman, 2006; Hirzel & Le Lay, 2008), describing the suitability of habitat to support a species. When used to predict in geographical space, they have been called, among other things, "predictive habitat distribution models" (Guisan & Zimmermann, 2000) and "spatially explicit habitat suitability models" (Rotenberry *et al.*, 2006).

The concept of habitat suitability is closely related to the idea of a resource selection function from wildlife biology (Manly *et al.*, 2002) as noted by Boyce *et al.* (2002), and can be applied to plants and animals. A resource selection function (RSF) is any function (for example, from a statistical model) that is proportional to the probability of habitat use by an organism (Manly *et al.*, 2002). If a resource selection function is proportional to the probability of use, then a SDM could be said to predict the likelihood that an event (species) occurs at a location – that is, *the probability of species presence.* Mackey and Lindenmeyer (2001) presented a very interesting analytical framework for modeling the spatial distribution of terrestrial vertebrates which tend to be mobile and may not occupy all suitable habitat (van Horne, 1983). They refer to the spatial occurrence and abundance of a species as its "distributional behavior" (discussed further in Chapter 3).

A species distribution model, when applied to maps of environmental variables, is said to predict the species potential geographical distribution (potential occurrence at a location); the resulting maps have been called ecological response surfaces (Lenihan, 1993); biogeographical models of species distributions (Guisan *et al.*, 2006), spatial predictions of species distribution (Austin, 2002), predictive maps (Franklin, 1995), predictions of occurrence (Rushton *et al.*, 2004), or predictive distribution maps (Rodriguéz *et al.*, 2007). According to Meyer and Thuiller (2006), one of the most important applications of RSFs is mapping species distributions.

Further, the same methods and spatial modeling principles apply whether the locations of plants, animals, microbes, community properties (e.g., diversity), ecosystem properties (e.g., productivity), or other response variables are being correlated with the mapped distributions of those factors hypothesized to control their geographical patterns. Some theory other than the species niche concept (Chapter 3) would have to support such analyses, for example an ecosystem productivity model or community diversity theory. These analyses might also have to rely on different predictors than those typically used in SDM (Chapter 5). Spatial overlay of geographical patterns to examine the suitability of a location for a particular use forms the historical foundation of geographical information science (McHarg, 1969).

The description of SDM that I have used before (Franklin, 1995), "geographical modeling of biospatial patterns in the relation to environmental gradients," while comprehensive, is cumbersome. In this book I will use the terms *species distribution model*, and *predictive [distribution] map*, for simplicity and generality.

1.3 Heightened interest in species distribution modeling

"The quantification of . . . species–environment relationships represents the core of predictive geographical modeling in ecology" (p. 148) (Guisan & Zimmermann, 2000). Species distribution modeling has its roots in ecological gradient analysis (Whittaker, 1960; Whittaker *et al.*, 1973), biogeography (Box, 1981), remote sensing and geographic information science (for review see Franklin, 1995). It has recently experienced explosive growth in the scientific literature, and in practice, especially by governmental and non-governmental organizations charged with biological resource assessment and conservation at larger spatial scales. The level of activity has been greatly enhanced by the development of digital databases for natural history collections, making species location information from specimen records and other sources more widely available (Graham *et al.*, 2004), for example via the Global Biodiversity Information Facility (http://www.gbif.org/). In 1995, I wrote a "comprehensive" review of the state of the art that described only 30 studies. SDM has since been the subject of a number international symposia and workshops resulting in an edited book (Scott *et al.*, 2002) and special features (collections of papers) in journals:

• *Ecological Modelling*, 2002, **157** (2–3) and 2006, **199** (2);
• *Biodiversity and Conservation*, 2002, **11** (12);
• *Journal of Applied Ecology*, 2006, **43** (3);
• *Diversity and Distributions*, 2007, **13** (3).

It is a topic that is, or has been, featured often in *Ecography*, *Ecological Applications*, *Journal of Vegetation Science*, and in a number of other scientific journals. Pick up any recent issue and you will find several

SDM studies. For example, the October 2008 issue of the journal *Ecological Applications* featured a study distinguishing the factors affecting the establishment piñon–juniper woodland before, versus after, Euro-American settlement in western North America using topo-climatic predictors (Jacobs *et al.*, 2008). Another study in the same issue showed that improved predictions of marine predator (bottlenose dolphin) distributions could be made using models of prey (fish) distributions (Torres *et al.*, 2008). Species distribution modeling methods have even been used to map the human "ecological" niche during the last glacial maximum in Europe (Banks *et al.*, 2008), and the "biophysical environmental space" of wildfire (Syphard *et al.*, 2008; Parisien & Moritz, 2009). These are just a few of the many fascinating SDM studies to pass my desk.

Several review papers, essays and editorials have summarized advances in different aspects of SDM. These advancements include conceptual issues such as a general framework for species distribution modeling (Franklin, 1995; Guisan & Zimmermann, 2000; Mackey & Lindenmayer, 2001; Guisan & Thuiller, 2005; Elith & Leathwick, 2009), links to ecological theory (Austin, 2002, 2007; Hirzel & Le Lay, 2008), modeling methods (Guisan *et al.*, 2002, 2006; Pearce & Boyce, 2006), data and scale issues and statistical model selection (Rushton *et al.*, 2004), use of natural history collections data (Graham *et al.*, 2004), modeling ecological communities (Ferrier & Guisan, 2006), and the influence of spatial autocorrelation on models of species distributions (Miller *et al.*, 2007). Other reviews have focused on applications, including the use of SDMs in land management under uncertainty (Burgman, 2005), and their applicability to conservation planning (Ferrier *et al.*, 2002a, b; Rodriguéz *et al.*, 2007). Each of these topics will be explored in this book.

Further, growing numbers of government agencies and non-governmental organizations have implemented ambitious, large-scale species distribution modeling programs (Iverson & Prasad, 1998; Ferrier, 2002). These projects usually involve modeling the distributions of hundreds (even thousands) of individual species or ecological communities over large regions. Often, these efforts are extensive and influential, affecting regional and global conservation decision making, but they are not necessarily reported in the scientific literature. For example, NatureServe, a non-profit conservation organization that represents an international network of biological inventories (http://www.natureserve.org/), calls their program predictive (or element) distribution modeling and they are using it to model species distributions in South America. They have run training workshops and provided tutorial materials (for developing the statistical models) on their website (http://www.natureserve.org/

prodServices/predictiveDistModeling.jsp).. The American Museum of Natural History, as part of their Remote Sensing and Geographic Information System Facility, has offered short courses in species distribution modeling for conservation biology (http://geospatial.amnh.org/). Clark Labs, in collaboration with Conservation International, an international non-governmental conservation organization, has developed specialized commercial GIS software to implement both species distribution models and project the impacts of land cover dynamics on biodiversity (http://www.clarklabs.org/). It is being used to model the distributions of 16 000 species in the Andes.

It is challenging to try and summarize the wealth of new information being produced in this exciting and dynamic field into a coherent narrative that can guide those who wish to use these methods in their work, but that is the goal of this book. Much recently published research in this area has focused on comparisons of different modeling methods and those comparisons have emphasized different measures of model performance or accuracy. While the findings of that important and useful work will be summarized, this book has a somewhat different emphasis. SDMs are empirical models and they start with data. I will first consider the qualities or characteristics of the data describing environment and species distributions in relation to the question being asked. This is in line with more recent studies that have emphasized the effect of species and environmental data properties on model performance. Trained as a geographer, I will emphasize the spatial nature of the data.

I will also emphasize how research informs the practice of SDM with reference to the published literature, as in the preceding paragraphs. Attribution is very important in science and scholarship, and I have tried to not only give credit where credit is due, but also provide signposts to where to look for those who want to find more detailed information about a particular topic.

1.4 What is species distribution modeling and how is this book organized?

This book is organized according to a framework for modeling species distributions presented by Austin (2002) that has three parts: the ecological, data, and statistical models (model here meaning framework or conceptual model). The ecological model includes the ecological theory applied or hypotheses tested in the study. The data model "consists of the decisions made regarding how the data are collected and how the data will be measured or estimated" (Austin, 2002, p. 102). The

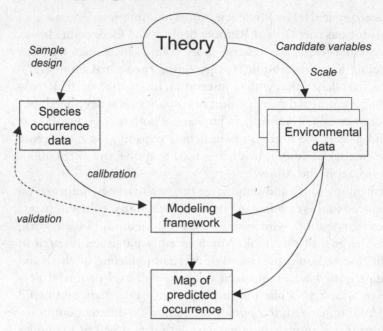

Fig. 1.2. Diagram showing the components of species distribution modeling. Biogeographical and ecological theory and concepts frame the problem, and identify the characteristics of the species and environmental data required to calibrate an appropriate empirical SDM and apply it to produce a map of predicted species occurrence or suitable habitat.

statistical model includes the choice of methods and decisions regarding implementation (calibration and validation).

The following elements are required for modeling and spatial prediction of species distributions and are described in the chapters of this book (Fig. 1.2):

- A theoretical or conceptual model of the abiotic and biotic factors controlling species distributions in space and time, and at difference scales, and the expected form of the response functions (Chapter 3);
- Data on species occurrence (location) in geographical space (a measure of presence, habitat use, abundance, or some other property) (Chapter 4), or expert knowledge about habitat requirements or preferences (Chapter 7);
- Digital maps of environmental variables representing those factors (or their surrogates) determining habitat quality, or correlated with it (Chapter 5). These are generally derived from remote sensing, from

spatial models of environmental processes, or from some other source, and stored in a GIS.

- A model linking habitat requirements or habitat use (species occurrence) to the environmental variables. The model can be statistical, descriptive, logical, or rule-based (Burgman *et al.*, 2005) (Chapters 6–8);
- Tools for applying the model (rules, thresholds, weights, coefficients) to the values of the mapped environmental variables to produce a new map of the metric of species occurrence – a geographic information system or GIS (Tobler, 1979);
- Data and criteria to validate the predictions and a way to interpret error or uncertainty in the analysis (Chapter 9).

Before discussing the theories supporting and tested using distribution models (Chapter 3), I review the growing number of ways that comprehensive geographical data on species distributions are being collected, assembled and displayed in Chapter 2. These do not all involve extrapolation using a statistical model. This will address the question "why do we need these correlative models at all?" and set the stage for the rest of the book.

The framework and procedures outlined are valid for spatial extrapolation of any ecological variable (e.g., biomass, species richness), and are often applied to spatial prediction in other domains, e.g., predicting the likelihood of deforestation (Ludeke *et al.*, 1990), urban growth, or fire risk, but the emphasis of this book is on individual species distributions. Many decisions regarding theory, data, models, and validation have to be made when developing species distribution models for a particular purpose. Often, the choice of a modeling framework is based on the experience of the practitioners (Burgman *et al.*, 2005; Austin *et al.*, 2006). I propose that the characteristics of the species (their life histories and distributions), the scale of the analysis, and the data available for modeling, determine the best modeling framework in any particular application. I will attempt to summarize this into a framework for implementation of SDM in Chapter 10.

1.5 Why model species distributions?

Why model habitat relations and make spatial predictions of species distributions? One reason is to *understand* the relationship between a species and its abiotic and biotic environment based on observations for the purpose of ecological inference, or to test ecological or biogeographical hypotheses about species distributions and ranges. Determining how pattern and scale influence the distribution and abundance of organisms is the focus of much of ecology (Elton, 1927; Scott *et al.*, 2002). When testing ecological (e.g., Jones *et al.*, 2007) and phylogenetic hypotheses (e.g., Peterson & Holt, 2003), the model describing species–habitat relationships or species niche may be the result of interest and spatial predictions from the model may not be needed. Even for applied problems, e.g., for wildlife management, ecological restoration (habitat manipulation, species translocation), species reintroductions or impact assessment, SDMs can be used to predict the value or suitability of habitat at discrete unsurveyed locations, or for a different time period. This does not necessarily require a map of predictions – a wall-to-wall spatial prediction for every location in a region.

However, species distribution models are now being widely used to interpolate or extrapolate from point observations over space, similar to forecasting through time, in order to *predict* the occurrence of a species for locations where survey data are lacking – that is, most locations on the earth's surface. This yields a predictive *map*. Maps of habitat suitability, or a predicted species distribution, are useful to test hypotheses about species range characteristics, niche partitioning or niche conservatism. Predictive distribution maps are also required for many aspects of resource management and conservation planning including biodiversity assessment, reserve design, habitat management and restoration, population, community and ecosystem modeling, ecological restoration, invasive species risk assessment, and predicting the effects of climate change on species and ecosystems. Further, a model calibrated for current conditions can be used to project potential species distributions at another point in time in order to predict the impacts of environmental change on species distributions. These are the kinds of potential uses that are usually enumerated at the outset of any one of the hundreds of papers now published on SDM. Examples of the primary applications are summarized here in order to provide a broad sense of the myriad uses of this methodology.

1.5.1 Reserve design and conservation planning

Regional to global scale conservation planning activities that require "biodiversity assessment," i.e., knowing the distribution of species, communities, species richness or other community attributes, on the landscape (e.g., Scott *et al.*, 1993; Li *et al.*, 1999; Rodrigues *et al.*, 2004;

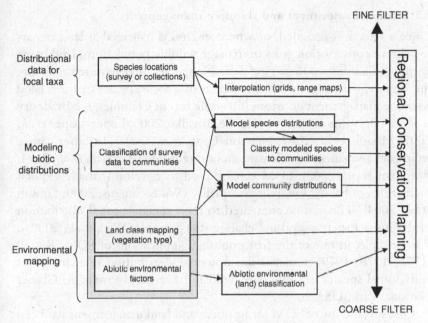

Fig. 1.3. Conceptual diagram (modified from Ferrier, 2002), showing how species distribution modeling, and specifically spatial prediction, supports regional conservation planning. Ferrier considers those entities (species, communities, other elements of biodiversity and the environment) for which good distributional data exist to be surrogates for the spatial distribution of biodiversity as a whole. These surrogates are used to ensure that conservation areas are designed to represent as many elements of biodiversity as possible. Species and community distribution modeling can be important tools in this framework, but so can direct interpolation of species locations (Chapter 2) and land classification systems, depending on data availability.

Kremen *et al.*, 2008; Thorn *et al.*, 2009), have made extensive use of SDM. Maps of species distributions or habitat suitability are required to prioritize or target areas for protected status, to assess threats to those areas, and to design reserves. To give just one of many examples, an impressive implementation was carried out by scientists at state agencies in Australia who developed models and spatial predictions for more than 4000 species of vascular plants, vertebrates, and ground-dwelling arthropods based on over a quarter of a million location records in a 100 000 km^2 region, and made important conceptual and methodological advancements along the way (Ferrier, 2002; Ferrier *et al.*, 2002b). Figure 1.3, modified from Ferrier (2002), illustrates how species distribution modeling supports regional biodiversity conservation planning.

1.5.2 Impact assessment and resource management

Once a reserve is designed, or wherever land is managed at least in part to achieve conservation goals or manage wildlife populations, land management planning, or managing desirable species, may require predicting the impact of an activity on the habitat of a focal species or for related resource management decisions (for some recent examples, see Bradbury et al., 2000; Nams et al., 2006; Traill & Bigalke, 2006; Lopez-Lopez et al., 2007; Massolo et al., 2007; Newton-Cross et al., 2007). In other words, an impact assessment or risk assessment is required (Rushton et al., 2004; Burgman et al., 2005). SDM can be used to develop spatially explicit predictions of habitat suitability or quality (Wu & Smeins, 2000; Gibson et al., 2004). This can be compared to the expected spatial distribution of the impact such as land use change (for example, Wisz et al., 2008a). For example, in one of the first published applications of SDM, Kessell (1976, 1978, 1979) modeled the potential distribution of many plant and animal species in support of wildland fire management in Glacier National Park, USA.

One specific use of SDM for resource and land management has been to designate so-called critical habitat for species that have legally protected status, e.g., under the US Endangered Species Act or similar legislation in Europe, Canada and elsewhere. In these studies, the results of an SDM are used to define the location and extent of habitat required for the protection or recovery of the focal species (Carroll et al., 1999; Turner et al., 2004; Carroll & Johnson, 2008; Thomaes et al., 2008). In aquatic or marine ecosystems SDMs have been used to designate essential fish habitat as legally mandated by environmental legislation (Valavanis et al., 2004).

1.5.3 Ecological restoration and ecological modeling

SDMs are increasingly being used to determine suitable locations for species reintroductions by associating maps of environmental factors with information on their historical ranges or habitat preferences using a model (Pearce & Lindenmayer, 1998; Schadt et al., 2002; Carroll et al., 2003; Hirzel et al., 2004; Martínez-Meyer et al., 2006).

In another exciting area of applied conservation biology, population viability analysis (PVA), a species-specific modeling framework used to forecast extinction risk (Beissinger & McCullough, 2002), has been widely used to predict the consequences of habitat loss and other threats for species of conservation or management concern (Brigham et al., 2003;

Akçakaya *et al.*, 2004; Henle *et al.*, 2004; Melbourne *et al.*, 2004). Population viability analysis often requires spatially explicit information about the distribution of habitat (location, size and quality of suitable habitat patches), and this can be derived using a SDM for the species under consideration (Akçakaya, 2000). PVA can incorporate landscape dynamics (Pulliam *et al.*, 1992; Lindenmayer & Possingham, 1996; Akçakaya & Atwood, 1997; Kindvall *et al.*, 2004), such as changing carrying capacities of habitat patches through time. SDMs may be used, in this case, to provide the initial conditions (spatial distribution of suitable habitat) or to provide maps of suitable habitat as different time steps whose changes are driven by landscape dynamics resulting from natural disturbance, land use change or climate change (Akcakaya *et al.*, 2004, 2005; Keith *et al.*, 2008).

Another form of ecological modeling, landscape modeling of plant community dynamics (e.g., forest succession), can require spatially explicit information on the distribution of potential habitat for the plant species comprising the community (He & Mladenoff, 1999; Mladenoff & Baker, 1999; Franklin, 2002; Franklin *et al.*, 2005). This information is typically generated using species distribution models. Again, this provides initial conditions for simulation models that explore the multiple impacts of natural and anthropogenic disturbance on ecological communities at large spatial scales (Syphard *et al.*, 2006; 2007; Xu *et al.*, 2007; Scheller *et al.*, 2008). For example, how do fire, logging, and climate change synergistically affect the distribution of late-successional forest patches on the landscape, and therefore on the distribution of old growth dependent species? Further, the initial distribution of species habitat provided by SDMs can then be modified by a landscape simulation model to provide dynamic habitat information to a PVA model (Akçakaya, 2001; Larson *et al.*, 2004) as described above.

1.5.4 Risk and impacts of invasive species including pathogens

Invasive species can have major economic and ecological impacts, and an important and growing use of SDMs is to determine locations where an invasive species is likely to establish (Peterson & Vieglais, 2001; Peterson, 2003; Andersen *et al.*, 2004). Two basic approaches are used – the first is to predict the geographic potential of an invasive species in a new region based on information about its habitat preferences in its native range (often using climate variables), and the second is to use information on where it has established in its new range to identify locations not yet invaded that may be at risk (Beerling *et al.*, 1995; Peterson & Robins, 2003; Underwood *et al.*, 2004; Mau-Crimmins *et al.*, 2006; Mohamed *et al.*, 2006; Roura-Pascual *et al.*, 2006; Herborg *et al.*, 2007; Loo *et al.*, 2007). Careful use of both kinds of distribution data (native range and invaded areas) can be very informative, indicating the degree to which invading species are restricted to the environmental conditions in which they are found in their native range during invasion (e.g., Broennimann *et al.*, 2007; Broennimann & Guisan, 2008). Some studies have specifically modeled factors associated with fitness of the invading species (Ficetola *et al.*, 2009). Multiscale environmental variables, including land use and other predictors in addition to climate, have been found to be important in predicting potential invasions (Meentemeyer *et al.*, 2008a; Ibáñez *et al.*, 2009).

In an inverted approach, Richardson and Thuiller (2007) modeled the current climatic distributions of South Africa's biomes, and then predicted the global distributions of regions with similar climates, thus identifying potential source areas for species likely to invade South Africa's biomes. While there are many factors involved in making an exotic, introduced species invasive (Rejmanek & Richardson, 1996), spatial prediction of the invasive species potential distribution based on environmental correlates with its established distribution is one tool for determining risk and targeting areas for monitoring and management.

If the species whose potential habitat is being predicted is a pest or disease organism that affects plants, animals or humans, or its vector or host, then spatial prediction of its potential distribution serves public health goals and supports epidemiological studies. This is another growing application of SDM (Kelly & Meentemeyer, 2002; Levine *et al.*, 2004; Saathoff *et al.*, 2005; Peterson, 2006; Raso *et al.*, 2006; Fleming *et al.*, 2007; Zeilhofer *et al.*, 2007; Menke *et al.*, 2007).

Predicting the potential location of alien species invasion, including pathogens, presents special challenges for SDM methods because the historical range or current distribution of the taxon may not fully represent the environments into which it might spread. An invasive species (a) can occupy environments outside its native range owing to competitive release, facilitation or genetic adaptations; or (b) may occupy a more restricted set of environmental conditions in a new location due to competition or dispersal limitations. Because of the ecological and economic importance of invasive species, the application of SDM and other approaches (such as genetic analysis or explicit consideration of dispersal) to risk analysis is an active area of research.

1.5.5 Effects of global warming on biodiversity and ecosystems

Changes in natural systems attributable to anthropogenic climate change are now well documented (Walther et al., 2002; Root et al., 2003; Parmesan, 2006; Rosenzweig et al., 2008). Species distribution modeling has been used to project the potential effects of anthropogenic global warming on species distributions and ecosystem properties for more than a decade (Neilson et al., 1992; Lenihan & Neilson, 1993; Mackey & Sims, 1993; Iverson & Prasad, 1998; Rehfeldt et al., 2006; Iverson et al., 2008). However, more recently the use of species distribution modeling for this purpose (Thomas et al., 2004) has received a lot of attention, as well as some criticism (Thuiller et al., 2004a; Akçakaya et al., 2006; Botkin et al., 2007). In this subsection I will expand more on this use of SDM than on previous applications because of the importance of, and growing interest in, understanding the effects of anthropogenic climate change on natural systems.

Although the relationship between climate and species ranges is well established (Woodward, 1987), as is the relationship between climate change and species range shifts based on paleoecological studies (Huntley & Webb, 1988; Webb, 1992), using SDMs to predict the impact of global warming on species distributions requires a number of assumptions. It must be assumed that species distributions tend to be in equilibrium with the climate and this assumption does not account for time lags (Lenihan, 1993). Species distribution modeling, as it is defined in this book, is a "static" approach to modeling species geographical response to climate change – it does not take into account species ability to move on the landscape (dispersal or migration), or typically does so in simple ways – usually assuming "all or nothing" dispersal or migration into new suitable habitat, or limited dispersal to contiguous suitable habitat (Peters et al., 2004; Araújo et al., 2006; Heikkinen et al., 2006; Midgley et al., 2006; Thuiller et al., 2006). Static SDM usually does not account for species interactions such as competition or predation, for evolutionary adaptation, or for a number of other potentially confounding factors, and thus could either over- or under-estimate species range shifts. Nevertheless, this tool for projecting the impacts of global warming in conjunction with other methods is continually being modified, improved and applied (Heikkinen et al., 2006; Pearson et al., 2006; Thuiller et al., 2006; Lawler et al., 2009).

However, it is worth discussing some of the controversy in the literature surrounding the use of SDMs to project the impacts of climate change on biodiversity. Controversy is a natural part of the scientific process, and it is important that practitioners state their assumptions and methods clearly so that their studies are replicable. In one study, already mentioned (Thomas et al., 2004), the authors developed species distribution models for over 1000 species in half a dozen regions of the world. They then estimated the proportion of those species "committed to extinction" by using the species area relationship and assuming that loss of habitat area would be related to extinction risk in a predictable way. They concluded that 15%–37% of the species they studied would be committed to extinction caused by anthropogenic global warming by 2050, depending on the climate projections they used (e.g., Gallagher et al., 2008) and the assumptions they made in their analysis. These estimates also varied by region and taxonomic group of species.

Thuiller et al. (2004a) responded that, although Thomas et al. had made an effort to measure the uncertainties in their estimates associated with different climate projections, there were additional sources of uncertainty in their approach that had not been accounted for (see also Botkin et al., 2007). Specifically, they noted that the original study had been based on projections from a variety of SDM modeling approaches applied to different subsets of the data (species within regions). Thuiller et al. conducted additional analyses, based on a different dataset, using four different SDM methods for each species, but with the same species-area approach and assumptions used in the Thomas et al. study. They demonstrated that, for the region and taxonomic group that they modeled (over 1000 plant species in Europe), uncertainty in the estimated proportion of species committed to extinction varied as much due to the SDM method used as to the different climate change projections.

Buckley and Roughgarden (2004) also questioned the findings of the Thomas et al. study because they contested that the species–area relationship had been incorrectly applied. They also argued that extinction risk is unevenly distributed with respect to range size. They concluded that, while species are vulnerable to extinction due to climate change, it is premature to conclude that the risk due to climate is equal to or greater than that due to land use change (as Thomas et al. stated). Botkin et al. (2007) also discussed the limitations to the use of the species-area method of forecasting changes in biodiversity under global warming. Additionally, Botkin et al. specifically questioned the ability of static, equilibrium SDMs to predict future distribution, positing that changing species interactions and adaptations would result in systematic underestimation of future distributions by SDMs. Akçakaya et al. (2006) further argued that the responses of most species to climate change are too poorly understood (e.g., Austin, 1992) to estimate extinction risks solely from SDMs applied to climate change scenarios. They recommend that using multiple models to address the interactions among potential habitat shifts, landscape structure (dispersal barriers caused by land use patterns, landscape patterning caused by altered disturbance regimes), and demography for a range of species functional groups is a way to develop guidelines for assigning degrees of threat to species (see Keith *et al.*, 2008). For example, SDMs can be combined with realistic rates (Fitzpatrick *et al.*, 2008), or simple models (Iverson *et al.*, 1999, 2004a), of species migration to evaluate the potential impacts of climate change. Transplant experiments (Ibáñez *et al.*, 2009) can be explicitly combined with an SDM approach to projecting species range changes under climate change scenarios.

At the same time, the assumptions and conclusions of the original study were being debated in the scientific literature, some scientists also expressed concern that misrepresentation of that study in the media could be damaging to biodiversity conservation in the political arena. Specifically, exaggeration of the threat of climate change to biodiversity could result in conservationists being accused of "crying wolf." Some media reports were highly inaccurate, stating that the study predicted that a million species could be extinct by 2050 (Ladle *et al.*, 2004). However, coauthors of the original study voiced the opinion that the publicity about the extinction risk posed by climate change was worth the inaccuracies in reporting in some outlets (Hannah & Phillips, 2004). They felt that the benefits of connecting species extinctions to climate change for the public and policy makers outweighed the costs of errors in reporting the details.

Another study modeling the response of just two species to climate change focused on the use of regional, rather than global climate models as a more appropriate data source for predicting shifts in species distributions, especially in mountainous terrain (Kueppers *et al.*, 2005). The species were endemic oaks in California and this brought a response from Carmel and Flather (2006) that ignoring dispersal or colonization rates might actually underestimate the negative impacts of climate change on species distributions. Specifically, those authors cited the documented lack of regeneration of oaks on abandoned agricultural land in California over the last 60 years. Kueppers *et al.* (2006) responded that the limitations of species distribution models that use a climatic envelope approach (described in Chapter 8) are well documented, but that they remain an important tool for estimating the first-order effects of climate change on species potential distributions (Pearson & Dawson, 2003).

I present these debates because I think they represent a healthy exchange of ideas in the scientific literature, and illustrate the importance of SDMs in the toolbox for predicting the impacts of land use and climate change on species distributions. I consider these arguments to be simply "growing pains" in the development of methodologies to address an important issue. One thing ecologists do agree on is that assessing the consequences of anthropogenic climate change for biodiversity is an important task on which scientific talent and resources should be focused (Thuiller, 2007).

2 · *Why do we need species distribution models?*

2.1 Introduction

There have been a number of recent large-scale programs to map species distributions from direct observations of occurrences at the national and regional scale in different parts of the world. If organizations are moving ahead to assemble large amounts of species location information globally, then why do we need species distribution modeling at all? Could species distribution modeling be considered a stop-gap approach, with a finite shelf-life? Will SDM become obsolete when all regions of the world have used global positioning systems (GPS), GIS and spatial analysis software to compile the comprehensive species distribution information that now only exists for a limited number of regions and taxa? Compiling and sharing comprehensive biodiversity data is, after all, the objective of ambitious projects like the Global Biodiversity Information Facility (GBIF, http://www.gbif.org/), whose database comprises more than 170 million records (accessed 11 March 2009). Aren't imperfect spatial predictions of species occurrence from models a poor substitute for comprehensive and spatially explicit compilation of direct observations? Perhaps an army of amateurs ("parataxonomists") with maps and GPS, and a good system for compiling and checking data, is where resources should be invested. E. O. Wilson has called for a digital "Encyclopedia of Life" (http://www.eol.org/; accessed 11 March 2009), and surely the encyclopedia would include spatially explicit information on species distributions. Even spatially incomplete species distribution data ("point maps") can be interpolated directly using so-called geostatistical and related spatial analysis techniques (see below and Chapter 6).

In this chapter, I will discuss some of the existing large-area mapping or atlas projects with particular attention to their scale, scope, and how they are being used. Methods that have been used to spatially interpolate species location data directly, without reference to environmental covariates, will be outlined. I will use the extent and availability of these data and methods as a benchmark against which to gauge the usefulness of the modeling approach that is the subject of the rest of this book. The use of old and new species atlas data for species distribution modeling will also be reviewed.

2.2 Mapping species – atlas projects and natural history collections

The purpose of SDM and, in particular, predictive mapping, is to provide spatially explicit information on species and other elements of biodiversity for conservation planning, risk assessment and resource management, as outlined in Chapter 1. Can existing species distribution maps be used for these purposes? A number of studies have directly used raw species distribution data (e.g., point maps of observations) or distribution maps that were derived from direct interpolation (Section 2.3), or some means other than species distribution modeling (e.g., range maps digitized from field guides), to evaluate patterns of biodiversity for conservation assessment (Rodrigues et al., 2004; Kremen et al., 2008), especially at broader spatial scales (large area, coarse resolution; see Fig. 1.3).

2.2.1 Grid-based atlases of species distributions

Since at least 1950 (Botanical Society of the British Isles), systematic grid-based surveys have been used to collect data on the distribution of different taxa, especially birds (Donald & Fuller, 1998). These surveys usually result in a published book of distribution maps – an atlas – and the distribution data may also be available in digital geographically referenced form in computer databases, accessed via the World Wide Web (WWW). The area covered by these atlases ranges from less than 100 km^2 to more than 10 million km^2 (Table 2.1). The survey data are often collected by dividing the landscape into equal-area grid cells at relatively coarse grain or low resolution – ranging from one to 10 000 km^2 in area (Fig. 2.1).

Typically, the larger is the extent of the area, the coarser the grain of the survey. This is presumably because finite resources are available for data collection (Donald & Fuller, 1998). These projects often rely on the volunteer efforts of hundreds or even thousands of amateur observers (Table 2.1). Many of these projects are frequently updated and continually improved. For example, in the months since this chapter was first drafted, several have started to provide maps that can be displayed using Google™ Earth. The Global Biodiversity Information Facility website

Database or publication	Date	Description and scope	url	Reference	Used in SDM?
Birds of Lesotho, South Africa	1989	Systematic observations in quarter-degree squares (QDS) ~24 × 27.5 km between 1986–1989			(Osborne & Tigar, 1992)
Atlas of the Continental Portuguese Herpetofauna	1999	Maps based on 9394 observations, 6485 from previous national atlases, 1790 from recently published work, 1119 new observations by authors from 1990 onward; 27 reptile and 17 amphibian species; 10-km UTM grid		(Godinho *et al.*, 1999)	(Segurado & Araújo, 2004)
European breeding birds	1997	Maps for 495 European bird species on 50-km grid, interactive atlas available on-line; 25 years of observations, thousands of volunteers	http://www.sovon.nl/ebcc/eoa/	(Hagemeijer & Blair, 1997) http://www.ebcc.info/atlas.html	(Araújo *et al.*, 2005)
Atlas of Amphibians and Reptiles in Europe	1997	Maps based on 85 067 records for 62 amphibian, and 123 reptile species; began in 1983 compiling all available data, 50-km UTM grid, about 4800 cells	http://www.seh-herpetology.org/atlas/atlas.htm	(Gasc *et al.* 1997)	(Araújo *et al.*, 2005)

(*cont.*)

Table 2.1. (cont.)

Database or publication	Date	Description and scope	url	Reference	Used in SDM?
Atlas of European Mammals	1999	Maps for 194 species, based entirely on field observations. Over 93 000 records are mapped with separate symbols for data collected before or after 1970	http://www.european-mammals.org/php/mapmaker.php	(Mitchell-Jones *et al.*, 1999)	(Araújo *et al.*, 2005)
Atlas Florae Europaeae (AFE)	1972–2006	Distribution maps in volumes 1 to 12 made manually. Later the maps were scanned. 3556 maps. Thus far, families which include more than 20% of the European flora	http://www.fmnh.helsinki.fi/english/botany/afe/index.htm	http://www.fmnh.helsinki.fi/english/botany/afe/publishing/index.htm	(Thuiller *et al.*, 2004b)
The Millennium Atlas of Butterflies in Britain and Ireland	2001	99% coverage of 10-km grids, >90% records with 1-km locational precision, presented on 10-km British National Grid. 1.6 m records 1995–1999 BNM survey, 10 000 volunteers		(Asher *et al.*, 2001)	
The Biological Records Centre (BRC)		The national focus in the UK for terrestrial and freshwater species recording (other than birds). Over 15 million records of more than 12000 species records on 10-, 2- or 1-km British National Grids, published distribution maps and atlases	http://www.brc.ac.uk/	http://www.brc.ac.uk/publications.htm	

New Atlas of the British and Irish Flora, Botanical Society of the British Isles	2002, ongoing	The BSBI Maps Scheme showing the distribution of vascular plants in the British Isles; dot maps, updated weekly. Precision of records based on the 10-km squares of the national grid, with 2412 dot maps. 9 million entries, 5 million gathered in intensive post-1987 fieldwork as of 2002. Currently mapping on 2-km grid system.	http://www.bsbi.org.uk/html/atlas.html	(Preston *et al.*, 2002)
San Diego County Bird Atlas (California, USA)	2007	492 natives, migrants, and well-established exotics. 479 grid squares (1 mile square), nearly 400 000 records. 400 volunteer observers, over 55 000 hours in the field from 1997 to 2002.	http://www.sdnhm.org/research/birdatlas/	(Unitt, 2004)

This is by no means a comprehensive list, but illustrates a range of projects at different scales, some that have been used as data sources for species distribution modeling. Websites accessed 11 March 2009.

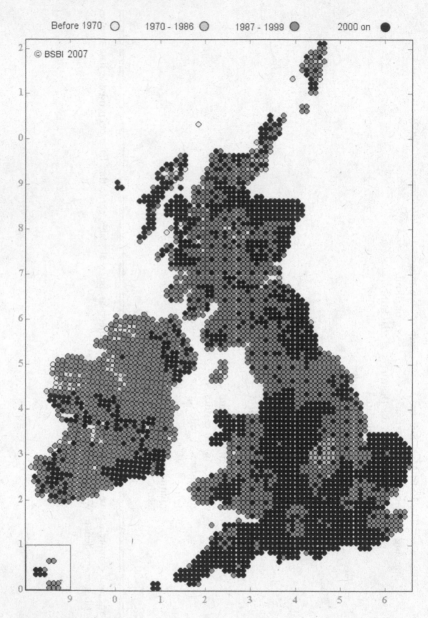

Before 1970 ○ 1970 - 1986 ◔ 1987 - 1999 ◉ 2000 on ●

© BSBI 2007

Fig. 2.1. Example of "dot map" of a vascular plant species, *Amsinckia micrantha* (Common Fiddleneck) in Britain and Ireland from the Botanical Society of the British Isles Maps Scheme (http://www.bsbi.org.uk/html/atlas.html), on the 10-km National Grid system. Dot colors indicate when observations were made: cyan = pre-1930 record; yellow = 1930–1969; green = 1970–1986; magenta = 1987–1999; blue = 2000 or later. Copyright © Botanical Society of the British Isles. Used with permission.

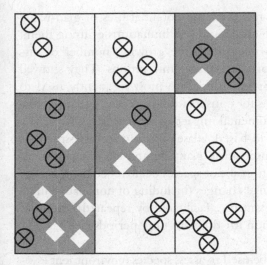

Fig. 2.2. Illustration of how species maps are typically compiled for atlas projects based on gridding point data from natural history collection records and other opportunistic observations. Open diamonds represent records of the target species occurrence, circled X's represent observation of other species. Gray grid cells represent target species presence in the resulting grid map and white grid cells absence. The larger the grid cells are, the greater the sampling effort per cell even with opportunistic observations, and the greater the confidence that lack of an observation of the target species implies true absence in a grid cell. However, information about species locations below the resolution of the map is obviously lost. The resolution of the species data is degraded.

now provides the ability to create "niche models" (SDMs using global-scale climate variables) on the fly.

Prior to these systematic efforts in recent decades, species distribution (range) maps and atlases were compiled from whatever observations were available including specimen records from natural history collections. These observations were usually highly non-random in space and time and of inconsistent locational precision, and therefore tend to more accurately reflect species distributions when compiled at coarser scales (Fig. 2.2). This spatial generalization evens out observer effort in sample units (grid cells) and gives more reliable evidence of species absence (at least for conspicuous, well-studied taxa in well-searched regions), but provides no information about what portion of the large grid cell a species actually utilizes (see McPherson *et al.*, 2006). Grid-based surveys have the advantage of using standardized, or at least known, protocols for making observations systematically in space, yielding contemporary distribution data.

Donald and Fuller (1998) reviewed the potential uses of grid-based ornithological atlases, and the challenges and limitations of using them, and their comments are also relevant to the growing number of distribution mapping projects for other taxonomic groups. They showed that bird atlas data for the British Isles were most frequently used to document species distributions for conservation purposes – that is, conservation planning and environmental impact assessment (see Chapter 1). However, a limitation of grid-based atlases for impact assessment is their coarse grain – often conservation planning and land management require finer-grained information about species occurrence (McPherson *et al.*, 2006). Documenting range changes (including of non-native invasive species) is another important use, facilitated by repeated systematic observations or data compilation for different time periods (see Fig. 2.1 and 2.3).

Grid-based atlases can also be used to assess species–environment associations – in other words, data from these relatively coarse grids can be used to develop species distribution models. Why would a species distribution model be needed if a species distribution map is already available? Understanding the environmental factors determining broad-scale species range and occurrence might be the objective (rather than predictive mapping). A growing number of SDM studies have used published coarse-grained atlas data to explore species–environment correlations, test hypotheses regarding species richness, ranges and equilibrium with climate, and even to test different SDM methods (Cumming, 2000b; Segurado & Araújo, 2004; Thuiller *et al.*, 2004b; Araújo *et al.*, 2005). Some of these studies have digitized species range maps from many sources to compile coarse-scale continental and global datasets that are not publicly available (Orme *et al.*, 2005).

Another reason for using grid-based atlas data to develop a SDM is to refine species range maps when some parts of the region were not surveyed as rigorously as others in a contemporary grid-based survey. Or, grid-based atlases may need to be refined when the range maps are based on natural history collections or a mixture of data sources including opportunistic observations with uneven effort due to accessibility (Osborne & Tigar, 1992; Godinho *et al.*, 1999; Cumming, 2000a). In these cases, absence of evidence (of species occurrence) cannot be taken for good evidence of absence. As noted above, the coarse resolution of the grids used to map distributions in atlas projects can help overcome sparse observations using spatial generalization (Fig. 2.2).

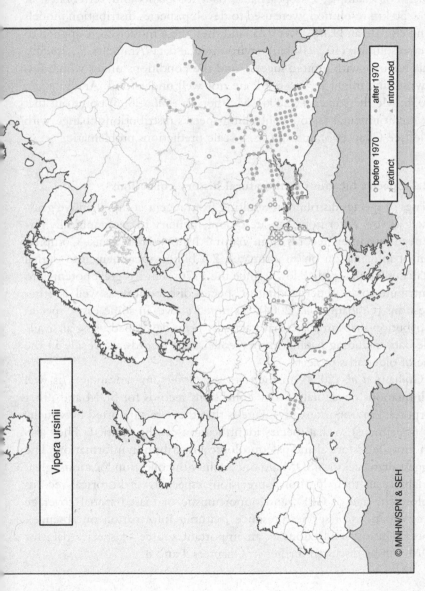

Fig. 2.3. Example of "dot map" for *Vipera ursinii* from *The Atlas of Amphibians and Reptiles in Europe* (Gasc et al. 1997) (http://www.seh-herpetology.org/). Each dot represents a 50 × 50 km UTM grid square. Copyright © Muséum National d'Histoire Naturelle & Service du Petrimone Naturel and Societas Europaea Herpetologica. Used with permission.

Vipera ursinii

before 1970   after 1970
extinct   introduced

© MNHN/SPN & SEH

There have been a number of attempts to address the limitation of coarse-scale atlas data via "downscaling" using species distribution modeling. For example, European atlas data for plants and vertebrates, at 50 × 50 km resolution, were used to develop species distribution models using climate and land cover data, and then these models were applied to finer-resolution (10 × 10 km) environmental data (Araújo *et al.*, 2005b). This has met with limited success, and the conditions under which this downscaling might be successful are not well understood (Araújo *et al.*, 2005b; McPherson *et al.*, 2006). McPherson *et al.* (2006) point out that the environmental factors controlling species distributions change with scale (see Chapter 3), making cross-scale predictions problematic.

2.2.2 Species locations from natural history collections

Atlases of species distribution resulting from recent comprehensive grid-based surveys differ from, but are complementary to geographically referenced specimen records from natural history collections. Compilation of information on an estimated 2.5 billion specimens into digital databases has proceeded slowly since the 1970s, but has great potential to contribute to an understanding of species distributions as well as other questions (Graham *et al.*, 2004). As noted above, in some cases, species distribution maps in atlases have actually been developed using all available data including natural history collections records, especially in the case of older atlas projects.

Graham *et al.* (2004), in their review, note some advantages as well as limitations of natural history collections records for SDM and other purposes. An important advantage is that they are supported by vouchers (specimens) so that species identification can be verified. They also can provide historical and paleontological distribution information. The biggest drawbacks for SDM are uncertainty in collection locations, great variability in their locational precision, especially of historical records (before the age of GPS), and opportunistic or biased spatial coverage of observations of species presence (without information on absence). Natural history collections as an important source of species data for SDM will be discussed further in Chapters 4 and 8.

2.3 Direct interpolation of species data

Some have taken the approach of directly interpolating species location data, usually using what are known as optimal interpolation (geostatistical) methods such as kriging, with limited investigation of environmental correlates. Interestingly, several of these recent studies are for the marine environment (Monestiez *et al.*, 2006; Sundermeyer *et al.*, 2006). *Interpolation* is the estimation of values for unobserved locations based on values at nearby locations and is an important tool in the cartographer's toolbox. The simplest form of spatial interpolation estimates the value of an unsampled location as the average of some number of its neighbors weighted by their inverse distance from the new location. Direct interpolation methods, by definition, assume that distance from observed values is the most important predictor of the variable of interest, and these methods have very limited capabilities to include environmental covariates as predictors (Bailey & Gatrell, 1995; Clarke, 1997; Burrough & McDonnell, 1998; Fortin & Dale, 2005).

Even though interpolation methods are usually applied to continuous variables (abundance or biomass, for example), other methods can be applied to point maps of species occurrence such as kernel density estimators (Fortin *et al.*, 2005). Kernel estimators convert point data to continuous surfaces showing point density – conceptually, they show a moving average of density (Nelson & Boots, 2008). Kernel estimators have been used to delineate animal home ranges, and can be used to enhance or detect edges or boundaries. Other boundary delineating methods (called hulls or spatial envelopes) have also been used for drawing range boundaries around species point location data (Burgman & Fox, 2003). A variety of spatial analysis methods are now available for delineating boundaries or interpolating surfaces (Fortin & Dale, 2005). Therefore, even if mapped environmental covariates are not available, there are a number of rigorous approaches that can be applied to get the most out of species location data.

Although they have been widely used in other fields of study, geostatistical and related interpolation methods have seen only limited applications in species distribution modeling (Bolstad & Lillesand, 1992; Araújo & Williams, 2000; Hershey, 2000; Segurado & Araújo, 2004; Fortin *et al.*, 2005), and there have been even fewer direct comparisons of interpolation versus multiple regression-type approaches (in the broad sense used in Chapter 6) – they have been reviewed by Miller *et al.* (2007). When space (location) alone is used to interpolate species distributions, patterns of environmental factors cannot be used to extrapolate distributions in time or space to similar environments. Further, environmental factors are unavailable to refine estimates of the extent of occurrence (the region encompassing all species localities) to the area of occupancy, those portions of a species range that are actually occupied by the species because they comprise suitable habitat (e.g., Elith & Leathwick, 2009). The concept of species distributions being strongly determined by multiple environmental factors (Chapter 3), and the characterization of species–environment relationships using quantitative models (Chapters 6–8), dominate in the myriad applications of SDM.

Alternatively, when species location data are lacking but environmental maps exist, it is possible to use simple decision rules to defined species habitat requirements derived from expert opinion and published literature (Chapter 8) and apply them in order to develop habitat suitability maps. For example, in the USA, the Gap Analysis Programs for various states used different approaches to develop species distribution maps, and composite maps of species richness, including the application of existing rule-based habitat suitability models to environmental maps (Scott et al., 1993; Davis et al., 1995). In the following chapters we will examine and compare these alternative approaches and try to develop guidelines for deciding under what circumstances each might be most effective.

2.4 Summary – what do we really want?

With an international organization like the Global Biodiversity Information Facility (http://www.gbif.org/) that is working to make the world's biodiversity data globally accessible via the internet and data sharing protocols, it seems that SDM may ultimately become obsolete for providing species location information. However, a quick perusal of the GBIF databases shows many species for which there are currently very little or no location information included. While recent large-area mapping efforts based on grid surveys are extremely useful, they may only be feasible at certain scales – typically coarse grain over large extents, or relatively finer grained over smaller extents. The atlas projects have yielded invaluable data that have been used to describe species ranges and address biogeographical research questions at broad spatial scales. However, the greatest demand for these atlas data actually appears to be for conservation planning and environmental impact analysis – and the coarse resolution of the atlas data are a great limitation to their use in this capacity. Often, conservation planning and resource management is carried out at smaller scales. SDM provides an alternative approach that can help extend the usefulness of direct observations and improve our interpretation and understanding of species distributions.

Further, it may only be possible to repeat large-scale systematic surveys at infrequent intervals. This is another important limitation of the current atlas data, or of direct spatial interpolation methods. Even a perfect and complete map of a species current distribution, like the "Marauder's Map" of Hogwart's Castle in the Harry Potter book series by J. K. Rowling – a magical, real-time, animated map showing where every individual in an area is (if you know the spell) – would not allow predictions of how species distributions or the location of suitable habitat might be altered in the face of environmental change, such as global warming or land cover change. A species distribution model does support these predictions. That is why SDMs are so important and receiving so much attention – interest in SDMs is driven by our need to forecast the impact of management actions or environmental change on patterns of biodiversity from local to global scales.

3 · Ecological understanding of species distributions

3.1 Introduction

Austin (2002) presented a framework for spatial prediction of species distributions (as outlined in Chapter 1) that links ecological theory to implementation (statistical modeling). This chapter discussed the ecological model portion of that framework – those ecological and biogeographical concepts and theories that are needed to frame the empirical modeling of species distributions. The ecological model is required in order to identify the characteristics of species occurrence data that are appropriate for modeling, select explanatory variables or their surrogates, specify appropriate scale(s) of analysis, hypothesize the nature or form of the species–environment relationship (the shape of the response curve), and select an effective modeling method. In this chapter, I will review the niche concept, and related to it, factors limiting species distributions, environmental gradients and species response functions. Finally, conceptual models of the environmental factors that control species distributions at hierarchical spatial and temporal scales that are particularly relevant to SDMs will be described.

3.2 The species niche concept

A number of ecological theories related to causes of species distributions, species diversity, and community structure may be relevant to SDM, but a lot of recent discussion has focused on the *species niche concept* as it relates to SDM (Austin & Smith, 1989; Austin, 2002; Guisan & Thuiller, 2005; Araújo & Guisan, 2006; Kearney, 2006; Soberón, 2007; Hirzel & Le Lay, 2008; Jiménez-Valverde *et al.*, 2008). The species niche concept is central to ecology, and the history of its development is discussed in most ecological textbooks and more specialized books (for example, Shugart, 1998; Chase & Leibold, 2003).

3.2.1 The species niche in environmental and geographical space

The concept of species niche has evolved over time and has several interpretations (Chase & Leibold, 2003). While Elton (1927) defined the niche as a species functional role in the biotic community, Grinnell (1917a) referred to those factors that influence where one would find a species (what it needs). Hutchinson, in a now-classic reference (1987), defined the niche as " . . . the hypervolume defined by the environmental dimensions within which that species can survive and reproduce." Hutchinson further distinguished the *fundamental* (physiological or potential) niche, defined as the response of species to environment (resources) in absence of biotic interactions (competition, predation, facilitation), from the *realized* (ecological, actual) niche – the environmental dimensions in which species can survive and reproduce, including the effects of biotic interactions (Austin & Smith, 1989; Austin, 2002; Chase & Leibold, 2003), depicted in Fig. 3.1a. He defined niche as a property of a species, not the environment (as discussed by Pulliam, 2000). SDM has relied on the niche concept that emphasizes species requirements, particularly (but not exclusively) representations of abiotic factors controlling species distributions. Hutchinson's hypervolume suggests multiple causal factors of species distributions and this has also been emphasized operationally in SDM, where multiple predictors are typically used. In contrast, theoretical treatments of the niche sometimes emphasize only a few resources, requirements, or limiting factors.

A recent synthesis links these concepts of species Eltonian impacts and Grinnellian requirements (Chase & Leibold, 2003, p. 15), and they define niche as "the joint description of the environmental conditions that allow a species to satisfy its minimum requirements so that . . ." there is net positive growth in the local population when the per capita effects of the species on the environment are taken into account. Pulliam (2000) also emphasized the importance of measuring fitness, specifically population growth rate, in order to identify a species niche (where growth rate is positive). He proposed that Hutchinson's niche concept, metapopulation theory (Hanski, 1999), and source-sink theory (Pulliam, 1988) together can help explain the relationship between the distribution of species and suitable habitat.

Pulliam (1988) differentiated source habitat, where local reproduction exceeds mortality, from "sink" habitat where individuals are found but do not contribute to net population growth. The metapopulation framework (Hanski, 1999) is used to describe and model populations as

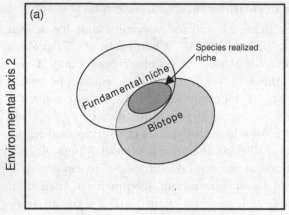

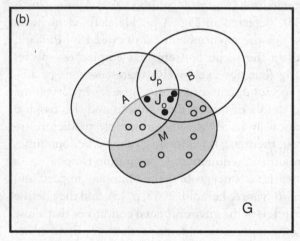

Fig. 3.1. The niche concept of Hutchinson. (a) The biotope encompasses the range of environmental conditions that occur in an area. The species fundamental niche represents its range of physiological tolerances in the absence of interactions with other species (competition, predation). The realized niche is the environmental space where a species actually occurs, accounting both for the availability of environmental conditions and for biotic interactions affecting a species distribution; after Shugart (1998) Fig. 5.7, with modifications. (b) representing geographical space, G, rather than environmental gradients, after Soberón (2007; Fig. 1). A = the geographical area where intrinsic growth rate is positive, B = the area where a species can coexist with competitors, M = the area that is accessible to a species within a given time frame given its dispersal ability. J_o is the occupied area and J_p is the potentially occupied area. Closed circles represent species occurrences in source habitat, open circles are occurrences in sink habitat (see text).

a group of spatially discrete subpopulations. A species can be absent from a patch of suitable habitat due to local extirpation resulting from population dynamics, or because of dispersal limitations. Taken together, the concepts show that a species might be found in unsuitable (sink) habitat, or might be absent from suitable habitat. This current understanding of the niche concept is important to keep in mind because most SDM studies rely on observations of species occurrence (Chapter 4), or sometimes abundance – fitness is rarely addressed directly, with a few exceptions (see Aldridge & Boyce, 2007). Equating habitat occupancy with habitat suitability may be an oversimplification.

The realized niche is often depicted as a subset of the fundamental niche (Fig. 3.1a), although it is not necessarily so, e.g., in the case of "sink" habitat or dispersal limitation, or in the case of positive biotic interactions (facilitation, symbiosis) where the occurrence of another species is a requirement. Figure 3.1b (modified from Soberón, 2007) depicts the concepts laid out by Pulliam in geographical space rather than environmental space. In this figure, A is the geographical area where intrinsic population growth rate is positive (corresponding to the fundamental niche); B is the area where a species can coexist with, or outcompete, competitors; M is the area that is accessible to a species within a given time frame, given its dispersal ability; J_o is the occupied "source" habitat and J_p is suitable habitat that is unoccupied due to dispersal limitations. If the species is found in portions of M and/or B that do not intersect A, or portions of A that do not intersect B, this would represent sink habitat.

Ackerly (2003) asserted that an assumption implicit in the Hutchinsonian niche concept is that all combinations of the relevant causal factors exist in the landscape. This is generally not the case, and he recounts the concept of "realized environment" proposed by Jackson and Overpeck (2000), who define "potential niche" as the intersection of the realized niche and the realized environment. This is an unnecessary elaboration on Hutchinson's concepts. Hutchinson defined the "biotope" as the realized environment in the sense used above, and the realized niche as the intersection of the biotope and species fundamental niche (Fig. 3.1a), as described by Shugart (1998, p. 121).

There has been a great deal of recent discussion about the relationship between the species niche concepts described above and species distribution modeling. Returning to a question posed in Chapter 1, what is actually being modeled in SDM, the fundamental species niche, the realized niche, or the probability of habitat use? Most studies identify a description of the realized niche as the outcome of SDM (Austin, 2002; Thuiller *et al.*, 2004b; Guisan & Thuiller, 2005), because data on actual species occurrence are usually used in modeling, and so the model extrapolates in geographical space those conditions associated with species abundance or occurrence (Chapter 4) in the environmental "hypervolume." When this realized niche described by the statistical model is mapped in geographical space it represents the potential distribution or habitat suitability (Araújo & Guisan, 2006; Soberón, 2007).

Others strictly prefer the term habitat model or distribution model for correlative SDMs (Jiménez-Valverde *et al.*, 2008), both for the model itself and the spatial prediction, and reserve the term niche model for mechanistic analyses of factors limiting species fitness (Kearney, 2006). I have used the term distribution model throughout this book to reflect the fact that these models are conditioned on the distributions of species, and predict the potential geographical distribution of species or suitable habitat. However, in SDM the connection to an underlying species niche concept should be made as explicitly as possible in the choice of predictors, interactions between predictors, response functions, model type, and interpretation of the resulting predictions. For example, it has been suggested that environmental envelope-type models using presence-only data (Chapter 8) tend to depict potential distributions (suitable habitat), and are more suitable for extrapolation, while more complex models that discriminate presence from absence (Chapters 6–7) tend to predict realized distributions (occupied habitat), and are more suitable for interpolation (Jiménez-Valverde *et al.*, 2008; Hirzel & Le Lay, 2008). Alternatively, Soberón and Peterson (2005) argue that SDM based on coarse-scale climate variables (bioclimatic niche modeling) describes the species fundamental niche. This concept is elaborated by Hirzel and Le Lay (2008) who noted that biotic interactions (competition, predation) tend to occur at short distances, and that dispersal limitations and fine-scale environmental heterogeneity allow inferior competitors to evade negative interactions by persisting in competitor-free locations. Thus, they conclude, the realized and fundamental niche may not differ that much in practice, especially when predicted from coarser-scale environmental factors such as climate. We return to this issue in Section 3.5.

When applying the niche concept in static (statistical) species distribution modeling, that is, without including dynamic elements, an underlying assumption is that species are in (quasi-) equilibrium with contemporary environmental conditions, and that observed distribution and abundance is indicative of environmental tolerances and resource requirements. The limitations of this assumption have been discussed extensively (Guisan & Zimmermann, 2000; Austin, 2002) and should be explicitly considered in each specific SDM study. For example, some species may still be spreading into suitable habitat following the last Glacial Maximum (Leathwick, 1998).

3.2.2 The species niche in evolutionary time

SDM is being increasingly used to test hypotheses in evolutionary biology regarding niche and geographic range as species traits (e.g., Ree et al., 2005; Peterson & Nyari, 2008). Do closely related taxa share similar climatic tolerances even if their current distributions are disjunct (Kozak & Wiens, 2006)? This has been taken as evidence of so-called "phylogenetic niche conservatism" (or ecological niche conservatism) – the tendency of species to retain characteristics of their fundamental niche over evolutionary time via stabilizing natural selection (Wiens, 2004; Wiens & Graham, 2005). For example, Huntley et al. (1989) studied the distribution of pollen of the tree genus *Fagus* (beech) in Europe and North America with respect to climate and concluded that the distribution of that taxon is in equilibrium with current climate, and further, that their similar ranges of tolerance to climate factors on the two continents (in spite of long separation) implied that those physiological tolerances had been evolutionarily conserved. While similarity of climatic tolerances has been used as a line of evidence for niche conservatism in relation to physiological tolerances in this and other studies, the alternative scenario, adaptive evolutionary shifts in environmental tolerances, certainly also occurs, especially when the species ability to disperse or migrate (track environmental change) is limited (Ackerly, 2003).

Further, it is important to acknowledge that niche conservatism is an assumption, rather than a hypothesis, in a growing number of studies that reconstruct contemporary and paleo-distributions in order to delimit species, especially in the case of morphologically cryptic species (Pearson et al., 2007; Rissler & Apodaca, 2007; Bond & Stockman, 2008). Niche conservatism is also assumed when using SDM to reconstruct paleo-distributions in order to examine other research questions in phylogeography (Hugall et al., 2002; Carstens & Richards, 2007; Knowles et al., 2007).

It is important to keep in mind that empirical SDM carried out to answer these phylogeographical questions has focused almost exclusively on coarse-scale climate variables as the only predictors and has even been called phyloclimatic modeling – "combining phylogenetics and bioclimatic modeling" (Yesson & Culham, 2006). In fact, specialized modeling software aimed at this application includes built-in global climate datasets to be used as environmental predictors (Hijmans *et al.*, 2001). I argue that the "climatic niche" actually encompasses only a limited set of dimensions in Hutchinson's "hypervolume" (see Chapter 5), and only at the broad scale described in the hierarchical framework in Section 3.5. If questions about speciation and phylogenetics are best addressed at the scale of species range limits, then broad-scale environmental predictors may be sufficient – but still, climate may not be the only factor circumscribing species ranges (e.g., Kozak *et al.*, 2008). These assumptions, and their limitations, should clearly be stated when using SDM in phylogenetic studies (Graham & Fine, in press).

While the growing use of SDM in phylogenetic research is an interesting development, the question of whether species niches are stable or not, over short (thousands of years) or long (millions of years) time periods, also has very practical implications for the use of SDM to predict the impacts of climate change on species. In particular, species whose climatic niche appears to have been conserved, at least over shorter time periods, may be the best candidates for reliable predictions of the impacts of future climate change (Pearman *et al.*, 2008).

3.2.3 Niche or resource selection function?

Niche is not the only ecological concept relating species to environment that has been used to frame SDM studies. The concept of habitat or resource selection in wildlife biology describes selection of resources by animals and is closely aligned to the quantitative methods developed to estimate it. Resource selection functions (see also Chapters 1, 4, and 8) are various types of statistical models that predict the probability of use of a resource unit, and RSF methods overlap substantially with SDM methods (Boyce *et al.*, 2002; Manly *et al.*, 2002; Boyce, 2006; Johnson *et al.*, 2006). The concepts of niche and resource selection are similar – sometimes niche is thought of in terms of the boundaries of the hypervolume where a species can exist, but these boundaries or limits simply represent habitat selection at the broadest scale – the species range. RSFs are thought of in terms of the probability of resource use relative to its availability, that is, the selection of home range or habitat patches by mobile organisms. But species response curves in the Hutchinsonian hypervolume also suggest probability of resource use, so the concepts are not really that different.

3.3 Factors controlling species distributions

Austin (2002) described the types of factors that affect species distributions, and distinguished *proximal* (causal) factors from *distal* (proxy or surrogate) factors. Distal factors are related to resources or regulators (proximal factors), and therefore correlated with species distributions, but may be easier to measure or observe than the proximal factors themselves. The example Austin gives is that the availability of a certain mineral nutrient in the root zone of a plant might represent a proximal factor related to growth and survival, while distal factors related to the availability of that nutrient might include soil texture, pH or even general soil type. Austin (1980) further distinguished *direct* factor gradients, those with a direct physiological effect (temperature, pH), from *resource* gradients, variables that are consumed or used up (water, light, nutrients, prey, suitable nest sites). *Indirect* factor gradients have no direct effect on species distribution or abundance, and so always are distal variables. Examples are latitude, longitude, elevation, slope angle (steepness) and aspect (exposure). Ideally, variables describing direct and resource gradients would always be used as predictors in SDM. However, when only variables describing indirect gradients are available, it is important not to extrapolate the model results beyond the range of conditions used to develop the model. Further, there is no theoretical expectation for the shape of a species response curve (see next section) on an indirect gradient (Austin, 2007).

3.4 Environmental gradients and species response functions

Whittaker (1956, 1960, 1967), developed key ideas for the analysis and interpretation of the distribution of species abundance along environmental {factor}gradients. Those *response curves*, often depicted graphically, are often called species response functions (for plants) or resource selection functions (animals) (Austin, 2002, p. 103). They depict a function describing the relationship of species abundance (or fitness, occurrence, or resource selection) in relation to values of an environmental variable (Fig. 3.2).

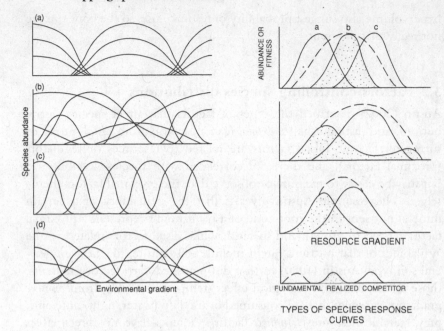

Fig. 3.2. Species distributions along an environmental gradient. Figures on the left (from Austin, 1985; Copyright © Annual Reviews; Used with permission): (a) community concept; (b) individualistic continuum; (c) competition and resource partitioning; (d) resource partitioning within strata or guilds, individualistic between strata. Figures on the right (Austin & Smith, 1989; Copyright © Kluwer Academic Publishers; Used with permission): top graph expresses same concept as (c) showing hypothetical response function in absence of competition (dashed line), middle graph showing Ellenberg's bimodal response with species restricted to extreme conditions under competition, next is species restricted to one extreme only (stippled areas). Bottom graph is the figure legend. These figures were themselves modified from those found in (Whittaker, 1970) and original sources quoted therein.

The individualistic continuum concept (Gleason, 1926) predicts species abundance optima and their limits to be independently distributed on environmental gradients (Fig. 3.2b). Alternatively, it has been hypothesized that species response functions in response to environmental factors are bell-shaped (Gaussian), equally spaced and of equal amplitude, with their width restricted by competition (e.g., Gause, 1934; Tilman, 1982), as in Fig. 3.2c (Gauch & Whittaker, 1972). However, symmetrical response functions are not usually found when they are carefully estimated using data. Unimodal, skewed responses to resource and direct

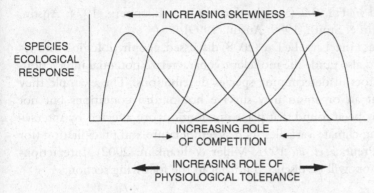

Fig. 3.3. Expected skewness of species response curves along environmental gradients. From Austin (1990). Copyright © Academic Press. Used with permission.

gradients are common. Skewed response curves may be expected (Austin & Smith, 1989), at least for certain resource gradients such as temperature, with physiological stress limiting species abundance or fitness at the "harsh" end of the gradient and competition limiting it at the benign end, as in Fig. 3.3 (Austin, 1990). Bimodal or multimodal response curves (Fig. 3.2) have also been hypothesized to result from competition (Whittaker, 1960, 1967; Mueller-Dombois & Ellenberg, 1974).

However (as reviewed by Austin 2007), Liebig's law of the minimum suggests that the true response of species to one (regulator or resource) factor can only be detected when all other factors occur at non-limiting levels (Huston, 2002, pp. 12–14). In reality, it is not usually possible to observe this without experimental data, and factors tend to covary in nature. Quantile regression, modeling the upper and lower bounds of the data, has been suggested and tested as an alternative approach to this problem (Huston, 2002; Guisan *et al.*, 2006; Austin, 2007; Vaz *et al.*, 2008), but has not been widely used in SDM (Chapter 6).

Mortality-causing disturbances (herbivory, fire) can create complex responses of species to environmental gradients (Huston, 2002), but responses to indirect gradients can also take complex forms. That form depends on the nature of the relationship between the indirect gradient and the "real" underlying causal factor (Austin, 2002). In fact, a number of alternative forms of the species response curves have been suggested that address the effects of interspecific competition, independent strata (layers or functional groups) in a plant community, and so forth, as depicted in Fig. 3.2 (Mueller-Dombois & Ellenberg, 1974; Austin, 1985; Austin & Smith, 1989; Austin, 1999).

Finally, as Hirzel and Le Lay (2008) discussed, ecophysiological knowledge should also guide the modeler to characterize potential interactions among factors in determining species distributions. The example they give is that an organism may survive hot or dry conditions but not both. It has been found that plant distributions are affected by interactions among climate variables such as temperature and precipitation (for example, Prentice *et al.*, 1991; Miller & Franklin, 2002). Interactions among factors will be discussed further in the following section.

3.5 Conceptual models of environmental factors controlling species distributions

The previous sections described in general terms the kinds of factors (abiotic and biotic, resource, direct and indirect) affecting species distributions, and the form those response functions are expected to have along a resource gradient, but did not explicitly define *what* variables are expected to determine species distributions. "One of the first steps in building predictive distribution models is to assume a conceptual model of the expected species-environment relationships..." (Guisan *et al.*, 2006a, p. 387). Selection of candidate variables is often made based on expert knowledge (Hirzel & Le Lay, 2008). Must the factors or variables controlling species distribution and abundance always be defined on a case by case (species by species) basis? Or, are there some generalizations that can guide empirical SDM for all species, or for different taxonomic groups?

3.5.1 Heat, moisture, light, nutrients, and the distribution of plants

Climatic conditions have long been observed to play a primary role in limiting species distributions and vegetation patterns (Grinnell, 1917b; Köppen, 1923; Holdridge, 1947; Gaston, 2003), dating back to the work of von Humboldt and de Candolle in the eighteenth and nineteenth centuries. Woodward (1987; see also Woodward & Williams, 1987) discussed the physiological effects of climate on plant growth and survival via climatic controls on the light, moisture and temperature regimes. He emphasizes the lethal effects of low temperatures in setting range limits of plant species at higher latitudes and altitudes. But, as Gaston (2003) points out, there are hundreds of possible climate parameters that can be derived from records of temperature, precipitation, humidity, evapotranspiration, solar radiation, cloudiness, etc., for any potential period of time, within and between years, for which records exist. So, connecting theory to data continues to be challenging (Chapter 5).

Several conceptual models describe the multiple environmental factors that control plant distributions. Mackey (1993a) described these factors as the *primary environmental regimes* of radiation, thermal, moisture and mineral nutrients. Most vascular plants are primary producers (photosynthetic autotrophs) and so perhaps developing a general model of environmental factors controlling plant distributions is easier than for animals of varying trophic levels, guilds, taxonomic groups, and so forth. Guisan and Zimmermann (2000) considered the resource gradients of nutrients, water, photosynthetically active radiation (light), and the direct gradient of heat sum to be the four gradients controlling plant distributions (see their Fig. 3). Franklin (1995) also described a conceptual model of plant distributions in which the environmental factors of climate, topography, and geologic substrate interact to control the primary environmental regimes in complex ways, for example, through evapotranspiration, soil development, and the subsurface moisture regime. I have modified that graphical model here to indicate that the four primary environmental regimes control the geographical distribution of the fundamental niche (potential distribution). Biotic processes and interactions – dispersal, competition, predation, facilitation, and mutualism – determine the realized niche, and additionally disturbance influences the actual distribution of species (Fig. 3.4).

Similar schemes have been developed in other fields such as ecohydrology (Wilby & Schimel, 1999) and vegetation science (Franklin & Woodcock, 1997). An important conceptual model of soil formation was developed by Jenny (1941), who proposed a functional factorial mode where soil properties in a location are a function of regional climate, parent material, topography, biota and time (for review see Scull et al., 2003, in the context of predictive soil mapping). Austin (1999), in discussing the contributions of plant ecology to biodiversity research, praised the early work or Perring (1958, 1959, 1960), who showed that in chalk grasslands the distribution of plant species on topographic gradients were contingent on climate.

The graphical model shown in Fig. 3.4 also shares some elements with the conceptual model of SDM presented by Guisan and Thuiller (2005) (see Fig. 3.5) – the potential "bioclimatic" range of a species based on its requirements for, or tolerance to, heat, moisture and light

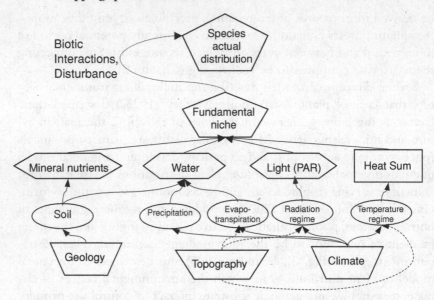

Fig. 3.4. Conceptual model of environmental factors controlling the primary environmental regimes circumscribing the fundamental species niche, further modified by biotic factors and temporal dynamics, resulting in observed species distribution. After Franklin (1995) (with inspiration from Guisan & Zimmermann, 2000).

regimes (controlled by climate at broad spatial scales) is then filtered by dispersal, disturbance and resource factors, and to tease apart their effects on the realized distribution requires dynamic modeling. This appears to limit the role of SDM to identifying the fundamental niche or potential distribution. However, SDMs, even though they are static and based on correlations, have been shown to be useful in identifying more than the "bioclimatic" niche or species range at broad spatial scales. I hope this book will convince you of this.

3.5.2 Hierarchical and nested scales of factors affecting species distributions

For animals, classically, wildlife habitat comprises cover, food and water (Morrison *et al.*, 1998, p. 41), but those authors emphasized the central role of vegetation in defining habitat conditions for wildlife. A more general, hierarchical framework for modeling the spatial distribution of animal species was presented by Mackey and Lindenmayer (2001). I think

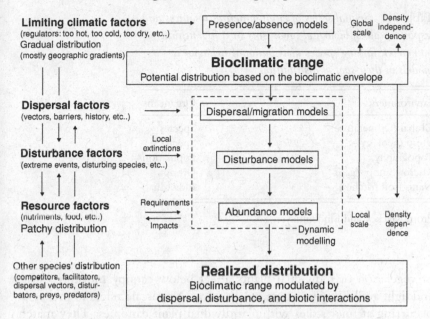

Limiting climatic factors
(regulators: too hot, too cold, too dry, etc..)
Gradual distribution
(mostly geographic gradients)

Presence/absence models

Global scale

Density independence

Bioclimatic range
Potential distribution based on the bioclimatic envelope

Dispersal factors
(vectors, barriers, history, etc..)

Dispersal/migration models

Local extinctions

Disturbance factors
(extreme events, disturbing species, etc..)

Disturbance models

Resource factors
(nutriments, food, etc..)
Patchy distribution

Requirements

Impacts

Abundance models

Local scale

Density dependence

Dynamic modelling

Other species' distribution
(competitors, facilitators, dispersal vectors, disturbators, preys, predators)

Realized distribution
Bioclimatic range modulated by
dispersal, disturbance, and biotic interactions

Fig. 3.5. A process hierarchy showing the relationship between static species distribution modeling and the dynamic population and community processes of dispersal, population dynamics, and disturbance. From Guisan and Thuiller (2005). Copyright © Wiley-Blackwell. Used with permission.

it is generally applicable to plants and animals and it directly and explicitly addresses spatial scale, which is important for SDM and lacking in the conceptual models just discussed.

Mackey and Lindenmayer built their framework on an ecological hierarchy of processes (Holling, 1992), and identified a five-level hierarchy of spatial scales (Table 3.1) defining natural breaks in the availability and distribution of the primary environmental resources (heat, light, water, and mineral nutrients, as discussed above). They define spatially nested global, topo-, meso-, micro-, and nanoscales. At the global scale, latitudinal and seasonal variation in incoming solar radiation drive atmospheric circulation and therefore regional weather and climate. At the meso-scale, the interaction of the synoptic weather patterns with the elevation of the land surface affects precipitation, temperature and radiation regimes, while underlying geology affects mineral nutrient availability. At the toposcale, local topography controls the distribution of water and radiation on the landscape via slope aspect, angle, slope length, and hillslope position. They define microscale as the scale at which patches

Table 3.1. *Spatial scales of environmental resources and response scale of biological organisms in a hierarchical framework linking species distributional behavior to models of their spatial distribution*

Environment	Organisms
Global (temperature)	Species
Meso (geology)	Species
Topography	Population
Micro (canopy/gap)	Group
Nano (soil patches)	Individual

From Mackey and Lindenmayer, 2001.

of vegetation or individual plants affect below-canopy moisture, heat, and light availability in a terrestrial setting. Nanoscale refers to processes occurring at finer scales within individual plant canopies. They match these five scales to appropriate biological levels of integration with regard to organisms – species, populations, groups, and individuals (Table 3.1). This conceptual model is similar in some ways to Fig. 3.4 but it describes an explicitly nested spatial hierarchy. The spatial occurrence of a focal taxon at each of these biological levels is referred to as its "distributional behavior."

The concept of spatially nested hierarchical orders of habitat selection is well developed in the wildlife ecology literature. Johnson (1980) described four levels of habitat selection that share some characteristics with the system just outlined, and apply, at least, to vertebrates. The first level is the species geographic range, the second is selection of the home range, the third is selection of resource patches within the home range, and the fourth is selection of food items within the patches. Scale-dependent wildlife habitat selection has been examined in SDM through varying both the grain or resolution of environmental maps (Nams *et al.*, 2006; Lopez-Lopez *et al.*, 2007) and the extent (Ciarniello *et al.*, 2007) of the study area. Animals do not respond to different scales (Hobbs, 2003), but rather the ecological patterns perceived depend on the scale at which they are investigated (Weins, 1989; Levin, 1992).

There are two components of spatial scale, grain, and extent. Grain is defined as sample resolution or the size of single observation. Extent refers to the size of the study area. In the previous paragraphs, scale referred to grain (as in Table 3.1). Spatial scale and habitat selection is discussed again in Chapter 4 with reference to collecting species data.

Pearson and Dawson (2003) presented another hierarchical framework for the scales at which environmental factors affect species distributions, and theirs specifically suggests spatial extents at which these factors operate with reference to SDM (Table 3.2). It is implicit that factors operating over larger extents typically vary slowly in space are therefore mapped at coarser grain, and vice versa. So, for example, they suggest that models based on climate variables may only be good at describing species distributions very generally at the macro-scale, and so broadly predict the potential future distribution when applied to climate change scenarios. But, these models based on macro-scale predictors do not perform well at reconstructing fine-scale distributions when those factors acting at the finer scale, such as land cover, are not included (Pearson et al., 2002). For example, because of the importance of terrain in modifying the heat and moisture balance at a site, and the role of substrate (soil) in determining gradients of mineral nutrient availability, species distribution models that include edaphic and terrain factors reconstruct species distributions better than those based on climate factors alone (e.g., Coudun et al., 2006; Coudun & Gégout, 2007).

Few studies have directly and comprehensively addressed the issue of selecting appropriate, spatially scaled and nested predictors of species habitat for SDM. However, one recent meta-analysis of 123 animal SDM studies did indeed find that those models that included environmental predictors from more than one hierarchical scale tended to yield more accurate predictions (Meyer & Thuiller, 2006). Specifically, they found that SDMs that included predictors at more than one grain size (local, landscape, and regional) were more accurate, even if the hierarchical relationship of those factors (finer nested within coarser) was not explicitly characterized in the model. They concluded that the majority of species were responding to habitat patterns at more than one grain (scale).

3.5.3 Environmental factors affecting species diversity and life form

Although the primary focus of this book is on modeling distributions of individual species, a number of related predictive mapping studies have used measures of species diversity, such as species richness (number of species in an area; for example, Pausas & Carreras, 1995; Waser et al., 2004; Carlson et al., 2007), or turnover (differences in species composition among sites; Ferrier et al., 2007), as the response variable. Other studies have used the concept of species functional type (Smith et al., 1997),

Table 3.2. *Hierarchical framework of factors affecting species distributions in space and their scale domains, expressed as extent*

Environmental driver	Global > 10 000 km	Continental 2000–10 000 km	Regional 200–2000 km	Landscape 10–200 km	Local 1–10 km	Site 10–1000 m	Micro < 10 m
			Extent range				
Climate	×	×	×				
Topography			×	×	×	×	
Land use				×	×	×	
Soil type					×	×	
Biotic interaction					×	×	×

Based on Pearson & Dawson, 2003, Fig. 5.

such as plant life form, operationally in SDM studies (Pausas & Carreras, 1995; Thuiller *et al.*, 2006; Syphard & Franklin, in press-a). Predictive mapping of species richness often uses remotely sensed variables (Chapter 5) as predictors (Kerr *et al.*, 2001; Saatchi *et al.*, 2008). Ecological theories other than niche theory are relevant to these efforts.

Austin (1999) reviewed studies that addressed the hypotheses regarding the latitudinal gradient in species richness (Pielou, 1975), namely the energy diversity hypothesis, predicting that more species are found in lower latitudes due to greater available energy. Several studies have shown that broad-scale patterns of species richness are related to potential evapotranspiration or primary productivity, both indicators of energy availability (e.g., Currie, 1991). Global scale species richness has also been related to actual evapotranspiration, and to topographic heterogeneity (see also Pausas & Austin, 2001; Pausas *et al.*, 2003). At the regional scale, species diversity may be the results of biotic processes such as competition and predation (Huston, 2002), and historical processes including dispersal and speciation. However, as reviewed by Austin (1999), several studies have shown that regional-scale patterns in species richness are correlated with the same factors that control species distributions: climate (temperature, precipitation, and their interaction) and terrain (Margules *et al.*, 1987; Austin *et al.*, 1996).

Using an ecophysiological framework, the global distribution of plant life forms (Box, 1981) has been related to bioclimatic variables and soil properties, for example, minimum temperature, growing degree days, actual evapotranspiration, and potential evapotranspiration, which vary as a function of soil depth and texture (Prentice *et al.*, 1992). This framework forms the basis for modeling the response of vegetation to climate change, and describing feedbacks from vegetation to climate (global dynamic vegetation modeling; Neilson *et al.*, 1992; Neilson & Running, 1996; Cramer *et al.*, 2001; Sitch *et al.*, 2003).

Perhaps unsurprisingly, many of the same environmental factors that control species distributions also affect patterns of species richness and plant growth form. If an SDM framework and methods are used to model biotic or ecological response variables other than species, ecological theory apart from niche theory should be used to inform the formulation of these models.

3.6 Summary

Niche theory underpins predictive models of species distribution; the environmental factors assumed to be limiting, and represent niche hyperspace for a species, as well as the form of the response function that is assumed or hypothesized, determine important choices about the data (sampling, variable selection, Chapters 4 and 5) and the selection and calibration of a model (Chapter 6–8). A hierarchical framework for modeling species distributions allows spatial scale to be explicitly considered when selecting environmental predictors for spatial prediction, and interpreting the resulting models and maps. Species distributions are determined by fine-scale as well as broad-scale environmental predictors, and SDMs that incorporate predictors at multiple scales, e.g., terrain, soil properties, or vegetation structure, as well as climate, more accurately describe species distributions.

General models of the factors affecting species distributions, as in Fig. 3.4, are useful as a starting point in any SDM study. However, it may be necessary to develop more detailed conceptual models linking species occurrence with environmental drivers on a case-by-case basis (for an example see Barrows *et al.*, 2005). However, even a well-developed conceptual model does not assure that data for the most important and proximal factors determining species distribution will be available, or in mapped form. But a conceptual model can help identify the most appropriate predictors, or their surrogates. For example, a model of topographically distributed solar radiation may be closer to a proximal driver of species distribution than are simple terrain variables such as slope and aspect (Chapter 5). Further, the expected form of the response function may be different depending on whether a measure of a direct or indirect gradient is used. Finally, the model may be more transferable in space or time if proximal predictors are used.

Part II

The data needed for modeling species distributions

This section addresses the data model – the second part of the framework presented by Austin (2002), outlined in Chapter 1 and used as an organizing principle for this book. Austin's data model encompasses theory and decisions about how the data are sampled, and measured or estimated. When making spatial prediction of species distributions, spatial aspects of both the species (Chapter 4) and environmental (Chapter 5) data are important.

Because species data (Chapter 4) are linked, in species distribution modeling, to digital maps of environmental variables (Chapter 5) used for spatial prediction, a conceptual model of geographical data will help to frame the material discussed in this section. Goodchild (1992, 1994) described two mental models of real geography relevant to species distribution mapping: the field and the entity. In the "field" view of the world, geographical variables can be categorical or continuous but they can be measured (have a value) at every location, so, mathematically, geography is a multivariate vector field. Examples are elevation (a continuous variable) and vegetation type (a categorical variable), which can be defined exhaustively for every point on the earth's surface. In the "entity" view, there are discrete geographic objects scattered in geographical space that is otherwise empty. Examples are records of species presence, or maps of trees, roads or fire perimeters.

In species distribution modeling we are frequently linking species location data (entities) with environmental maps (fields) to produce a new field-view of species occurrence, the distribution map. Often, the modeling techniques yield a predictive distribution map that depicts a continuous variable, the probability of occurrence or suitability of habitat (discussed in the next section), even if the response variable used to train the model is discrete (a record of species presence). Frequently, the resulting prediction is then made into a discrete suitable/unsuitable habitat map by imposing a threshold on that continuous prediction of probability suitability. A specific digital data format is usually most convenient for implementing predictions – the raster. It is important to think carefully about the characteristics of the species data, and about what the predictions represent; these issues will be emphasized in this section.

4 · Data for species distribution models: the biological data

4.1 Introduction – the species data model

This chapter discusses the species (or biological) data used to develop distribution models – their spatial, temporal and measurement scales and characteristics. I will summarize what is known about the effect of the spatial and temporal sampling of species occurrence on SDM performance using information gained from a growing number of studies on the topic. This will help guide species distribution modelers to select or collect appropriate data for modeling, and to understand the limitations of the SDMs they produce given certain characteristics of the training data such as spatial or environmental bias or small sample size.

4.2 Spatial prediction of species distributions: what is being predicted?

If a model of a geographical distribution is conditioned on a continuous ecological variable, such as biomass, species richness, or species abundance (for example, Meentemeyer *et al.*, 2001; Cumming *et al.*, 2000b; Thogmartin *et al.*, 2004; Bellis *et al.*, 2008), then that "dependent variable" is the attribute being predicted. The resulting prediction is in units of grams per m^2, species per km^2 or individuals per km^2, for example. However, models conditioned on observations that a species was present in a location – it was growing someplace or sighted somewhere or recorded utilizing a habitat (foraging, nesting) – are different. Even when biological surveys record some measure of abundance, they may be simplified to presence versus absence for reasons discussed below. When species abundance data are available for modeling, but the data include a lot of zeroes (locations where abundance is zero), a recommended procedure is to model in two stages (Austin & Cunningham, 1981), " . . . first modeling the association between the presence and absence of a species and the available covariates and second, modeling the relationship between abundance and the covariates, conditional on the organism being present" (Barry & Welsh, 2002, p. 179).

Revisiting the question introduced in Section 1.2 and discussed in Section 3.2, if the presence or absence of a species at a location is used to condition a model, what is being predicted? Because many of the statistical models used in SDM (Chapter 6), such as logistic regression, predict a probability of a binary condition, these predictions have been variously interpreted as:

- the probability of species presence or the potential species distribution (Scott *et al.*, 2002);
- potentially occupied habitat (Guisan & Zimmermann, 2000; Edwards *et al.*, 2006);
- habitat suitability or quality (Hirzel & Guisan, 2002; Hirzel *et al.*, 2002; Gibson *et al.*, 2004);
- the conditional probability of habitat use or ranked habitat suitability (Keating & Cherry, 2004);
- the species geographic range (Dunk *et al.*, 2004);
- a resource selection function – a value proportional to the probability of use of a resource unit (Boyce *et al.*, 2002; Manly *et al.*, 2002);
- the realized ecological niche (Peterson *et al.*, 2003), or,
- the fundamental niche (Soberón & Peterson, 2004).

I argued that spatial predictions, made by applying the model to maps of the predictors, depict a species or habitat distribution, not the niche, which exists in environmental space, but not geographical space (Chapter 3).

Although Keating and Cherry (2004) caution that both the nature of the data and the model can determine the appropriate interpretation of the prediction in habitat-selection studies, in general, there are two basic kinds of predictions. On the one hand, the SDM is used to *interpolate* or fill in the gaps in geographical information about where a species occurs (for a species whose distribution is in equilibrium with environmental conditions). On the other hand, the model is used to predict suitable conditions for species survival (for more mobile species or those that are not in equilibrium with environment, e.g., a recently introduced species), that is, to *extrapolate* areas where the species could occur based on known locations, either in the present or in some other time period (discussed by Chefaoui & Lobo, 2008). The nature of the species occurrence data determines whether SDM is suitable for interpolation versus extrapolation, and affects the interpretation of the resulting SDM and spatial prediction.

4.3 Scale concepts related to species data

Spatial and measurement scale principles related to species distribution models were discussed in an excellent paper by Huston (2002), and general discussions of scale and spatial data can be found in introductory texts on geographic information science (Clarke, 1997; Burrough & McDonnell, 1998; DeMers, 2009). Fundamental concepts are summarized here. Spatial scale has two components, grain and extent (Turner *et al.*, 2001). *Grain* or sample resolution refers to the size of single observation, for example, a 50-m radius "point count" used to survey birds, or a 0.1-ha plot used to record forest plants. *Extent* refers to the size of the study area and in SDM defines the area over which the model is used to extrapolate from data. Extent also determines the total range of environmental conditions that can be sampled. Its boundaries might be determined by practical constraints, for example, a National Park might require species distribution models for planning and management. Wouldn't it make more sense if the extent encompasses the range of the species being modeled? This may be desirable but is not always practical, especially for very wide-ranging species.

Sample density is defined as the ratio of the total sampled area to the extent. It is typically very low for species and for some environmental data sources (see Chapter 5) such as climate stations (0.01, 0.001), reasonably dense for others (elevation). Sample density can be very high for remotely sensed data (up to 1.0) and its derivatives (digital elevation models, geology, land cover) when there is more or less direct observation of every "pixel" or ground resolution element.

Measurement scale refers to the kind of variable being measured: continuous (interval or ratio, for example density, count or cover), ordinal (ordered classes such as low, medium and high density), or categorical (species). Species surveys are also characterized by temporal scale, both resolution (interval) and extent (time period, duration) that must be considered even though in species distribution modeling we frequently make the simplifying assumption that those distributions are static in time and space.

Salvadora hexalepis

Rhus integrifolia

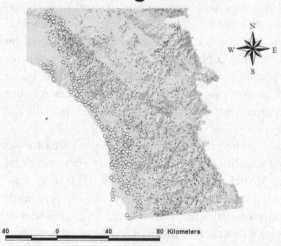

Fig. 4.1. Examples of species presence (open circle) and absence (dot) data from two surveys in the same region. Upper map: A purposive sample of clustered locations (arrays of pitfall traps), located on a gradient of human disturbance, used to survey herpetofauna (Fisher *et al.*, 2002). Map shows the occurrence of *Salvadora hexalepis* (Western patch-nosed snake) found in 12% of the sites. Lower map: an undesigned purposive survey of plant communities conducted in the 1930s in California (Kelly *et al.*, 2005). Map shows the occurrence of the shrub *Rhus integrifolia* (see Fig. 9.4) in 11% of 1481 vegetation plots in the southwest Ecoregion of California (see Fig. 5.1). Although locations were subjectively chosen, this survey is similar in appearance to a geographically stratified sample, with plots well distributed throughout the study area, because locations were deliberately located to represent contiguous map polygons.

Another measurement scale to consider is the scale of biological organization. Although species distributions are the focus of this book, the modeling framework described here is broadly applicable to biological and geographical variables that can be functionally related to other environmental maps via process models or theory. When species, other taxonomic units, communities or community properties such as species richness are the subject of interest, organisms are generally the basic units of observation (Huston, 2002, p. 11). Typically, primary data for geographically referenced species distribution models come from two kinds of sources, surveys based on probability sample designs, or undesigned purposive (Fig. 4.1) or opportunistic samples (sets of observations), especially natural history collections records (museums and herbaria). A third source, atlas data from systematic wall-to-wall surveys of a system of grids, was discussed in Chapter 2.

4.4 Spatial sampling design issues related to species data

Sample design principles are discussed thoroughly in a number of excellent texts (for example, Cochran, 1977; Thompson, 1992; Barnett, 2002). The statistical objective of sampling is estimation – sampling is the process by which units are selected from a population for measurement in order to estimate the population mean of a variable (abundance, biomass), or, more relevant to this discussion, estimate the extent (range) of a variable in space, or its value at unmeasured locations (spatial interpolation). Another objective might be to detect all species in an area (estimate species richness). Five aspects of sampling in space will be considered in the following subsections: probability sample design, sample size, species prevalence, sample resolution (grain), and survey area extent. Implications of using data from undesigned surveys, particularly presence-only data, will also be discussed. Then several important issues related to temporal aspects of species sampling are discussed in Section 4.5.

Sample size refers to the number of observations in the sample, and as sample size increases for a given area (extent), sample density increases. The objective of allocating an efficient sample is to collect an adequate minimum set of data for accurate estimation or interpolation, thus optimizing sampling effort. With respect to SDMs, *prevalence* is a term that has been widely used to refer to the frequency (proportion) of events, records of species presence, in a presence/absence sample.

4.4.1 Probability sample designs

Probability sample designs allocate sample locations in space according to data required for estimation. Random sampling, e.g., simple random sampling without replacement, can provide unbiased estimates of a population mean and variance. All possible locations (at a given grain) are treated as the sampling frame. For example, grid cells from a raster environmental map could be randomly chosen as locations for species survey. Pseudo-random number generators (PRNGs) available in widely used software, such as spreadsheet or GIS, are used for this purpose. It is important to note that a recent study has shown that pseudo-random number generators can produce spatially biased samples, and that study recommended particular PRNGs for generating spatial samples (Van Niel & Laffan, 2003).

However, the objective in species distribution modeling is usually to detect species–habitat associations, not to estimate population size. Stratification partitions a population spatially or thematically so that units within strata are similar (Bourgeron *et al.*, 2001), and random stratified sampling is often used in environmental and geographical survey (Theobald, 2007). Strata for species surveys are usually defined using those mapped environmental variables that are expected to be related to species distributions and that will also be used as predictors in the modeling. More than one variable can be combined to define classes for sampling. For example, in the Mojave Desert in California broad scale climate and geology maps and fine scale topographic maps were used to allocate a two-stage sample for vegetation survey of a large ecological region (Franklin *et al.*, 2001). The broad-scale climate and fine-scale terrain variables shown in Fig. 5.1 could be used in a similar way (see also Cawsey *et al.*, 2002).

Samples can also be geographically stratified so that all regions are sampled adequately, or geographical and environmental criteria can be combined. Predicted habitat quality from a preliminary habitat model for a target species, or a more common but similar species, has also been used successfully as a basis for stratified sampling (Engler *et al.*, 2004; Edwards *et al.*, 2005). A random-stratified sample can use *proportional* allocation of observations among strata (proportional to the area of the stratum when larger strata are expected to have greater variance). However, a proportional random-stratified sample will approach a random sample if the sample size is large and there are no very small classes. If rare strata are as "important" as common ones (e.g., if detecting all species in an area is important), then *equal* random-stratified sampling is more appropriate (equal allocation of effort to all strata without regard to size). For example, if rare strata may support rare species, equal random-stratified may provide more data for modeling the distributions of all species in a survey, yielding some minimum number of presence observations even for rare species.

A specific implementation of random stratified sampling was developed for large-scale (regional) biodiversity inventory in areas where little survey data existed. Called "gradsect sampling," it concentrated sampling in a few subareas of a study region ("gradient directed transects" or gradsects) that were both accessible and found to be representative by comparing the environmental characteristics of the gradsect to the entire area (Gillison & Brewer, 1985; Austin & Heyligers, 1989, 1991; Margules & Austin, 1994; Wessels & van Jaarsveld, 1998). This approach is based on the assumption that stratifying a sample of geographic locations across environmental gradients will locate the most species, species assemblages, and vegetation communities, and yield an adequate number of observations of each of them. In gradsect sampling, all environmental strata are sampled at least once, while larger strata are sampled proportionally. However, sampling rare entities adequately may require additional allocation of samples to rare strata, as noted above. A variation on gradsect sampling, called SR^3 (which stands for stratification, representation replication and randomization), uses map sheets instead of transects for initial geographical stratification. This reduces bias and is possible when access to all parts of the study area is possible (Cawsey *et al.*, 2002).

Is systematic, random, or random stratified sampling best for SDM? One study found that systematic survey design yielded the greatest accuracy of resulting species distributions models (Hirzel *et al.*, 2001). Another using simulated data found that random or (equal) random stratified designs yielded the best results, but their systematic design had lower sample size which could account for lower accuracy (Reese *et al.*, 2005). A rigorous comparison, that used independent data for evaluation and controlled for sample size, found that a designed sample gave consistently and substantially better species distribution predictions than a purposive sample (Edwards *et al.*, 2006). In that study the designed sample was a systematic sample with a random start, and the four species tested were moderately prevalent in each sample (15–45%).

Sample design considerations are somewhat different when sampling is carried out to support models that rely partially or completely on location or proximity as a predictor (Chapters 2 and 6). For spatial interpolation (e.g., linear spatial prediction or kriging), systematic (regularly spaced) sample designs are best because they minimize the distance from any point in the study area to a sample point. For estimating spatial dependence (the variogram or correlogram, e.g., Burrough & McDonnell, 1998; Fortin & Dale, 2005) systematic samples are often used (leading to decreasing numbers of observations at greater lagged distances), but other sample designs can be used, e.g., random (leading to decreased number of observations at short lags). Species distributions are not usually modeled strictly by spatial interpolation (but see Chapter 2 and Bolstad & Lillesand, 1992; Hershey, 2000; Fortin *et al.*, 2005). However, it has been shown in some instances, as noted above, that systematic samples also provide better data than random samples or random stratified samples for estimating species distribution models based on species–environment relationships (Hirzel & Guisan, 2002).

Adaptive cluster sampling aims to estimate unbiased parameters for rare, clustered populations (Thompson, 1992). Is this a promising sampling method for biological surveys to gather data for rare species distribution models? Adaptive cluster sampling increases the prevalence (frequency) of species presences in the sample. This results in SDMs with lower omission errors (false negative predictions), but higher commission errors (false positive predictions; see Chapter 9 for a thorough discussion of SDM validation). Adaptive cluster sample-based SDMs are also very sensitive to sample size – increasing the overall sample size really brings the commission errors down (Reese *et al.*, 2005). What really seemed to improve SDM predictive accuracy in that particular study was having a sample where the species prevalence is about 50% (see Section 4.4.3).

4.4.2 How many observations?

What is an adequate or optimum sample size and sample density for species distribution modeling? Is more data always better? Sample size has been found to be positively related to performance of species distribution models (Cumming, 2000a; Hirzel & Guisan, 2002; Reese *et al.*, 2005). However, some studies have found that only 50–100 observations of species presence were required to achieve acceptable performance (Stockwell & Peterson, 2002; Kadmon *et al.*, 2003; Loiselle *et al.*, 2008). Some methods seem to require even fewer observations, as few as 30, to produce acceptable results (Elith *et al.*, 2006; Kremen *et al.*, 2008; Wisz *et al.*, 2008b). These particular comparative studies examining the minimum sample size required for SDM focused on the types of models that require "presence-only" data (Chapter 8). Other studies that examined both sample size and prevalence got decent predictions with sample sizes of 100–500. This is of practical importance because, although some datasets used in SDM include thousands of observations, often the number of presence observations for any single species is much smaller, and this is discussed in the next section (Section 4.4.3). Coudon and Gégout (2007) carried out a modeling experiment using artificial data and concluded that, as a rule of thumb, at least 50 observations of species occurrence are needed to accurately estimate species response functions with logistic regression.

SDMs are often developed using a multiple regression statistical framework (Chapter 6), or something conceptually similar to that, and there are formal methods, power analysis, as well as heuristics ("rules of thumb"), for determining an appropriate sample size for multiple regression models. This topic is covered in many statistics textbooks (for example, Quinn & Keough, 2002). One rule of thumb is that there must be 20 times as many observations as predictor variables, or 40 times in the case of stepwise variable selection in regression. Of course, larger samples should be used when the effect size is small (which it could very well be in the case of SDM), if there is measurement error in the predictor variables (again, likely in SDM) and if training data are being cross-validated to test data – as is frequently done in SDM! A statistical power calculation, for example, yields a sample size, N, of 76 if alpha is set to 0.05, desired statistical power is (by convention) 0.8, effect size is moderate (0.15) and the number of predictors is 3. However, for comparison, if the effects size were small (0.02) the N should be 542. If the number of predictors were 10 instead of 3, N should be 118, and if predictors were 10 and effects size were small, N should be 818.

The absolute number of observations may be less important than having observations of a species that are well-distributed throughout the environmental space that it occupies (Fig. 3.1). This is the definition of representativeness and is the objective of stratified sampling. Kadmon *et al.* (2003) referred to how well the species occurrence data covers the environmental space that a species occupies as "completeness," and this is discussed further in Section 4.4.5.

4.4.3 Species prevalence

What constitutes an adequate overall sample size depends on the rarity of the species. A large random sample of survey sites where the species is mostly absent may not give enough observations of presence. This occurs when species prevalence in the sample (the frequency or proportion of locations or observations in which species is present) is allowed to vary with a species "natural" prevalence in the surveyed area – the proportion of sites it occupies, reflecting species actual commonness or rarity (Manel *et al.*, 2001). In trying to distinguish habitat preference based on presence and absence data, what is the optimal ratio of presence to absence (prevalence or frequency) in the sample? Some have suggested 1:1 (Fielding & Bell, 1997; Reese *et al.*, 2005).

A recent study showed that a 1:1 ratio of presence to absence observations yields an optimal balance between omission and commission errors in model predictions, and they concluded that perceived relationships between range size and model accuracy might simply be an artifact of species prevalence in the sample (McPherson *et al.*, 2004). However, it has been found in other studies that model accuracy increases with species rarity (Segurado & Araújo, 2004), independent of manipulated sample prevalence (Franklin *et al.*, 2009), suggesting that rarer species may be niche specialists with more predictable patterns of occurrence relative to environmental gradients (see Chapter 10).

In contrast with the effect of a species natural prevalence on presence/ absence modeling, the effect of prevalence on presence-only modeling methods (Chapter 8) may be different. In the case of presence-only data, the sample prevalence is under the control of the modeler and is a function of the number of "available" or background locations that are designated. When generating a random sample of background locations to complement presence-only species data (Section 4.4.6), a very large number of background sites may be required to fully describe variation in the environment – several orders of magnitude larger than the number of presences (Manly *et al.*, 2002; Phillips & Dudík, 2008).

4.4.4 Sample resolution

Species are commonly surveyed over areas (quadrats, transects, traps, point counts, specimen collection locations, radio tracking locations, Breeding Bird Survey routes) that are small relative to the extent of the survey area and the predictive distribution map (Fig. 4.1). They are typically treated as points, both conceptually (recall the entity concept of geographical objects) and in practice (in a geographic information system the survey locations are recorded as x, y locations, without area). However, it may be more appropriate (true to the data) to treat them as areas when their size is greater than the resolution of environmental maps

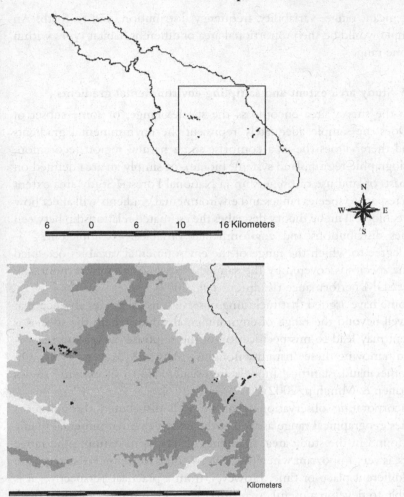

Fig. 4.2. Example of species observation recorded in areas of variable sizes. Stands of the endemic Island Oak (*Quercus tomentosa*) on Catalina Island, Channel Islands, southern California (data courtesy of the Catalina Island Conservancy), show as black polygons, with reference to the extent of the island (upper map), and inset enlarges these stands and shows with a raster map of elevation (light gray = 0–212 m; medium gray = 212–424 m; dark gray = 424–636 m). See also Fig. 9.5.

used for modeling. Some examples are large home ranges mapped from radio or satellite telemetry (mountain lions, porpoises), floras (lists of plant species in relatively large but well-delimited locations such as an island or national park), or variable-sized polygons delimiting stands of a rare plant (Fig. 4.2). Then, a number of observations from the environmental maps might be necessary to characterize the species observation using their mean, range, variability, frequency distribution, and so forth. An example would be the proportional area of different habitat types within a home range.

4.4.5 Study area extent and sampling environmental gradients

Does the survey area encompass the species range, or some subset of it? Does the sample adequately represent the environmental gradients found there? Does the area comprise some natural region (ecoregion, physiographic region, land system, biome), or simply an area defined on the basis of land use or ownership (a National Forest)? Study area extent with respect to species ranges and environmental gradients will affect how well ecological niche theory describes the estimated relationship between species distributions and environmental variables. "Completeness" – the degree to which the range of the environmental variables occupied by the species is covered by the sample – has been shown to positively affect SDM performance (Kadmon et al., 2003).

Some have argued that including observations of species absence that are well beyond the range of environmental values where the species is present may lead to misspecification of the response curve, and recommend removing these "naughty noughts" (Austin & Meyers, 1996). On the other hand, statistical models are usually able to fit extreme zeroes (Oksanen & Minchin, 2002).

Incorporating observations that are well distributed throughout a species geographical range as well as the extent of environmental gradients found in the study area, including observations defining the range limits, is very important when the SDM is going to be used to extrapolate to a different place or time. However, from a practical perspective, it is possible to develop a useful correlative SDM even if it is calibrated from only a portion of the species range if the main purpose is interpolation (predicting habitat suitability of unsurveyed locations within the study area). This type of model and its projections might be needed for impact assessment or conservation planning. In this case, however, the form of the response curve would not be expected to match what would be predicted from niche theory.

4.4.6 Using existing data for modeling

As a generalization, designed biological survey data that have been carried out recently tend to be of known and uniform sample area, accurately

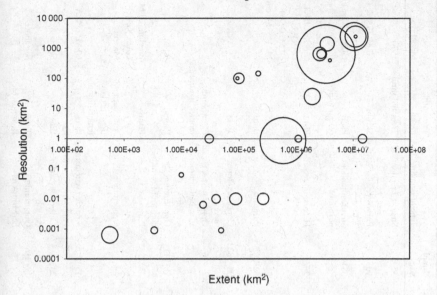

Fig. 4.3. The extent and resolution of the studies listed in Table 4.1, shown on a log–log scale, with the size of the circle proportional to the number of species modeled in the study, showing ecological-scale, high–resolution studies (lower left) versus very-large area, low resolution, biogeographical-scale studies (upper right), with some studies beginning to bridge the scale gap.

located (using GPS), based on probability sample designs, and provide information on species presence, absence and even detectability (see Section 4.5). On the other hand, specimen collection locations, historical surveys, and other opportunistic sources of species observations (e.g., fisheries data) tend to have lower, more variable, or unknown location accuracy, to be based on undesigned, opportunistic, or purposive observations, and to come from multiple sources of varying or unknown integrity. Further, they often provide information only on species presence. Spatial errors in species location data degrade the performance of SDMs, although some modeling methods are more robust than others are to this source of noise (Graham *et al.*, 2008).

Species distribution modeling studies fall roughly into two broad groups based on these two types of data (Table 4.1; Fig. 4.3). I will refer to these as biogeographical versus ecological scales. In ecological-scale SDM studies using data from surveys that are based on designed probability samples, typically the resolution of the species observations

Table 4.1. *Twenty-five species distribution modeling studies, arranged by increasing spatial resolution (grain) of study*

Study	Emphasis	Location	Extent (km²)	Resolution (km²)	N species	Sample size	Type of sampling	Taxonomic group	Model type
(Drake et al., 2006)	Test modeling method	Switzerland	564	0.000625	106	550	Stratified random	Trees and shrubs	SVM
(Franklin, 1998)	Modeling methods comparison	Southern California, USA	3383	0.0009	20	906	Purposive	Shrubs	GLM, GAM, CT
(Miller & Franklin, 2006)	Include spatial dependence in SDMs	Mojave desert, California, USA	50370	0.0009	11	3819	Stratified random, purposive	Plant communities	GLM, CT, spatial (Kriging, simulation)
(Elith et al., 2006)	Best methods for p-only data	Australia wet tropics	24000	0.0064	20	9–74	Herbarium records	Plants	Bioclim, BRT, GARP, GAM, GLM, GDM, MARS, Maxent
(Elith et al., 2006)	Best methods for p-only data	Wet tropics, Australia	24000	0.0064	20	32–265	Purposive	Birds	"
(Elith et al., 2006)	Best methods for p-only data	Switzerland	41000	0.01	30	36–5822	Unplanned forest survey	Trees	"
(Elith et al., 2006)	Best methods for p-only data	New South Wales, Australia	89000	0.01	54	2–426	Opportunistic, faunal atlas, herbarium records	Plants, birds, reptiles, mammals	"
(Elith et al., 2006)	Best methods for p-only data	New Zealand	265000	0.01	52	18–211	Opportunistic, herbarium records	Plants	"
(Elith & Burgman, 2002)	Accuracy vs extent, validation methods	Victoria, Australia	10000	0.0625	8	3522	Purposive	Plants	Bioclim, GLM, GAM, GARP
(Kremen et al., 2008)	Conservation planning	Madagascar	587040	0.86	829	8+	Opportunistic	Ants, butterflies, frogs, geckos, lemurs and plants	Maxent
(Brotons et al., 2004)	prevalence, PA versus p-only	Catalan, Spain	31000	1	30	1500	Stratified systematic, Purposive	Birds	GLM, ENFA
(Elith et al., 2006)	Best methods for p-only data	Ontario, Canada	1 088 000	1	20	255	Collections and nest records	Birds	See above
(Elith et al., 2006)	Best methods for p-only data	South America	14 653 000	1	30	17–216	Opportunistic, herbarium records	Trees, Bignoniaceae	See above

(Stockwell & Peterson, 2002)	Effect of sample of size and species range on performance	Mexico	1 923 040	25	103	Collection records, atlas	Birds	GARP, GLM
(Thuiller et al., 2003)	Effect of extent and resolution on model performance	Portugal	93 300	100	4	Database project	Trees	GLM, GAM, CT
(Segurado & Araújo, 2004)	Prevalence vs. type model	Portugal	99 300	100	44	Occurrence records, atlas	Reptiles and amphibians	Gower, ENFA, CT, ANN, GLM, GAM, Spatial interpolation non-linear DA
(McPherson et al., 2006)	Downscaling SDMs	Uganda	215 784	144	9	Atlas, point observations, opportunistic	Birds	non-linear DA
(Prasad et al., 2006)	Model comparison	Eastern USA	3 912 800	400	4	Forest Inventory Analysis	Trees	RT, BT, RF, MARS
(Cumming, 2000b)	Map regional diversity	Southern Africa	2 604 960	648	74	Collection records	Ticks	GLM
(McPherson et al., 2004)	Effect of sample size vs. prevalence model performance	Southern Africa	2 770 200	648	32	Atlas	Birds	GLM, DA
(McPherson & Jetz, 2007)	Effect of species traits on model performance	Southern Africa	33 47 568	648	1329	Atlas	Birds	GLM, AR
(Iverson & Prasad, 1998)	Effect of climate change on abundance	Eastern USA	3 500 000	1400	80	Forest Inventory Analysis	Trees	RT
(Huntley et al., 2004)	Modeling from climate data	Europe	10 390 000	2500	306	Mean prevalence 0.00169	Plants, insects, birds	Locally weighted regression
(Thuiller et al., 2003)	Effect of extent and resolution on model performance	Europe	11 047 500	2500	4	Atlas	Trees	GLM, GAM, CT
(Araújo & Williams, 2000)	Model comparison, conservation planning	Europe	11 047 500	2500	174	Atlas	Trees	GLM, spatial interpolation

This is by no means a complete, or even a representative sample of the hundreds of recent SDM studies. Rather, these were selected to illustrate the broad range of extents, resolutions (Fig. 4.3), data sources, number of species, taxonomic groups modeled, model types (see Chapters 6–8), and purposes of these studies.

(plots, transects, point counts) are on the order of 100–2500 m², approximately 1/10 to 1/25 the size of the prediction resolution (30- to 250-m pixels or 900–62 500 m²), so it is unusual for more than one species observation to occur in a grid cell (no aggregation takes place). These data may have been collected specifically for the purpose of SDM or for a related purpose such as inventory and monitoring of vegetation or fauna.

At the biogeographical scale, very large pre-existing species datasets such as 34 000 museum collection records for ticks in Africa (Cumming, 2000a, 2000b, 2002), 300 000 collection records for birds in Mexico (Stockwell & Peterson, 2002), or 100 000 forest inventory plots (Iverson & Prasad, 1998), covering very large extents (subcontinental to global), are typically aggregated to a coarse grain. For example, if observations such as natural history collection records are aggregated to 10 × 10 km (the species is considered present if any record of it occurs in a cell), to 1/16 square degree (about 25 × 25 km), and then to 1 degree of latitude and longitude, the total number of observations decreases as each cell becomes an observation (see Fig. 2.2). The area represented by the original observation is generally about three orders of magnitude smaller than the aggregation scale. This spatial aggregation has the effect of increasing the reliability of "presence" observations. However, these coarse grid cells are environmentally heterogeneous, and environmental factors that are associated with species observations at this resolution may not correspond to the environment within the grid cell that the species experiences.

Aggregation also has the effect of generating absence data where none were explicitly collected (e.g., collection records), because the larger a cell containing no observation of presence for a species (assuming good coverage of observations in space, even if the locations are biased), the more likely it represents an area where that species is truly absent. Some studies have addressed this directly by only including cells as representing the "absence" of Species A if there are a minimum number of total collections of any species there. This would imply that the area has been well-searched and the lack of any record of Species A means real absence rather than absence of evidence (McPherson *et al.*, 2004). If this is the case, however, one could argue that a coarse-scale map of the species distribution has now been compiled, so there is no need for a model for spatial prediction. In fact, as noted in Chapter 2, sometimes models at this scale are generated using species distribution data from existing species atlases. However, an inferential model may still be useful to understand species–environment relationships at coarse scale (to test biogeographical hypotheses), to improve distribution maps in areas that were less-well surveyed, or to predict temporal change, e.g., under climate change or other large-scale environmental change (land use).

Some recent studies attempt to develop species distribution information for large extents at relatively high resolution (1 km^2), but these still generally depend on collections-based or opportunistic "presence-only" observations (Table 4.1, Fig. 4.3). Instead of relying on spatial aggregation to generate absence information, they use modeling techniques that require only presence data.

Opportunistic data collected without a probability sampling design may be spatially biased. A biased survey design (e.g., a sample collected along roads) or purposive sample (the species is searched for based on expert knowledge, an approach often used for rare or cryptic species) does result in SDMs that perform more poorly (Kadmon *et al.*, 2004; Reese *et al.*, 2005; Edwards *et al.*, 2006). However, existing data can be evaluated for bias – how much it deviates from random, or how adequately it samples defined strata – and then supplemented with additional survey locations based on an unbiased design (Nelder *et al.*, 1995). Further, a recent study showed that a biased sample of environmental gradients produced acceptable SDMs, and small sample size degraded model performance more than sample bias (Loiselle *et al.*, 2008).

4.4.7 Species presence-only data

Several of the data sources discussed consist only of records of species presence, or were collected using a method that only documents species presence (radio telemetry), rather than distinguishing presence and absence. Much of the data used in large-scale SDM efforts (those involving lots of species and large geographical extents) can be characterized as "presence-only" data. Specialized modeling methods that associate species presence data with environmental variables can be applied to presence-only data (Chapter 8). Some of these methods require a sample of observations that characterize the available environment, describing the range of conditions found in the study area (Phillips *et al.*, 2009). A random sample, random stratified sample, all non-presence locations, or all locations (grid cells) can be used to represent the available environment, referred to as the background. This sample of background locations can also be used in any of the model types suitable for binary response variables (Chapters 6, 7) to discriminate used from available habitat (Boyce *et al.*, 2002; Meyer

& Thuiller, 2006). These background locations are often referred to as "pseudo-absences" in the literature because it is not actually known if the species does or could occur there. However, the terms background or available sites more accurately describe the outcome of SDM based on presence-only data – a model distinguishing occupied versus available habitat.

When generating background data to use with presence-only modeling methods, the background points could hypothetically come from anywhere within the study extent. Studies have shown that survey designs other than random, e.g., environmentally weighted, may be better for selecting these background sites (Zaniewski *et al.*, 2002). Excluding locations where a species was documented to have historically existed from candidate background locations has also been shown to improve SDM performance (Lutolf *et al.*, 2006).

It has been suggested that including background observations from beyond the species range, outside the limits of its hypothesized distribution, or those that are environmentally dissimilar from the presence locations, yields more accurate or plausible SDMs (Chefaoui & Lobo, 2008; Le Maitre *et al.*, 2008), especially when the purpose of modeling is to project species invasion or range changes under climate change scenarios. This strategy is also said to better estimates response curves (including the tails) (Thuiller *et al.*, 2004b). A preliminary SDM can be developed based on a presence-only method, and then background points can be selected from only those areas falling below some habitat suitability threshold predicted from the preliminary model. However, recent studies have strongly cautioned that background data should be located in areas that could reasonably have been sampled for species presence/absence, not in unlikely areas (Elith & Leathwick, 2009; Phillips *et al.*, 2009). Selecting pseudo-absence points that are environmentally dissimilar from presences results in models that predict a greater areal extent to be suitable, in other words, "over-prediction" (Chefaoui & Lobo, 2008). This is discussed further in Chapter 8.

Generalizing from these findings, a random or random stratified sample of background locations results in a model that reflects resource selection or extent of occupancy more than species range, and in a predicted distribution map that is more geographically restricted. Using some means to exclude locations from the absence sample that are environmentally similar to the presences results in a more spatially extensive prediction of suitable habitat, that may, in fact, represent "over-prediction."

4.5 Temporal sampling issues and species data

4.5.1 Species detectability

The frequency, timing, and duration of species surveys depend on the main objectives of the survey. In species distribution models the "response variable" is some measure of habitat use, species occurrence, or abundance (see Section 4.3), and the objective is to model a relationship between this and habitat variables. However, species distribution modeling often makes use of data collected for other purposes. Often the primary aim of biological surveys is population monitoring where the variable of interest is a population estimate, or "proportion of area occupied" by a species (e.g., Bailey et al., 2004). A major issue in designing surveys to monitor wildlife populations is "detection probability" and the problem of "false negatives" – that is, distinguishing unsuitable from unoccupied habitat (MacKenzie et al., 2002, 2005). The more mobile or cryptic the species, the greater is the detection problem. Even "static" species distribution models need to take into account species movement or detectability (Tyre et al., 2003; Gu & Swihart, 2004; Wintle et al., 2004). This issue has mainly been considered for animals, but some cryptic plants, plants in certain life stages (a soil seed bank), or other types or organisms, could have detection probabilities <1 (Dreisbach et al., 2002; Edwards et al., 2005; Wollan et al., 2008). Generally, estimating detection probabilities can improve SDMs. Conducting some minimum number of repeat site visits (three, for example) to confirm species presence, will eliminate bias in estimates of habitat effects. Six repeat visits will improve precision of those estimates; when false-negative error rates are <50% it is better to add sites, and when they are >50% it is better to add repeat visits (Tyre et al., 2003).

4.5.2 Historical species data

What if species distribution models are developed using existing data that were not all collected at the same time, but over a number of years, decades or even centuries, as is the case with museum collection specimen records? Typically these observations have become more spatially accurate (with the use of GPS) in recent years. Observations also may have gotten denser over time, but not always (some taxa may have been better collected during some early period of exploration). It may be necessary to delete older records if they contain imprecisely located observations, and reflect historic rather than present distributions, which, while it might be of great interest, may create problems in modeling. Many species distribution modelers have ignored "short-term" (twentieth century) fluctuations in species distributions when using natural history collections for SDM, or have simply excluded older records. However, in recognition of the effect of anthropogenic global change on species distributions over that very time period (Root *et al.*, 2003), some studies have used the interesting approach of matching species occurrence data to interpolated climate data and other maps (forest cover) from the same general period, e.g., 1930–1960 versus 1970–2000 (Kremen *et al.*, 2008).

4.6 Summary

What can we glean from this research on species data characteristics for SDM? What are the best survey design characteristics for habitat models? The spatial distribution of observations, sample size, scale of the observations, extent of the study area, evenness of the sample (species prevalence), species detectability, and temporal resolution of the species observations all need to be considered when collecting or using existing species distribution data for SDM.

Probability sample design in space is the sampling issue that has most often been addressed in comparative studies. A designed sample results in more accurate SDMs than a purposive or opportunistic sample. Systematic, random, or stratified survey designs are all likely to yield a proportional sample of habitats and species if neither are strongly patterned (that is, with periodicity matching the spacing of systematic sample), clustered or extremely rare. Environmentally-stratified sampling is often used when many species are being surveyed simultaneously in large areas without previous survey data. These surveys have the advantage of recording species co-occurrence in the same site, and also recording species absence. Adaptive cluster sampling is a useful approach to estimating parameters of rare populations such as density, but may also be useful for sampling rare species for distribution models. However, it is important, if implementing adaptive cluster sampling, to have an adequate sample of species absence also. Larger samples improve models and may overcome problems of noisy and biased data, but the balance between number of presences and absences in the sample is also important. Existing data can be examined for spatial bias, and supplemented with an additional unbiased or representative sample.

The size of surveyed area (plot, transect) is usually based on existing survey methods and has not been extensively evaluated with respect to habitat modeling (but see Reese *et al.*, 2005). Study area extent also affects the model that is estimated and the resulting predictions. If the extent is less than the species range, the species response functions may not have the form predicted by niche theory. However, a predictive model can still be estimated and may be useful for environmental impact assessment. Species detectability does affect habitat model estimation. Detectability should be estimated during biological surveys and an adequate number of repeat site visits will assure that modeling data are suitable. Often, estimating detectability is part of standard survey and monitoring protocols for wildlife (but not usually for plants).

Tremendous efforts are under way to archive, digitally, all kinds of species location information, especially from natural history collections and opportunistic observations, for many parts of the world. These data may have imprecise or inaccurate location information, they do not represent a probability designed sample and they only include presence information. However, their reliability can be improved through spatial aggregation, and then SDMs can be developed using moderate-resolution predictors (Chapter 5) and a growing number of methods optimized to perform well with only presence observations (Chapter 8). Observations generally exist throughout the species range and this may be an appropriate scale for asking biogeographical questions and studying global environmental change. However, important environmental determinants of species distributions that are heterogeneous at the resolution of these models cannot be characterized.

Biological surveys or natural resources inventories of impressive scope are carried out for a number of reasons and can yield high-quality data for SDM. These surveys have several advantages for SDM. They usually use a probability designed sample, they generate information on multiple species over a limited and well-defined area, and therefore include spatially explicit presence and absence information, and their precise locations are recorded. Therefore, relatively fine scale environmental maps can be used to develop SDMs. Sometimes, they are repeated regularly and provide information on different time periods, for example Breeding Bird Survey data.

Still, one of the greatest challenges is the lack of sufficient numbers of observations for SDM of many species on Earth. In an interesting recent example, data were painstakingly compiled for 2315 species across several taxonomic groups in Madagascar; over 60% of them had eight or fewer observations, and SDMs could not be developed (Kremen *et al.*, 2008).

5 · *Data for species distribution models: the environmental data*

5.1 Introduction

In general, the suitability of spatial (GIS) environmental data sources for analytical use in ecological modeling (Hunsaker *et al.*, 1993) has not been given as much attention as other aspect of the SDM problem, such as the species data (Chapter 4) and modeling methods (Chapters 6–8). A few studies have explicitly examined issues of spatial resolution and data quality of predictors on SDMs (Aspinall & Pearson, 1996), have shown that including satellite-derived climate variables with land cover as predictors improves SDMs (Suarez-Seoane *et al.*, 2004), and have illustrated the impact of predictors derived from digital elevation models on SDMs via error propagation (Van Niel & Austin, 2007). Daly (2006) provided guidelines for assessing the quality of interpolated climate maps used in environmental modeling. However, these efforts, while important, do not give comprehensive guidelines for selecting data to represent environmental gradients related to species distributions. Some software systems developed for SDM include global environmental datasets at 0.01– (mostly 0.5-) 1.0 degree resolution (Hijmans *et al.*, 2001; Stockwell, 2006), but this is the exception rather than the rule. Usually it is the job of the modeler to select and assemble appropriate predictor data, and consider issues of data quality and resolution; the effort involved is non-trivial.

Chapter 3 outlined conceptual models of factors driving species distributions, and in this chapter, I will describe the types of environmental data typically used in SDM to represent those factors, including climate, topography, substrate, land cover and vegetation, disturbance maps, remote sensing-based land surface characterizations, measures of landscape pattern, and information about other species (biotic interactions). I will discuss the assumptions and limitations of using these data, and in particular I will examine how directly the mapped factors are linked to causal factors, and consider issues of spatial scale.

Spatial prediction in SDM relies on the availability of environmental predictors in the form of maps or, more precisely, digital spatial data (Goodchild, 1996). As discussed in Chapter 1, environmental maps must be readily available and correlated with species distributions in order for SDM to be a practical endeavor – otherwise some form of spatial interpolation would be the only way to develop species distribution maps from point observations. There is a widely held believe that "GIS data" (digital environmental maps) are abundant and ubiquitous (Franklin, 2001). It is true that, for a growing number of places in the world, there are many geospatial datasets available, often free or nearly free from public agencies (developed at public expense) and private sources, much more so than when Estes and Mooneyhan (1994) wrote about the "mythical map." It is also true that the number and variety of remotely sensed data products is growing (Section 5.3.4). The challenge for the modeler is to identify environmental maps that represent resource gradients or other factors determining species distributions at an appropriate scale.

Further, while it is often assumed that environmental variables measured *in situ*, that is, in the field, concurrent with species observations, will more accurately characterize species distributions than GIS surrogates, this has rarely been tested (Thomas *et al.*, 2002) and is not necessarily the case. For example, a recent study showed that SDMs for wildlife species were slightly more accurate when based on GIS-derived predictors (land cover, soil, topography) than on field-measured environmental covariates (Newton-Cross *et al.*, 2007).

5.2 Spatial data representing primary environmental regimes

5.2.1 Climate maps

As discussed in Chapter 3, although the primacy of climate in controlling the thermal, moisture and light regimes and in determining species range limits at larger spatial scales is widely acknowledged, theoretical studies do not specifically address *which* environmental parameters represent these primary environmental regimes.

However, in an early example of SDM, mean annual temperature and rainfall, as well as a radiation index, were related to tree species distribution (Austin *et al.*, 1984). A group of scientists in Australia defined a comprehensive set of climatic parameters that they expected to be relevant to species distributions at broad spatial scales (Busby, 1986;

Wiegland *et al.*, 1986; Busby, 1991). The modeling procedure they used was called bioclimatic ("BIOCLIM") or environmental envelop modeling, which referred to both the driving gradients (bioclimatic variables) and the method ("envelope" – a boxcar or parallelepiped classifier – described later in Chapter 8). BIOCLIM was also the name given to the software and data system they implemented. They used long-term (30-year) monthly averages of precipitation, maximum daily and minimum daily temperature (36 variables) to derive 16 climate parameter indicative of average, seasonal and extreme conditions (Table 5.1). Subsequently, several other researchers used similar bioclimatic variables, or a subset of them, for SDM (Walker & Moore, 1988; Walker, 1990; Walker & Cocks, 1991; Mackey, 1993a). Today, virtually all SDM studies that encompass regional to global scales use climate parameters (for examples see Table 10.2). Species modeling at smaller spatial extents often uses topographic variables (next section) as surrogates for radiation, temperature and moisture gradients.

Integral to the BIOCLIM system were the methods developed and used to create climate maps by spatial interpolation of climate station data (Hutchinson, 1987; Hutchinson & Gessler, 1994; Hutchinson, 1996). Currently, maps of basic climate parameters (usually monthly precipitation, and maximum and minimum temperature), based on interpolated climate station data (e.g., Daly *et al.*, 1994; Daly & Taylor, 1996; Hijmans *et al.*, 2005; Daly *et al.*, 2008), including retrospective data (Di Luzio *et al.*, 2008), are available for many parts of the world. From them these bioclimatic variables (Table 5.1) could be derived. Generally, seasonal means and extremes of precipitation and temperature are more strongly related to species distributions, theoretically and empirically, than annual averages (e.g., Stockwell, 2006), owing to interactions of temperature and moisture availability. Figure 5.1 shows an example of a map of mean minimum temperature of the coolest month, along with annual average precipitation.

Other variables measured at climate stations can be interpolated and used in modeling, for example incoming solar radiation, relative humidity or cloudiness. However, there are fewer examples of this. Mackey (1993a) used various indicators of the radiation regime (annual, monthly and seasonal minima and maxima) to predict rain forest plant functional groups. Annual and seasonal measures of radiation, interpolated from climate station data, have also been used as predictors in other SDM studies (Leathwick, 1998; Elith *et al.*, 2006). Methods have been proposed for interpolating photosynthetically active radiation (PAR) from measurements of shortwave radiation that also account for terrain effects

Table 5.1. *Bioclimatic parameters used in a selection of species distribution modeling studies*

Bioclimatic variables	AWT	CAN	NSW	NZ	SA	SWI	(Phillips et al., 2006)	(Leathwick, 1998)	(Iverson et al., 2008)	(Luoto et al., 2005)	(Guisan et al., 2007)	(Hernández et al., 2006)	(Loiselle et al., 2008)	(Skov & Svenning, 2004)	(Araújo & Williams, 2000)	(Araújo et al., 2005b)	(Brotons et al., 2004)	(Thuiller et al., 2003)	(Vayssières et al., 2000)
Temperature																			
*Annual mean temperature**	×	×	×	×	×	×	×	×	×				×		×	×			×
Mean annual maximum temperature																		×	
Mean annual minimum temperature			×																
*Minimum temperature coolest month** (A = average temp. coolest month)	×				×	×	×		×	A	A				A	×		×, A	×, A, max
Absolute minimum temperature																			
Departure minimum temperature of the coolest month from annual								×						×					
*Maximum temperature warmest month** (A = average temp. warmest month)	×	×		×	×		×		×		A	×	×		A	×		×, A	×, A, min
*Annual temperature range** (Seasonality, CV, S.D.)	×						×												×
Mean difference temperature of coolest and warmest month									×				×						
Mean diurnal temperature range					×														

(cont.)

Table 5.1. (cont.)

Bioclimatic variables	AWT	CAN	NSW	NZ	SA	SWI	(Phillips et al., 2006)	(Leathwick, 1998)	(Iverson et al., 2008)	(Luoto et al., 2005)	(Guisan et al., 2007)	(Hernández et al., 2006)	(Loiselle et al., 2008)	(Skov & Svenning, 2004)	(Araújo & Williams, 2000)	(Araújo et al., 2005b)	(Brotons et al., 2004)	(Thuiller et al., 2003)	(Vayssières et al., 2000)
Isothermality (mean diurnal range / annual temperature range)	X											X	X						
Mean temperature coolest quarter*																			
Mean temperature warmest quarter*					X												X		
Mean temperature wettest quarter*																			
Mean temperature of driest quarter*																			
Mean temperature April									X										
Average number of days > 28 °F		X																	
Growing degree days (GDD) > 0 °C														X					X
Growing degree days (GDD) > 5 °C																X			X
Annual temperature sum > 5 °C										X	X								
Mean annual frost frequency						X	X				X								
Average number of days of summer frost																			
Precipitation																			
Annual mean precipitation*	X	X	X	X	X	X	X		X	X	X	X	X			X		X	X
Precipitation of wettest month*																			

*Precipitation of driest month**	x								
*Coefficient of variation of monthly precipitation * (Seasonality)*	x	x							
*Precipitation of wettest quarter**	x								
*Precipitation of driest quarter** (M = driest month)	x	x			M				
*Precipitation of coolest quarter**	x								
*Precipitation of warmest quarter**	x	x	x		x				
Winter precipitation			x				x	x	
Spring precipitation			x				x	x	
Summer precipitation			x					x	x
Fall precipitation			x					x	x
Mean number days precipitation > 1 mm			x					x	
Solar radiation		x							
Annual mean solar radiation	x	x	x	x	x				
Departure of winter solar radiation from annual			x	x					
Solar radiation seasonality (annual range)	x		x						
Annual cloud cover		x							
Available moisture									
Moisture Index seasonality	x								
Moisture Index lowest quarter	x								
Coefficient of Variation monthly relative humidity			x						
Mean annual vapor pressure deficit (O = October VPD)	x, O			x					
Mean monthly ratio precipitation to potential evapotranspiration (PET) (A = annual)	x				A				

(cont.)

Table 5.1. (cont.)

Bioclimatic variables	AWT	CAN	NSW	NZ	SA	SWI	(Phillips et al., 2006)	(Leathwick, 1998)	(Iverson et al., 2008)	(Luoto et al., 2005)	(Guisan et al., 2007)	(Hernández et al., 2006)	(Loiselle et al., 2008)	(Skov & Svenning, 2004)	(Araújo & Williams, 2000)	(Araújo et al., 2005b)	(Brotons et al., 2004)	(Thuiller et al., 2003)	(Vayssières et al., 2000)
Annual water deficit (precipitation, temperature, radiation)								×											
Water Balance (sum monthly difference precipitation and PET)														×					
Site water balance (precipitation, temperature, soil)																			
Mean annual ratio actual to potential ET (AET/PET)																×			
Mean annual PET											×								
Mean annual AET																			×
Mean PET coolest month															×			×	×
Mean PET warmest month															×			×	
Minimum mean monthly humidity								×											

The first five columns (AWT, CAN, NSW, NZ, SA and SWI) refer to the five data sets in Elith et al. (2006). Asterisk (*) and italics show the 16 original variables used in BIOCLIM (Busby, 1991).

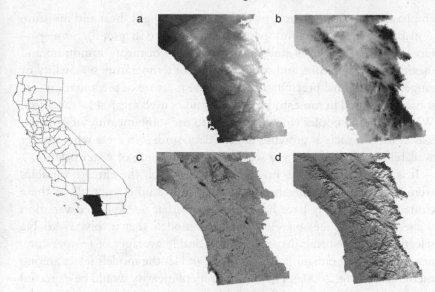

Fig. 5.1. For a 16 079-km² portion of the southwestern ecoregion of coastal southern California, in the southwestern United States, broad scale (1 km² resolution) climate variables are shown: (a) temperature (average January minimum) and (b) precipitation (average annual). Fine scale (900 m² resolution) secondary terrain attributes are also shown: (c) the topographic wetness index, and (d) estimated potential winter insolation (solar radiation at the surface), influenced by slope aspect. In all maps darker colors indicate higher values. No scale bars are given so these maps represent relative values.

(Wang *et al.*, 2006), and PAR is also estimated from satellite data (Section 5.3.3). However, topographically-distributed shortwave radiation (Section 5.2.2) is also frequently used in SDM.

Surprisingly, although predicting the response of species distributions to future anthropogenic climate change has become a major application of species distribution models (Chapter 1), even recent reviews of this topic have emphasized the modeling methods (Heikkinen *et al.*, 2006) rather than the climate variables used as predictors, the data sources, and their quality. Table 5.1 shows the range of variables derived from interpolation of climate station records used in a number of SDM studies. This is not a comprehensive list of all climate variables used in SDM studies. These studies are featured because they emphasized climate predictors, modeled a large number of species, or both (see also Table 10.1). The variables selected or derived from monthly averages and extremes of temperature, precipitation, radiation, humidity, cloud cover, and so forth, emphasize the seasonal patterns and extremes of light, heat and moisture availability expected to limit species distributions (Chapter 3). Some patterns emerge. Almost all studies used measures of mean, minimum and maximum temperature, and some measure of temperature seasonality or range. Mean annual precipitation and some measure of precipitation seasonality was used in most studies. Many studies used mean solar radiation. While studies in cooler climates tended to use some measure of growing season length, such as growing degree days, studies in areas where water availability is limited in some seasons used a measure of water balance.

It is important to note, however, that although the climate variables were indicated as having been used in these studies, not all of these climate variables may have been important (that is, explained variation is species occurrence) or selected in the models that involved variable selection. Other studies have used all monthly averages of temperature, precipitation and radiation, for example, and let the models select among them (Cumming, 2000a). However, multicollinearity would be expected among monthly averages of a variable, as well as between many of the climate variables discussed. Therefore, climate predictors used in SDM should be screened for multicollinearity (Chapter 6).

5.2.2 Digital terrain maps

Just as coarse-scale species distribution models rely, sometimes exclusively, on climate maps as environmental predictors, many landscape-scale species distribution models use topographic or terrain variables, derived from digital elevation grids, for prediction. The comprehensive review by Moore *et al.* (1991b) provided a summary of hydrological, geomorphological and biological applications of digital terrain modeling. Subsequently, a book was published that includes a more detailed treatment of digital elevation data and the derivation of terrain attributes (Wilson & Gallant, 2000). In this section, I will focus on terrain variables used in SDM that are related to direct and resource gradients (Chapter 3) through the effect of elevation on temperature and precipitation; of slope characteristics on radiation regime and moisture availability; and of landform, hillslope position and catchment position on soil moisture and on erosion and deposition, which in turn affect soil development and properties (Franklin, 1995).

All of the terrain variables described here are derived from a digital elevation model (DEM). A DEM is an ordered array of numbers (for example grids, contours, or a "triangular irregular network") representing the spatial distribution of elevations in a landscape (see a review on the production of DEMs by Florinsky, 1998). For more detailed information on the data structures used to represent spatial data in geographic information systems, I recommend Burrough and McDonnell (1998). Derivation of terrain attributes from DEMs, and use of terrain attributes in landscape analyses were among the first challenges tackled in geographic information science (Mark, 1979; Marks *et al.*, 1984; Franklin *et al.*, 1986; Goodchild & Tate, 1992). Note that, sometimes a DEM is called a digital terrain model (DTM), while in other sources DTM is used to refer to the terrain attribute maps derived from a DEM.

Moore *et al.* (1991b) distinguished primary terrain attributes, derived directly from the DEM, from secondary attributes, which are calculated from a combination of primary attributes to represent physically-based processes that vary systematically with terrain. Terrain attributes are generally related to landform (geomorphology) and to hydrology – the routing of water flow on the landscape (Lane *et al.*, 1998). The following primary attributes are related directly or indirectly to environmental gradients affecting species distributions:

1. *Elevation* systematically affects temperature regime, and it affects precipitation via orographic lifting.
2. *Slope* angle or gradient is the steepness of the slope (i.e., the rate of change in elevation over some distance), and is related to overland and subsurface flow of water, and therefore it affects potential soil moisture (higher on gentler slopes) and soil characteristics (texture, moisture regime, development).
3. *Slope aspect*, the direction the slope is facing, affects the amount of solar radiation received on the slope and the seasonal and annual patterns of solar insolation (radiation). This in turn affects soil moisture availability. There is greater insolation, and therefore evaporative demand, on equatorward-facing slopes.
4. *Specific catchment area* (upslope area or drainage basin position per unit width of contour) is related to runoff volume and rate (Band, 1986; Quinn *et al.*, 1991), and therefore soil moisture availability, soil development and soil characteristics.
5. *Slope curvature*, the second derivative of elevation, is the degree of concavity or convexness of a slope and can be measured across-slope

or down-slope (or averaged). Related variables are the terrain shape index (mean difference between the center of a plot and its boundary) and land form index (the average vertical gradient to the topographic horizon; McNab, 1989, 1993; Fels, 1994). Slope curvature affects subsurface water flow, and therefore soil moisture content, litter accumulation, erosion and deposition rates, which in turn are related to soil texture, depth and nutrient availability.

6. *Hillslope position* or "terrain position" (Skidmore, 1990) is the relative distance between the nearest ridge and stream, and in geomorphology and soil geography is usually assigned to ordered categories (ridge, upper slope, midslope, toeslope, etc.). Methods for automatically deriving hillslope position are usually similar to those procedures used to derive stream networks and delineate drainage basins from DEMs (Jenson & Domingue, 1988; Skidmore, 1990). A related measure is the landscape position index, the average inverse-distance weighted difference in elevation between a grid cell and surrounding cells within a specified radius (Fels, 1994). It has been suggested that hillslope position may be a better indicator of geomorphic processes, and thus soil properties, than upslope catchment area and slope curvature, which are more related to hillslope hydrology (Swanson *et al.*, 1988).

There are many more primary terrain attributes – Moore *et al.* (1991b) noted that Speight (1974) described more than 20 topographic attributes that are related to landform – but the ones listed above are most frequently used in SDM. Another non-local terrain attribute that has been used in SDM is topographic complexity or heterogeneity, used to represent habitat heterogeneity at a variety of scales (see Section 5.3.3). It is usually quantified as the variation in elevation values within some window (Jetz & Rahbek, 2002; Shriner *et al.*, 2002).

Secondary or compound terrain attributes, derived from process-based models or indices, quantify the role played by terrain in redistributing water on the landscape and in modifying the amount of solar radiation received at the surface. The main ones used in SDM are topographic wetness, temperature and radiation indices (see Table 1.2 in Wilson and Gallant, 2000).

Topographic wetness or moisture indices describe the spatial distribution and extent of saturation for runoff generation as a function of upslope contributing area, slope angle, and soil transmissivity (Moore *et al.*, 1988). The topographic wetness index, also called the topographic moisture index, is

$$W_T = \ln\left(\frac{A_T}{T \tan \beta}\right),$$

where T is soil transmissivity, A is upslope contributing area, and β is slope angle. Often when calculating W_T for each grid cell in a DEM (Fig. 5.1), transmissivity is assumed to be constant and equal to unity (indicating steady state hydrological conditions and uniform soil properties), and T drops out of the equation.

Solar radiation or insolation for SDMs is often calculated, using simulation models, as the shortwave solar irradiance at the top of the atmosphere for some time period (daily, monthly, annual) as an approximation of radiation incident on the surface. More detailed models calculate shortwave radiation incident on the earth's surface by accounting for atmospheric transmission, direct and diffuse radiation, and terrain shading (Dozier, 1980; Dozier & Frew, 1990; Dubayah & Rich, 1995; 1996). In SDM, if relative rather than absolute values are required, daily insolation can be calculated for certain days (1 day per month, or solstices and equinoxes). Those daily values can then be used directly as indices of season or annual insolation, or scaled up to approximate to monthly, seasonal or annual by multiplying calculated values by the appropriate number of days (Fig. 5.1). Alternatively, approximate solar radiation indices can be calculated using slope, aspect and, in some cases, latitude (McCune, 2007).

Temperature maps that are spatially interpolated from climate station data (see Chapter 5.2.1) are sometimes refined, using a DEM, to account for the effect of elevation on temperature via the lapse rate, and also slope-aspect effects on radiation (e.g., Daly *et al.*, 2008).

Terrain attributes are usually derived from elevation data structured as a grid in a GIS. Slope and aspect are "local" attributes in the sense that they are calculated from the elevations of a target grid cell and its immediate neighbors, while hillslope position and upslope contributing area are nonlocal (Florinsky, 1998). I refer you to the references mentioned in this section (e.g., Moore *et al.*, 1991b; Florinsky, 1998; Wilson & Gallant, 2000) for more details about the algorithms used to derive terrain attributes from DEMs. It is worth investing some time to learn more about this when using terrain attributes in species distribution modeling because, although most GIS packages provide tools for terrain analysis, they may not clearly state what algorithms are used, or provide the user with parameter options. As a simple example, slope can be calculated as the "steepest descent" (based on the greatest elevation difference between a grid cell and any of its eight neighbors), or based on the averaged finite differences in elevation between the grid cells' north and south neighbor, and its east and west neighbor. The second method gives slightly smaller average slopes. Slope curvature (the second order finite difference in elevation) can be calculated for the same local neighborhood, or for a larger window, and it can be calculated across slope (plan curvature), down slope (profile curvature), or the two can be averaged. It is possible that these differences could affect the results of SDM, and at the very least the methods used to derive terrain attributes should be reported so that comparisons across studies can be correctly interpreted.

DEMs also contain systematic and non-systematic errors (Weibel and Heller, 1991; Florinsky, 1998). Both these errors and the resolution (scale) of the elevation grid affect the derivation of primary and secondary terrain attributes, and this has been studied extensively (e.g., Brown & Bara, 1994; Hunter & Goodchild, 1997; Deng *et al.*, 2007). These errors can propagate to species distribution models sometimes in surprising ways, a topic that deserves much greater consideration. For example, Van Niel *et al.* (2004; see also Van Niel & Austin, 2007) found that, by simulating DEM errors in a landscape of relatively gentle topographic relief, the compound terrain attribute solar radiation (representing a resource gradient), was less affected by DEM error than the simple attributes, slope and aspect, which represent indirect gradients (as discussed in Chapter 3). However, topographic position was less affected by DEM error than the topographic wetness index. These error patterns might be different in an area of steeper topography or greater relief. Another study showed that some temperature variables (seasonal minima) were correlated with elevation at a specific (finer scale) location, while others (seasonal maxima) were correlated with elevation averaged over a larger area, and this has implications for the scale at which terrain variables should be derived for SDM (Ashcroft, 2006).

DEMs for much of the world are available at 25- to 50-m resolution (e.g., 25 × 25-m to 50 × 50-m grid cells), and global topographic data are available at 30-arc second resolution (about 250 m). Some studies have suggested that higher (10- to 20-m) resolution DEMs are required to accurately model topo-hydrological variables that are frequently used in SDM (Hutchinson, 1996; Hutchinson & Gallant, 2000). Very high resolution DEMs (up to 1 m resolution) are now being produced for some regions, and a recent study showed that slope originally calculated from 1-m data but using a window (neighborhood) of 100 × 100 m showed the greatest predictive ability in SDMs of alpine plants (Lassueur *et al.*, 2006). Aspect was optimized at 20 × 20 m, suggesting that moderately high resolution (10–20 m) DEMs are probably of sufficient resolution to estimate the simple topographic variables that are related to direct and indirect environmental drivers of many species distributions.

In addition to elevation data, other information found on topographic maps, such as streams, water bodies, and roads, are available in digital vector and raster form (for example, from the US Geological Survey in the USA). These have been used to derive predictors for species distribution modeling. For example, proximity to streams and rivers can be an important habitat requirement for both animals and plants (Franklin *et al.*, 2000; Fleishman *et al.*, 2002; van Manen *et al.*, 2002; see Section 5.3.4).

5.2.3 Soil factors and geology maps

Nutrient availability has been defined as one of the primary environmental regimes determining the distribution of plants (Chapter 3), and given the direct or indirect effect of substrate (parent material and soil) on nutrient and water availability, it is not surprising that digital geology and soil maps have been used to derive chemical and physical input data for GIS-based ecological models (Burrough & McDonnell, 1998), including SDMs. Further, animal distributions may be tied to soil properties indirectly, through the effect of soils on plant community composition and structure, or directly, for example when the distribution of animals is tied to a soil type or texture required for burrowing, movement or camouflage (Barrows, 1997; Fisher *et al.*, 2002). There are two issues to consider when incorporating mapped soil or substrate data into a SDM: 1) what are the soil factors that proximally determine the species distribution in question, and 2) what is the relationship between available mapped data and those factors?

Traditionally, soil surveys qualitatively relate soil and substrate properties to landscape and environmental factors, and therefore using soil survey maps in SDM is asking quite a lot of them (Scull *et al.*, 2003). As I noted in Franklin (2001), digitizing geology, soil or vegetation maps for quantitative modeling can be problematic because those thematic maps may have been developed using a cartographic approach and a "communicative paradigm" (DeMers, 1991) – they were intended to present a visual model of a spatial distribution at a particular scale (Goodchild, 1988). Although in recent decades geology and soil survey maps are being produced in digital form from the outset, and based on remote sensing (image classification and interpretation), or even on predictive modeling (Scull *et al.*, 2003), they should be used with careful regard for the scale at which they were produced. Further, the mapped categories may not bear a one-to-one relationship with the environmental factors driving species distributions. In spite of these challenges, SDM studies have derived nutrient or fertility indices, texture, and other attributes from thematic maps of soil type or geology and used them in modeling (for example, Austin, 1992; Mackey, 1993a; 1993b; Cawsey *et al.*, 2002).

In the USA, the US Department of Agriculture's agency that is responsible for mapping soil resources (formerly the Soil Conservation Service, now called the Natural Resources Conservation Service or NRCS) has developed several digital soil databases, including the State Soil Geographic Database (STATSGO). These data have coarse spatial resolution (map units drawn on 1:250 000 base maps with approximately a 625-ha minimum mapping unit), many components (subunits that are not spatially explicit and correspond to phases of soil series) and multiple attributes. Each component can have up to six layers (soil horizons) and each layer can have up to 28 properties – for example, percent clay, cation exchange capacity, permeability, bulk density and pH (Lytle, 1993; Lytle *et al.*, 1996). Although the complex data structure makes it challenging to use STATSGO data, they have been increasingly used in quantitative models of productivity, crop yield, hydrology, water quality and soil carbon dynamics (Wilson *et al.*, 1996; Abdulla & Lettenmaier, 1997; Epstein *et al.*, 1997; Homann *et al.*, 1998; Hernandez *et al.*, 2000; Rosenblatt *et al.*, 2001; Heath *et al.*, 2002; Niu & Duiker, 2006; Rasmussen, 2006). STATSGO data have also been used in SDM and related forms of ecological modeling (Lathrop *et al.*, 1995; Mann *et al.*, 1999; Hathaway, 2000; Strickland & Demarais, 2006). The coarse spatial resolution (coarser than a traditional County soil survey, which is typically mapped at 1:10 000 to 1:30 000 scale), and correspondingly large and heterogeneous map units intended for multicounty, state and regional planning, mean that STATSGO data are not always of appropriate scale for SDM.

In an interesting approach, extensive plant community survey data have been used, in France, to infer certain soil properties (pH, nitrogen conditions and base saturation), based on published tolerances of plant species to those properties (Coudun *et al.*, 2006; Cord, 2008). Over 3000 relevés from the EcoPlant database (Gégout *et al.*, 2005) were used to impute soil properties based on Ellenberg indicator values (Ellenberg,

1988). Ellenberg *et al.* (1992) derived preference values for moisture, light, temperature, nutrients and soil properties for thousands of plant species in central Europe (see also Ertsen *et al.*, 1988; Hill *et al.*, 2000). Soil properties estimated from Ellenberg indicator values for species in sites were then spatially interpolated to maps. These maps were used to improve predictions of the distribution of another plant species based on an independent set of presence/absence data (88 000 relevés from the French Forest National Inventory).

There are no guidelines on how to define the relationship between mapped soil types or properties and the environmental gradients driving species distributions, but we can look at examples and try to draw some generalizations. Leathwick (1998) used maps of geologic substrate classes to predict the distribution of tree species in New Zealand, but grouped those classes according to their age, weatherability and chemical composition. These factors would presumably be related both to soil texture and water holding capacity, and also to nutrient availability. In the rain forests of Queensland, Australia, Mackey (1993b) used mapped parent material rock type as a surrogate for soil information in order to predict the distribution of forest types (see also Austin, 1992). Lithology classes were ranked according to their potential mineral nutrient supply (into ordinal numerical classes) and so, for example, dolorite had a high value and beach sand a low value. Franklin *et al.* (2001) used coarse-scale geology maps aggregated from 22 to eight classes thought best to represent gradients of the availability of water and nutrients related to plant distributions in the Mojave Desert of California, USA. When landform maps became available based on contemporary image interpretation, these 29 classes were used instead of the more general geology classes in vegetation modeling (Miller and Franklin, 2002).

5.3 Other environmental data for SDM

5.3.1 Vegetation maps

Paraphrasing Köchler (1973; Köchler and Zonnèveld, 1998), vegetation is the tangible integrated expression of the entire ecosystem. Categories of vegetation or plant communities represent habitat classes and habitat structure for animal as well as plant species and have been widely used as predictors in SDM (for example, Leyequien *et al.*, 2007). Traditionally, wildlife habitat suitability models rely on vegetation categories as their main drivers (Morrison *et al.*, 1998). For example, the Wildlife Habitat

Relationships models developed by the State of California's Department of Fish and Game (Mayer & Laudenslayer, 1988) classifies habitat suitability for all vertebrate species based on broad vegetation (habitat) categories and, in some cases, general forest cover classes (habitat structure). In another example, a distribution model was developed for a vegetation type and used to predict the distribution of key habitat for a spatially dynamic and threatened butterfly species (Early *et al.*, 2008). Vegetation maps, like most thematic maps, are available at low resolution for large areas (continents) or high spatial and categorical detail for smaller areas (national forests, habitat reserves, counties) (Franklin & Woodcock, 1997). Annual land cover and vegetation cover data products are available globally from NASA and could be used in coarse-scale, large area SDM (see Section 5.3.3).

More typically, SDM relies on vegetation mapping carried out at the scale of a national park, forest, or reserve, or a political unit such as a county or state. As with mapped soil data, it can be conceptually and methodologically challenging to describe a meaningful relationship between numerous vegetation categories mapped for multiple land management purposes and habitat suitability for a target species. This is compounded when the vegetation data must be compiled from multiple sources and with different classification schemes.

Recently, land cover maps are being compiled at regional to continental scales in order to monitor and study global environmental change (Franklin & Wulder, 2002). These new data products often contain information about vegetation type. For example, the land cover database of North America (Latifovic *et al.*, 2002) uses a modification of the National Vegetation Classification Standard (NVCS) for the USA (Grossman *et al.*, 1998) as a basis for regional land cover classes. Further, these NVCS classes can be cross-referenced with other major systems, such as the European CORINE (Coordination of Information for the Environment) Land Cover classification (Ahlqvist, 2005). CORINE digital land cover maps have been produced for 29 countries in Europe at a scale of 1:100 000, or roughly a 25-ha minimum mapping unit (Feranec *et al.*, 2007). CORINE Land Cover data are intended for many purposes, from water quality modeling to carbon accounting, and have been evaluated by comparison to other land cover data (Bach *et al.*, 2006; Schmit *et al.*, 2006; Waser & Schwarz, 2006; Smith & Wyatt, 2007). Their suitability for use in species distribution modeling has been tested, and in these published cases, these general purpose land cover maps were adequate to support habitat modeling for wetland birds, other bird species, and the brown bear (Posillico *et al.*, 2004; Suarez-Seoane *et al.*, 2004; Virkkala *et al.*, 2005).

Human land use, e.g. agriculture, forestry and urban development, is a major form of human disturbance affecting the distribution of suitable habitat and of species. Land use/land cover mapping is commonly carried out by planning agencies, and land cover data are becoming increasingly important for land surface characterization required for modeling global environmental change (Loveland & Ohlen, 1993; Steyaert, 1996; Defries & Townshend, 1999; Foley *et al.*, 2005). Land use is a driver of land cover change, although there is not always a simple, one-to-one relationship between land use and land cover (Hurtt *et al.*, 2001; Schmit *et al.*, 2006). Future projections of land use have been used to project the effect of land cover change on species distributions (Wisz *et al.*, 2008a). While Suarez-Seoane *et al.* (2004) concluded that there were limits to how much a more detailed land cover or vegetation map would improve SDMs, this depends both on the scale of the modeling and the taxa being modeled.

Sometimes, rather than plant community type (species composition), some aspect of vegetation structure such as cover, size distribution of plants, or vertical canopy structure, is related to habitat suitability. This information can be extracted from certain remotely sensed data products (Section 5.3.3). Alternatively, structure variables can be interpolated from forest inventory or similar data (Zielinski *et al.*, 2006) if they are dense enough for spatial interpolation (Ohmann & Gregory, 2002). In some cases, vegetation structural attributes have been predictively mapped using the very same methods described in this book (e.g., Moisen & Frescino, 2002).

5.3.2 Disturbance and disturbance history

Natural disturbance is an important driver of ecological succession and population dynamics (Pickett & White, 1985). Ecosystems are characterized by various disturbance regimes – the typical frequency, magnitude and spatial patterns of natural disturbances such as floods, fires, avalanches, hurricanes and tornadoes. It was noted early on in SDM (Lees & Ritman, 1991) that, while remote sensing-based mapping is particularly adept at identifying broad land cover types (including those affected by disturbances such as land use), SDM based on relatively stable features of the physical environment (climate, terrain, substrate) should yield a relationship between species occurrence and environment in the absence of disturbance. This argues for the combined use of remotely sensed data, revealing contemporary and dynamic land surface characteristics, together with variables related to the primary environmental regimes, in order to isolate the effects of disturbance on species distributions.

Spatially explicit records of the footprint of fire disturbance (fire history maps) exist for fire-prone landscapes in many parts of the world, and are being incorporated into digital databases. These time series usually only go back decades, perhaps to the early twentieth century, the period for which fire perimeters have been mapped on the ground, by eye from the air, or more recently using GPS or from aerial imagery. Fire history maps have also been reconstructed using fire scar mapping methods. In recent years, monthly maps of burned areas are provided globally from MODIS data (Section 5.3.3), at low spatial resolution. Fire history maps could be used to derive information about time since disturbance, or disturbance frequency, which might be useful for predicting the distributions of early-successional "pioneer" versus late successional species (e.g., Bazzaz & Pickett, 1980). For example, descriptors of fire disturbance were significantly related to the shifting distributions of early-successional (open habitat) birds in Catalonia, Spain (Vallecillo et al., 2009).

5.3.3 Remote sensing

In an earlier review, I discussed the initial development of SDM in the context of remote sensing, i.e., biophysical mapping of the earth' surface based on satellite and airborne imagery or aerial photography (Franklin, 1995). Why can't species distributions be mapped directly with remote sensing – at least large plant species of the upper canopy layer? Although advances are being made in this area, it is challenging to directly map species with remote sensing. This is because multispectral remote sensing is more effective for determining land surface characteristics and the structure and physiognomy of vegetation (Running et al., 1994; Wulder, 1998; Gamon & Qiu, 1999), and other biophysical properties of the earth's surface, than identifying individual species. Even when "hyperspectral" imaging spectrometer data are available, species do not necessarily have unique spectral signatures. Lees and Ritman (1991) recognized early on that both remotely sensed data and other geospatial data describing the landscape are required to make spatial predictions of species composition.

But, if we map plant communities in detail, using air photos or satellite imagery, then haven't we directly mapped the "communities" (plants) or "habitats" (animals) where species occur? This has traditionally been the basis of much wildlife habitat suitability modeling (Morrison *et al.*, 1998), but it is only effective if there is a known one-to-one relationship between vegetation community type and species, and if the communities are mapped without very much error. That is why, early on, remotely sensed images were combined with other data to develop spatially explicit information about species distributions. As I have previously quoted (Franklin, 1995), ". . . if {plant}species composition varies systematically with terrain, topographic variables. . . can be used to improve predictions of {forest} species composition through implicit or explicit use of an ecological model. . . The technique is a prototype for a set of tools that will become increasingly important as *geobased information systems* using {remotely sensed} and collateral data develop in the coming years" (Strahler, 1981) – emphasis added. Notice that this quote actually predates the widespread use of the term "geographic information system" (GIS). Digital environmental data layers (terrain, geology, soils) were incorporated as "ancillary" or "collateral" data into early efforts to develop regional forest maps from Landsat. Working together, multispectral and other digital mapped environmental data were much better able to accurately map plant species assemblages even 30 years ago (Hoffer, 1975; Strahler, 1981; Hutchinson, 1982).

What progress have we made in the last quarter century in using remote sensing to map biotic distributions? Remote sensing has seen the development and operational use of data and data products from a plethora of passive and active airborne and satellite-borne sensors, in particular hyperspectral (Collins *et al.*, 1993) and hyperspatial (Culvenor, 2002; Chen & Stow, 2003; Wulder *et al.*, 2004) imaging, as well as active sensors – LIDAR and RADAR instruments. These new data have been used to move far beyond general plant community mapping to describing biodiversity at multiple levels of ecological organization – individuals, species distributions, community composition (species assemblage), and diversity (species richness), from global to regional to landscape scales (Stoms & Estes, 1993; Nagendra, 2001; Turner *et al.*, 2003; Foody & Cutler, 2003). Many SDM efforts use habitat predictors derived from remotely sensed imagery (Rushton *et al.*, 2004; Bradley & Fleishman, 2008; Gillespie *et al.*, 2008). In this subsection I will give an overview of remotely sensed data and data products relevant to SDM, and then describe some recent applications of remote sensing to SDM.

While satellite and airborne remotely sensed image data have been applied to ecological studies for decades, since civilian remote sensing began (Tucker, 1979; Curran, 1980; Strahler, 1981; Hutchinson, 1982), these data are now finding their way into the hands of a growing community of increasingly sophistical data "consumers" to be used for ecological and other environmental applications (Gamon & Qiu, 1999; Pettorelli *et al.*, 2005). It still requires specialized knowledge and software to use these data effectively, although some standard data products are being made available (see below). Recent reviews and texts provide excellent summaries of the remotely sensed data, their sources, characteristics and relationships to ecosystem properties, with an emphasis on forested environments (Franklin *et al.* 2003), and biophysical characterization of species habitats (Kerr & Ostrovsky, 2003).

Satellite remotely sensed data are usually described in terms of their spatial resolution (pixel size), temporal resolution (frequency with which data are collected for the same place) and spectral resolution or range (number of bands and bandwidths). Optical sensor data are most often used to infer properties of the earth's surface in ecological applications – that is, sensors that measure upwelling (reflected or emitted) electromagnetic radiation in short (visible), near- and middle-infrared wavelength bands. Data from longer-wave passive (thermal) and active (microwave, LIDAR) sensors are less commonly but also used in ecological applications. Commonly-used datasets include those from the following sensors:

• Multispectral sensor data from the French SPOT satellites (2.5–20 m pixels),
• Landsat Enhanced Thematic Mapper (ETM+) with 15–120 m pixels,
• the US National Oceanic and Atmospheric Administration's Advanced Very High Resolution Radiometer (NOAA-AVHRR) with 1–4-km pixels,
• the Vegetation sensor on the SPOT-4 and -5 satellites (1-km pixels), and
• the US National Aeronautics and Space Administration's Moderate Resolution Imaging Spectroradiometer (NASA MODIS) sensor with 250–1000-m resolution.

A breadth of standard data products are being produced from MODIS data (Running *et al.*, 1994; Justice *et al.*, 1998), including daily surface reflectance, daily surface temperature, yearly land cover, vegetation (tree) continuous cover (VCF), spectral vegetation indices, surface albedo, Leaf Area Index (LAI), net photosynthesis, annual vegetation cover, and monthly burned areas (http://modis.gsfc.nasa.gov/). These data products are being incorporated into SDMs. For example, recent studies have used MODIS data products such as LAI and VCF, along with microwave backscatter from the space-borne QSCAT instrument and topography (elevation) measured from satellite-based RADAR, to develop species distribution models for species from several taxonomic groups (birds, mammals, trees) in South America (Buerrman et al., 2008; Prates-Clark et al., 2008). The microwave measurements are related to vegetation canopy roughness, moisture and deciduousness. Including these coarse-scale remotely sensed predictors improved SDMs over those developed using only climate variables as predictors.

A very widely used vegetation index, the Normalized Difference Vegetation Index (NDVI), contrasts the red (R) absorption and near-infrared (NIR) reflectance of healthy green vegetation using the simple formula (NIR-R)/(NIR+R) (Tucker, 1979). This index can be derived from broad-band red and NIR reflectance measured by all the sensors mentioned above as well as others. A number of studies modeling species distributions at larger spatial scales have used annual or monthly sums, or transformations (such as by principal components analysis) of NDVI from MODIS or AVHRR as measures of broad-scale primary productivity patterns and effective predictors of species distributions (Cumming, 2000b; Osborne et al., 2001; Cumming, 2002; Jetz & Rahbek, 2002; Zinner et al., 2002; McPherson et al., 2004; Suarez-Seoane et al., 2004).

Some of these same remotely sensed indices and data products have been related to measures of biodiversity (reviewed by Gillespie et al., 2008). Reflectance, surface temperature, NDVI summaries, and other spectral indices from a variety of sensors have been used to predict the species richness of different taxonomic groups, in particular, plants and birds (Oindo & Skidmore, 2002; Hurlbert & Haskell, 2003; Fairbanks & McGwire, 2004; Seto et al., 2004; Waring et al., 2006; Levin et al., 2007; Saatchi et al., 2008). Biochemical diversity measured from hyperspectral AVIRIS imagery (Carlson et al., 2007; Asner et al., 2009), and structural diversity and landscape patterns measured from multi-spectral ETM+ imagery (Gillespie, 2005) have also been linked to tree species richness. Habitat heterogeneity, as measured by spatial heterogeneity in spectral measures, has been correlated with species richness in a number of studies (Gould, 2000; Palmer et al., 2002; Rocchini et al., 2007).

Land cover or vegetation type is the thematic variable derived from remote sensing that is still the most commonly used in SDM (for some recent examples see Poulos et al., 2007; Zeilhofer et al., 2007). Gottschalk et al. (2005) reviewed over 120 studies that use remotely sensed data for mapping bird species habitat or distributions, and most of those studies derived land cover, vegetation or habitat classes from satellite imagery, Landsat in particular. They noted that, in some of these studies, the imagery was of insufficient spatial and spectral resolution to identify important habitat features or categories. When SDM requires land cover, landscape pattern (Section 5.3.4) or surface characterization at finer spatial resolution than is available from satellite-borne sensors such as Landsat, higher-resolution data such as SPOT may be more suitable. If these data are not available or affordable, high spatial resolution digital multispectral cameras may be useful. These instruments, mounted on aircraft and satellites, are blurring the traditional boundaries between digital imagery and air photos, and are increasingly used for regional land cover mapping and ecosystem characterization (Phinn *et al.*, 1996; Lefsky *et al.*, 2001; Chen & Stow, 2003; Thenkabail *et al.*, 2004; Greenberg *et al.*, 2006).

Hyperspectral (many narrow wavebands) and hyperspatial (high spatial resolution) images are now facilitating the mapping of individual trees and tree species (Culvenor, 2002; Clark *et al.*, 2005). Still other recent studies have emphasized the use of newer sensors and products such as LIDAR, ASTER and Quickbird to predict the distribution of suitable habitat for individual species (for example, Sellars & Jolls, 2007; Estes *et al.*, 2008; Stickler & Southworth, 2008). Light detection and ranging (LIDAR) is a remote sensing system using properties of scattered light to find the range of (e.g., distance to) a distant target. This allows canopy height, as well as topography, to be measured. Canopy heterogeneity from LIDAR has been used to predict bird species richness (Goetz *et al.*, 2007). ASTER (Advanced Spaceborne Thermal Emission and Reflection Radiometer) is a USA-Japan satellite-borne multispectral scanner that captures data in 14 bands, from the visible to the thermal infrared wavelengths at spatial resolutions of 15 to 90 m. Quickbird is a commercial satellite carrying a hyperspatial imaging system with sub-meter panchromatic resolution. In studies of individual species, the remotely sensed variables were often related to vegetation canopy structure measures, which, in turn, were determinants of habitat quality.

Another approach that has been used to map plant species, communities or vegetation types is the use of multitemporal imagery to differentiate species based on aspects of their phenology such as their differential time of leafing out or greening up (for example, Wolter *et al.*, 1995; Townsend & Walsh, 2001; Dechka *et al.*, 2002). This may be particularly useful for mapping the extent of non-native, invasive plant species whose phenology differs from native species (e.g., Hestir *et al.*, 2008; Huang & Geiger, 2000).

Some generalities emerge from these studies – remotely sensed variables that are related to productivity, vegetation structure, habitat heterogeneity and biochemical diversity are correlated with biodiversity measures, and with species habitat quality, at least for some species in some environments. When these measures are acquired regularly from satellite or airborne platforms they have the potential to track the dynamics of environmental change (Pettorelli *et al.*, 2005; Prates-Clark *et al.*, 2008), and this potential could be more extensively exploited for SDM. For example, an NDVI time series was used to characterize temporally dynamic habitat and generate a "temporal suitability map" for little bustards in Spain (Osborne & Suarez-Seoane, 2007), and has been used similarly to identify suitable habitat for other species (Mueller *et al.*, 2008; Wiegland *et al.*, 1986).

However, a comprehensive framework (e.g., Phinn, 1998; Phinn *et al.*, 2003) for the effective use of remotely sensed data in biodiversity mapping – one that matches the spatial, spectral and temporal characteristics of the sensor to specific biodiversity mapping goals – is still lacking. One recent study directly examined species traits that predict whether remotely sensed data are likely to improve spatial predictions of suitable habitat and found that remote sensing based predictors improved distribution models of rare, early-successional and broadleaf tree species (Zimmermann *et al.*, 2007). Studies like this will help to develop a framework for the effective use of remotely sensed data in SDM.

5.3.4 Landscape pattern

The spatial arrangements of land covers or habitat structure, or other spatial attributes of the landscape (Pickett & Cadenasso, 1995), can be described with landscape metrics (McGarigal & Marks, 1995; McGarigal, 2002) or other measures of spatial heterogeneity (Turner & Gardner, 1991; Plotnick *et al.*, 1993; Gustafson, 1998). A number of SDM studies have shown that measures of landscape (land cover) composition, heterogeneity, or proximity to certain landscape features, such as streams, forest edges or roads, are important predictors of habitat suitability, primarily for animal species (Pearson, 1993; Lindenmayer *et al.*, 2000; Kerr *et al.*, 2001; Saveraid *et al.*, 2001; Fleishman *et al.*, 2002; Johnson & Krohn, 2002; McAlpine *et al.*, 2002; van Manen *et al.*, 2002; Vernier *et al.*, 2002;

Heikkinen *et al.*, 2004; Murphy & Lovett-Doust, 2004; Shriner *et al.*, 2002; Warren *et al.*, 2005; Ciarniello *et al.*, 2007; Lassau & Hochuli, 2007; McAlpine *et al.*, 2008).

In all of these investigations, landscape composition and pattern were significant and effective predictors of species distributions. For example, in one study the distribution of shorebirds was predicted from landscape variables including distance to channels and area of oyster beds (Granadeiro *et al.*, 2004). Land cover composition has also been shown to be an important predictor of the distribution of invasive plant species (Ibáñez *et al.*, 2009). Habitat heterogeneity has been estimated from satellite imagery using image texture (local variance) measures, and correlated with habitat suitability for the Greater Rhea (*Rhea americana*; Bellis *et al.*, 2008). Many of these studies emphasize habitat selection at different scales – patch versus landscape (Graham & Blake, 2001; Lee *et al.*, 2002; McAlpine *et al.*, 2008), as discussed in Chapter 3.

Often measures of landscape pattern and habitat fragmentation used in biodiversity studies are derived from remotely sensed imagery, either directly, or from thematic land cover maps derived through image classification (Cohen & Goward, 2004). There are also theoretical reasons for expecting habitat heterogeneity to be related to species diversity (Chapter 3). Land cover diversity has been correlated with butterfly species richness (Kerr *et al.*, 2001). Habitat heterogeneity estimated from local variation in canopy structure, elevation, substrate type, or climate, has been used to predict species richness at various scales (Jetz & Rahbek, 2002; Pausas *et al.*, 2003; Goetz *et al.*, 2007).

5.3.5 The distributions of other species

It is often lamented that correlative SDMs based on predictors representing the physical environment and landscape cannot truly describe the species niche because they do not explicitly consider how the distribution of other species (facilitators, competitors, predators, prey) affect the distribution of the species being modeled. It is generally assumed that SDMs estimated from distribution data approximate the realized niche, because observed distributions integrate any effects of species interactions on distributions. However, these effects are thought to be predictable only if they covary with one of the mapped environmental covariates. Otherwise, these effects are relegated to unexplained variance. But, in fact, the positive or negative effects of other species can be and have been taken into account in SDMs. If species affect each other's occurrence with feedback (as in predator-prey interactions), or if there are multiple and complex interaction among many species (as, of course, there are in food webs), this would be difficult to fully specify in a correlative SDM. However, if one species has a rather large (asymmetrical) affect on another species distribution, and its distribution is known or could be modeled, this is more tractable.

Plant competition is often asymmetrical between species (Zobel, 1992). Therefore, the distribution of the superior competitor could be modeled assuming no effect of other species. Then, this predicted distribution map can be used as a predictor in a model of the inferior competitor (as in Austin, 1971; Leathwick & Austin, 2001; Meentemeyer et al., 2001). Spatial co-occurrence of predators and their prey is of great interest in wildlife biology (Alexander et al., 2006). As mentioned in Chapter 1, improved predictions of marine predator (bottlenose dolphin) distributions were made using models of prey (fish) distributions (Torres et al., 2008). A species distribution model for a host plant was used to improve a distribution model for an endangered butterfly species (Preston et al., 2008). Species interactions, including competition, predation and facilitation (Callaway & Pennings, 2000), generally take place at local scales (Pearson & Dawson, 2003). Even at macro scales, however, models of species distributions have been improved by including the modeled distributions of other species with expected positive impacts (facilitation) as predictors (Araújo & Luoto, 2007; Heikkinen et al., 2007).

If the distribution of other species (competitors, prey, predators, pollinators, etc.) is not known, or if the biotic interaction cannot be assumed to be asymmetrical, then some other statistical modeling approach, such as structural equation modeling, may be able to estimate the effect of biotic interaction on species distributions (Austin, 2007).

5.4 Environmental data for aquatic and marine species

SDM has also been applied to marine species. Although different environmental predictors are obviously required, these are becoming more readily available owing to an established archive of remotely sensed data for the oceans and growing development of marine GIS, both for deep ocean (Wright & Goodchild, 1997; Wright, 1999) and coastal (Stanbury & Starr, 1999) environments. Further, while some variables like bathymetry (bottom depth) have terrestrial analogs (digital elevation models), others factors (e.g., sea surface temperature) tend to be more temporally dynamic than many terrestrial predictors because of the fluid nature of the oceans (although primary productivity, estimated from satellite remote sensing, would be a seasonally dynamic feature of both marine and terrestrial environments). Therefore, with the environment and/or species in motion, environmental correlations with marine species distributions can be difficult to establish (Sundermeyer et al., 2006).

In spite of these challenges, habitat suitability for aquatic and marine species has been predicted based on a number of variables. Many of these are derived from remote sensing and mapped at coarse spatial scales, which is unsurprising given the vast expanse and dynamic nature of the oceans and large river systems. For example, sea surface temperature, chlorophyll-a concentration, depth, and distance to the shore were used to model habitat suitability for a threatened Mediterranean seabird (Louzao et al., 2006). Sea surface temperature was based on nighttime AVHRR imagery (Section 5.3.3), chlorophyll-a from the Sea-viewing Wide Field-of-view Sensor (SeaWiFS), and bathymetric data from NOAA's ETOPO five minute gridded elevation data (hwww.ngdc.noaa.gov/mgg/global/seltopo.html), all available on-line. Demersal fish species richness in the oceans surrounding New Zealand was predicted using remote-sensing based on estimation of chlorophyll-a concentration and spatial gradients of annual sea surface temperature, among other predictors (Leathwick et al., 2006a). Variables describing climate, the physical nature of the river basin, and historical events such as past glaciation, were used to model the distributions of diadromous fish among 196 river basins in Europe, North Africa and the Middle East for conservation planning purposes (Lassalle et al., 2008).

In another recent example, temperature, salinity and depth were examined for correlations with fish stocks, but in this case, temperature was derived from ship-based observations, and the bathymetry data used were higher-resolution (15-second) from the US Geological Survey (Sundermeyer et al., 2006). The ship-based CTD (conductivity, temperature, depth) profiles were spatially interpolated to derive surface and bottom temperature maps for the time period of the study. A map of bottom sediment type was available from a published source. There are growing efforts to map near-shore marine habitats (substrate type) using technologies such as side-scanning sonar (McRea et al., 1999). At an even finer scale, substrate type (from existing maps), depth (100-m resolution) and salinity (from a hydrodynamic model) were used to model habitat suitability for invasive non-native bivalve species in harbors (Inglis et al., 2006).

Some studies have successfully used "landscape" pattern variables as predictors in marine and aquatic species distribution modeling. Measures of habitat structure or the spatial arrangement of substrate types (Section 5.3.4) have been used to predict species distributions (Sheppard *et al.*, 2007; Whaley *et al.*, 2007), as have other landscape variables such as distance to an estuary or land-sea interface, and exposure (Beger and Possingham, 2008).

Although there are fewer published studies of marine than terrestrial SDM, these examples show that potential environmental variables are numerous because of extensive remote sensing of the oceans for the purpose of studying ocean dynamics, and ocean–atmosphere processes. However, the dynamic nature of the predictors is the rule rather than the exception in marine studies, and this can pose significant challenges for marine SDM when using the correlative statistical models emphasized in this book.

A growing number of SDM studies for aquatic (freshwater) species use predictors that reflect the hydrological and chemical conditions and substrate quality of the stream segment, and the land characteristics of the watershed or catchment area, including land cover and land use (Manel *et al.*, 1999a; Leathwick *et al.*, 2006b; Lohse *et al.*, 2008).

5.5 Summary

Species distribution models should use variables related to the primary environmental regimes of heat, light, moisture, and nutrients (Chapter 3). These should include both broad-scale climate variables (Table 5.1), and finer-scale predictors that capture terrain-mediated variation in water, energy, and nutrient availability. Species distribution models that rely only on climate predictors can produce only a very rough approximation of those distributions.

Climate variables are generally derived from interpolated climate station records, and climate variables relevant to biotic distributions range beyond annual averages of precipitation and temperature. Seasonal extremes of temperature and moisture, and other variables (Growing Degree Days) are related more strongly to factors limiting growth and survival than are annual averages (Table 5.1). Finer-scale factors are generally derived from digital topographic data, as well as maps of soil or substrate types or properties. Physically based models or indices of topographically distributed energy and moisture balance are superior to indirect variables such as elevation and aspect. The propagation of DEM errors through these models, as well as through the SDM itself, should be considered. Variables representing the availability of mineral nutrients are rarely mapped explicitly, but for SDM are often derived from soil and geology maps and databases.

In addition to these primary environmental regimes, SDMs for animal species should include predictors related to the availability of food, cover and water. Often, maps of vegetation type or structure, and measures of landscape structure (such as proximity to surface water, or habitat composition within a given area), are important predictors in wildlife habitat modeling. These predictors can be derived from remote sensing of land cover (via thematic mapping), or by using biophysical indices based on remotely sensed data as measures of productivity, seasonal patterns of productivity, species diversity (via spectral diversity), habitat structure, or habitat heterogeneity. Including remotely sensed variables as predictors in SDMs improves their performance because these spectral measurements and their derivatives are directly related to biophysical properties of the earth's surface that, in turn, are linked to the primary environmental regimes and to habitat quality (productivity, vegetation structure, land cover type, available moisture).

In an evaluation using artificially generated species distribution data, Austin *et al.* (2006) found that the choice of environmental predictors had a large influence on SDM performance. Further, they emphasized that a process-based understanding of the relationship between direct gradients related to species distributions, and those indirect variables that are often used as predictors, is required. For example, understanding the form of the relationship between slope angle, slope aspect and solar radiation, or between elevation and temperature, helps the modeler interpret the response curves estimated during modeling if indirect variables are used as surrogate predictors.

In a very useful paper on error and uncertainty in SDMs, Barry and Elith (2006) pointed out that no species modeling study is likely to have a comprehensive set of proximate predictors in mapped form. They recommend that conceptual models linking species distributions to causal factors (Chapter 3) should be used to assemble the most appropriate variables from available data, but that a "sufficient" (rather than exhaustive) set of variables should be sought that allows models with acceptable levels of error to be specified. It is very likely that distal or indirect variables will have to be used in some cases.

Part III

An overview of the modeling methods

"Any mechanistic process model of ecosystem dynamics should be consistent with a static, quantitative and rigorous description of the same ecosystem"

(Austin 2002, p. 112)

This section addresses the third part of the framework presented by Austin (2002), outlined in Chapter 1 and used as an organizing principle for this book – the statistical part. Austin's statistical modeling framework includes the choice of modeling methods and decision regarding implementation (calibration and validation) of a model. Some appropriate and widely used methods in SDM are not statistical in the strict sense, and so we can more broadly refer to quantitative and rule-based empirical models. In any case the methods included are explicit and the modeling is repeatable.

Guisan and Zimmermann (2000) divided the statistical modeling portion of Austin's framework into four steps: (a) conceptual model formulation, (b) statistical model formulation, (c) calibration (fitting or estimation), and (d) evaluation. Those steps provide a useful outline for this section. Conceptual model formulation in species distribution modeling generally relies on a number of key ecological concepts and was described in Chapter 3. Guisan and Zimmermann emphasized that species distribution models are usually empirical or phenomenological models, designed to condense empirical facts, and are judged on their ability to predict, that is, judged on their precision and reality. This distinguishes them from distinct but complementary mechanistic (process) models, that aim to be general and realistic, and from analytical (theoretical) models, built for generality and precision. The complementarity of these three modeling approaches is the subject of Austin's comment, quoted above.

Guisan and Zimmermann define statistical model formulation as choosing a type of model that is suitable for the data – often, statistical models are specific to a type of response variable (to use traditional statistical terminology) and its probability distribution (Franklin, 1995). What is the measurement scale of the response variable: interval, ratio, ordinal, counts, or categories? While many of the methods considered in this book are remarkably flexible with regard to the measurement scale and distribution of the predictor and response variables, some have advantages over others when applied to certain kinds of data, and certain forms of the relationship between response and predictors, as summarized in Table III.1. For example, categorical variables can be used as predictors in regression modeling – each category is simply treated as a "dummy" variable, that is, with a value of zero or one for each observation. But, when there are many categories, for example, if soil series is being used to predict the occurrence of a plant species, it can be difficult to estimate and interpret meaningful parameters for all the categories that occur in the data. Decision trees (regression or classification trees) handle this problem very easily. In addition, decision trees automatically identify interactions between variables, while interactions must be specified in advance in regression modeling. Hopefully, these kinds of differences will become clear with more examples in the chapters of this section. Elith and Leathwick (2009; their Table 6.2) also provide a summary of the key features of modeling methods and associated software with respect to types of species data, complexity of fitted functions, and ability to estimate model uncertainty.

As was discussed in Part II (Chapters 4 and 5), the species occurrence and environmental data typically used in SDM have some particular characteristics. The observations are geographically referenced. Species occurrence, abundance, or some other measure, is recorded from surveys, sometimes as an ordinal (none, few, many) or binary (presence or absence) variable. Alternatively, records of species occurrences are available from natural history collections, atlases, or other kinds of observations that are not from probability-based samples, and that only provide information on species occurrence (and not species absence). If there are no geographically located observations, other kinds of information on habitat preference might be used to develop a model (Chapter 8). With multiple environmental predictors, collinearity is an issue to be concerned about, and environmental predictor variables often represent surrogates or indirect gradients, rather than proximal drivers of species abundance or fitness. Those mapped predictors used in SDM often include a mixture of categorical and continuous variables, the relationship between predictors and response is not expected to be linear, and interactions

Table III.1. *Modeling methods that can be applied to quantitative, categorical or binary (presence–absence) response variables*

Modeling method	Best for type of response	Predictors (covariates)	Response function
Statistical			
Discriminant function analysis	Categorical	Quantitative	Linear
Generalized linear model (GLM)	Quantitative, categorical binomial	Quantitative, categorical	Parametric – linear, polynomial, piecewise, interaction terms
Spatial autoregressive models	Quantitative, categorical binomial	Quantitative, categorical	Same as GLM. Includes autocovariate
Generalized additive model (GAM)	Quantitative, categorical binomial	Quantitative, categorical	Smoothing function, estimated using local regression, splines or other method
Multivariate adaptive regression splines (MARS)	Quantitative, categorical	Quantitative, categorical	Adaptive piecewise linear regression (combines splines and binary recursive partitioning)
Machine learning			
Decision tree (DT) – classification and regression trees	Quantitative, categorical multinomial	Quantitative, categorical	Divisive, monothetic decision rules (thresholds) from binary recursive partitioning
Ensemble trees (bagging, boosting, random forests)	Quantitative, categorical multinomial	Quantitative, categorical	Weighted and unweighted model averaging applied to decision trees
Artificial neural network (ANN)	Categorical multinomial	Quantitative, categorical	Non-linear decision boundaries in covariate space

(cont.)

Table III.1. (cont.)

Modeling method	Best for type of response	Predictors (covariates)	Response function
Hybrid methods (GARP)	Categorical binomial	Quantitative, categorical	Combines decision rules using a genetic algorithm (response functions not visualized)
Maximum entropy	Categorical binomial	Quantitative, categorical	Non-linear response functions can be described
Other			
Distance methods	Categorical "event only"	Quantitative	Similarity to conditions where event occurs: does not estimate response function or importance of predictors
Expert methods	Any	Quantitative, categorical	Response functions and weights defined by expert knowledge

The statistical methods GLM, GAM and MARS can handle dependent variables with Gaussian, binomial, Poisson, ordinal, or other distributions via a link function. Categorical variables can be binomial or multinomial. "Event-only" refers to data documenting the occurrence of a species only, and not its absence, that is, not resulting from a probability based sample design aimed at documenting occurrence and absence (Chapter 3). Multivariate "distance methods" are described in Chapter 8.

between predictors are expected. These are all factors to keep in mind when formulating a statistical model for SDM.

Model estimation or fitting (called calibration by some) is usually taken to mean estimation of model parameters, for example, the coefficients in a regression model, the splitting rules in a decision tree, or the weights in an artificial neural network. Model fit is evaluated, based on some measure of the agreement between the model and the data (Rykiel, 1996), such as variance reduction in traditional regression, deviance reduction in modern regression, or information criteria such as the Akaike Information Criterion (AIC, discussed in the following chapter). Model estimation is also frequently investigated by some kind of analysis of model residuals (the pattern of differences between the modeled cases and the data), including spatial analysis. By its very nature, SDM almost always examines multiple environmental variables (Chapters 3, 5) as potential predictors of species' distributions, and calibration usually involves selecting a subset of candidate explanatory variables. Model fitting also includes the transformation of variables, including polynomials of, and interactions between, candidate explanatory variables, with attention to the expected shape of the species' response curves (Chapter 3).

In the model evaluation step, many criteria could be used for validating the output of a model of species–habitat relations (Chapter 10 in Morrison et al., 1998). Evaluation is distinct from model fitting when the model is used to make predictions based on new or different data. If a strictly independent dataset with suitable attributes is not available, it is common to divide the dataset into "training" and "testing" data prior to modeling, or to use some kind of resampling method (such as bootstrapping) to estimate, from the training data, what the prediction accuracy of the model would be if it were applied to new data. This is the subject of Chapter 9.

My understanding of the quantitative empirical models used in SDM has been greatly influenced by Hastie et al. (2001), and some fundamental concepts from that excellent book are a good place to begin this section. (Note that, while I relied on the 2001 edition, a second edition was published in 2009). The organizing principle for their book is the idea of learning from data. In classical hypothesis testing in statistics, relatively small amounts of data result from controlled experiments that enforce certain restrictions in order to meet a set of criteria and assumptions. Although exploratory data analysis has long been part of scientific problem solving (Tukey, 1977), today more than ever many problems in science, engineering, business, marketing and medicine start with very large datasets – many observations and many variables. The objective is to learn from the data and build a model that can make accurate predictions, for example of the price of a stock, the chances a patient will have a heart attack, or the likelihood that a pixel in a satellite image belongs to a certain land cover class. Hastie *et al.* refer to this as the supervised learning problem (what some of us from remote sensing background call supervised classification). Observations of the outcome (response variable) are used to guide the learning – in other words, to estimate the rules or parameters –in order to calibrate the model. The observations of the predictor and response variables are called the training data, and as noted above, a separate set of test data (or a cleverly fabricated "independent" dataset) is used to validate the ability of the model to predict.

Because there have been such great changes in statistical and computational methods in recent decades, it is somewhat difficult to classify them or group them into chapters, as the distinction between classical statistical approaches, like regression, and various other methods of supervised "learning from data," has become blurred. In this section I have grouped the models into: statistical (emphasizing regression; Chapter 6), machine learning (Chapter 7), and classification and distance methods (Chapter 8). Chapter 8, in particular, addresses methods that are widely used for "presence-only" data on species occurrence. This grouping and progression makes sense to me – however, it is certainly not the only way that these methods can be related one to another. In each chapter an overview of model formulation and estimation is given. An alternative approach to mapping species distributions, direct interpolation of species data, is really a form of modeling done in geographical space, using only location, or proximity to species records, as predictors. This was discussed in Section 2.3, but is also relevant to issues considered in Chapter 8. In Chapter 9 I discuss model validation or verification, because most of the validation concepts and approaches used in SDM can be applied to all types of models.

There are now a number of both general purpose and dedicated software packages and systems available for SDM, as shown in Table III.2 (see also Table 3 in Guisan & Thuiller, 2005 and Table 4 in Elith *et al.*, 2006). Some are proprietary, some are open source, and some are free. Some include environmental data (primarily low-resolution global climate maps). A list is provided here with the caveat that it is undoubtedly incomplete and will be out of date by the time you read it. However, it will help in identifying software resources for SDM.

Table III.2. *Examples of stand-alone or dedicated software for species' distribution modeling*

Software	Algorithm family	URL, reference	Additional software
ANUCLIM	EE	http://cres.anu.edu.au/outputs/anuclim.php	–
BIOCLIM	EE	http://cres.anu.edu.au/outputs/anuclim/doc/bioclim.html; (Busby, 1991)	BIOCLIM ArcView ® extension; DIVA-GIS
BIOMAP	EE	http://cres.anu.edu.au/outputs/anuclim.php	–
BIOMAPPER	Ecological niche factor analysis (ENFA)	www.unil.ch/biomapper (Hirzel *et al.*, 2002)	–
BIOMOD	GLM, GAM, CART, ANN	At the discretion of the author (Thuiller *et al.*, 2003)	–
DIVA	EE	http://www.diva-gis.org (Hijmans *et al.*, 2001)	–
DOMAIN	Gower-similarity; multivariate distance	www.cifor.cgiar.org/scripts/default.asp? ref=research_tools/domain/index.htm (Carpenter *et al.*, 1993)	–
ECOSPAT	Logistic regression (GLM, GAM)	http://uiwadmnweb.uwyo.edu/wyndd/ http://www.ecospat.unil.ch; with permission of the developer	–
GARP	Genetic algorithms	http://lifemapper.org/desktopgarp (Stockwell & Peters, 1999)	–
GARP/ WhyWhere?	WhyWhere	http://landshape.org/enm/whywhere-22-download/ (Stockwell, 2006);	–
GRASP	Logistic regression (GLM, GAM)	https://www.unine.ch/cscf/grasp (Lehmann *et al.*, 2002)	S-Plus ® or R
HyperNiche	Non-parametric multiplicative regression	http://home.centurytel.net/~njm/hyperniche.htm (McCune, 2006)	–

(*cont.*)

Table III.2. (*cont.*)

Software	Algorithm family	URL, reference	Additional software
MARS	Multivariate adaptive regression splines	www.salford-systems.com/mars.php;	MDA package ®
MaxEnt	Maximum Entropy	http://www.cs.princeton.edu/~schapire/maxent/ (Phillips *et al.*, 2006)	–
NNETW	Artificial neural networks		S-plus ® w/libraries (nnet, NNETW)
OpenModeller	GARP, bioclimate envelopes	http://openmodeller.sourceforge.net/	–
PRESENCE	Logistic regression	http://www.mbr-pwrc.usgs.gov/software/presence.html (MacKenzie *et al.*, 2002)	–
Presence Absence	SDM performance evaluation (Kappa, AUC, etc.)	http://cran.r-project.org/web/packages/Presence Absence/index.html	R
SAM	OLS, Autoregression	http://www.ecoevol.ufg.br/sam/ (Rangel *et al.*, 2006)	–
STATMOD ZONE	CART	http://www.gis.usu.edu/~chrisg/avext/	ArcView and S-plus

Table modified from Guisan and Thuiller (2005), courtesy of J. Miller. Note that many methods, including GLM, GAM, CART, ANN, MARS, and Mahalanobis distance, are also available in proprietary or open source general purpose statistical and/or GIS software, either integrated into the software or available as additional modules or packages CART = classification and regression trees; EE = Environmental (climatic) envelope; GAM = generalized additive models; GLM = generalized linear models; OLS = ordinary least squares regression.

6 · *Statistical models – modern regression*

Janet Franklin and Jennifer A. Miller

6.1 Introduction

In this chapter we will review the linear (regression) model, the generalized linear model (flexible, modern regression) and related statistical models that are frequently used in species distribution modeling. We will also discuss the statistical treatment of spatial autocorrelation in SDM, and review some classic and recent applications of these modeling methods in SDM.

The methods discussed in this chapter include generalized linear models (GLM), generalized additive models (GAM), both used in SDM since the late 1980s and early 1990s, and more recently arrived on the scene, multivariate adaptive regression splines (MARS). Multivariate and Bayesian statistical approaches are also discussed. These abbreviations and names become more daunting and numerous as the years go by, and can make the choices of modeling methods seem overwhelming if they are new to you. Those of us who have been around for a while just learn to accept, even embrace, the funny names and try to rise to the challenge of understanding the nuances and strengths of new methods. Optimistically, these improvements offered by statisticians, econometricians, data miners, and engineers, may better fit our data and objectives than what we already have in our toolbox. However, these methods need continued testing in different problem domains and there may not be a magic bullet for all problems and data configurations. At the very least, we should be able to understand the method we use in relation to other commonly used alternatives and be able to justify our choice and apply the method correctly. We hope that this and the following chapters will help with that objective.

It is impossible to treat this topic comprehensively in a single chapter, and anyone doing much of this type of modeling will probably need to consult additional references. For example, Chapters 1, 2 and 4 in McCullagh and Nelder (1989), *Generalized Linear Models* (2nd edn), give a great overview of logistic regression. Two other often-cited reference books about logistic regression are *Applied Logistic Regression* by Hosmer & Lemeshow (2000), and *Categorical Data Analysis* by Agresti (1996). Already mentioned in the overview of Section III, *Elements of Statistical Learning* by Hastie *et al.* (2001; and the 2009 2nd edn) is a comprehensive treatment of the modern approach to statistical learning from data, and places GLMs within the context of other supervised and unsupervised learning methods. In addition, although they are tied to implementations of GLMs, in particular statistical software packages, namely R, S or S-Plus, Hastie and Pregibon's (1992) chapter in *Statistical Models in S* (Chambers & Hastie, 1992), and Chapter 6 in Venables and Ripley (1994) both contain fairly detailed descriptions of the modeling framework (see also more recent contributions by Maindonald & Braun, 2003; Crawley, 2005). Guisan *et al.* (2002) give a nice overview of these statistical models specifically with reference to species distribution modeling. Chapter 5 in Fortin and Dale (2005) provides a great general overview of how spatial autocorrelation is addressed in spatial analysis of ecological data. In addition to these resources, we encourage everyone to consult their own favorite statistical text for an explanation of the linear model, the generalized linear model, and the important issues of variable transformation, model fitting, regression diagnostics and model selection in multiple regression.

Once again, we turn to Hastie *et al.* (2001) to set the stage. They presented two alternative simple approaches to prediction, the linear model, fit by least squares, and the *k*-nearest-neighbor method. A local method, such as a *k*-nearest neighbor prediction rule, simply finds the *k* observations closest to a value of *x* and averages them, and is the least generalized and most data-driven approach to learning from data. It is rarely used in SDM, but some of the methods discussed in this section have some of the qualities of a *k*-nearest-neighbor classifier in that they estimate local, rather than global parameters or rules. The linear model is a mainstay of statistics, and can be used to predict quantitative outputs (regression) or qualitative, categorical outputs (linear discriminant analysis).

6.2 The linear model

The linear multiple regression model predicts the output or response variable (also referred to as the dependent variable), Y, from a vector of multiple inputs or predictor variables (also called explanatory variables, independent variables, or covariates), $X = (X_1, X_2, \ldots, X_p)$:

$$\hat{Y} = \hat{\beta}_0 + \sum_{j=1}^{p} X_j \hat{\beta}_j + \varepsilon \qquad 6.1$$

where $\hat{\beta}$ is the vector of estimated coefficients, and $\hat{\beta}_0$ is an estimated constant known as the intercept. It is assumed that the error term, ε, is normally distributed with zero mean and constant variance, and the variance of Y is constant across observations.

As Hastie *et al.* (2001) point out, the linear model makes a lot of assumptions about the structure of the data and yields stable but possibly inaccurate predictions –this global method uses all the data to estimate a linear relationship between the Xs and Y, and those authors would say that it has low variance but high bias. In other words, if you had another sample of the same population, you would tend to estimate the same model parameters (low variance). This is sometimes called a model-driven approach to modeling. In contrast, a local method, such as a k-nearest neighbor prediction rule, simply finds the k observations closest to a value of x and averages them. This more data–driven approach makes very few assumptions and can make accurate predictions for cases similar to the training data but they may be unstable when confronted with novel data. This approach has high variance but low bias, meaning that if you had another sample of the same population you would be likely to estimate a very different model. Many advanced methods of statistical learning fall between these two extremes, containing elements of both, for example kernel methods, local regression, and even neural network models. Some of these will be discussed in this section of the book.

6.3 Generalized linear models

Ecological data often violate the assumptions of the linear model. Generalized linear models (GLM) are extensions of linear models that can cope with non-normal distributions of the response variable (Venables & Ripley, 1994). GLMs provide an alternative to transforming the response variable and then applying the linear model. Distributions that can be used to characterize Ys in ecology, in addition to Gaussian, include Poisson, binomial, negative binomial, and gamma. These are collectively referred to as the exponential family of distributions. For example, if Y assumes values over a limited range, or if variance changes systematically with the value of Y, it may be characterized by a Poisson distribution. If Y is a binary variable (it can assume two states, such as dead/alive), it may be described by a binomial distribution. The linear model can be generalized using what is called a link function that describes how the mean of Y depends on linear predictors, and a variance function that describes how the variance of Y depends on its mean (Chambers & Hastie, 1992, p. 192). The generalized linear model can be expressed as:

$$g(E(Y)) = LP = \hat{\beta}_0 + \sum_{j=1}^{p} X_j \hat{\beta}_j + \varepsilon \qquad 6.2$$

where the predictor variables (far right side of the equation) are combined to produce a linear predictor, LP, and the expected value of Y, $E(Y)$, is related to the LP through the link function, $g()$. So, formulating a generalized linear model for SDM includes selecting the response distribution and the link function (together called the family of the GLM), the variance function, and the predictor.

The link function or inverse function describes how the mean of Y depends on the linear predictor, or "links" the expected values of Y to the predictors. For example, if the response is binary, the binomial distribution is used to describe the distribution of Y, and the logit link is used. This link function is very widely used in species distribution modeling because the species occurrence data often consist of presence and absence observations (a binary response), and so we will give some attention to this example. A GLM with a binomial family (logit link) is commonly called "logistic regression." The logit link is:

$$LP = \log\left(\frac{\mu}{1-\mu}\right)$$

or "logit" μ – the "log odds" ratio of the probabilities of classes 1 and 0 – where $\mu = E(Y)$ is the probability of class 1, and $(1 - \mu)$ is the probability of class zero, e.g., the probabilities of class 1 and class 0 must sum to one.

The coefficients (β) of the predictors in a logistic regression model can be conveniently interpreted, as in the following hypothetical example. If a predictor, such as average annual temperature, has a positive coefficient of 0.0249 in an estimated model of the occurrence of a species, this implies that a one unit increase in temperature results in an increase of $\exp(0.0249) = 1.025$ (the log-odds ratio), or 2.5%, in the probability of species presence.

The LP produces fitted values and predictions in units of that LP – for example, predicted values from the LP of a logistic regression are on the logit (log odds) scale. An inverse transform is then applied to the LP in order to predict the values on the response scale (in units of the response variable). In logistic regression we are often interested in the predicted probability of a class, μ (e.g., probability of species occurrence, or tree death, or heart disease, and so forth) rather than the log odds ratio, so by inverting the above equation we can solve for μ.

$$\mu = \left(\frac{e^{LP}}{1 + e^{LP}} \right)$$

The variance function describes how the variance of Y depends on the mean. For example, the binomial variance function for the logistic regression model is:

$$V(\mu) = \mu(1 - \mu).$$

For another example, the Poisson link,

$$LP = \log(\mu).$$

Inverting this, $\mu = (e^{LP})$, and the variance function is $V(\mu) = \mu$. Some distributions can have more than one link. Note that the Gaussian family is a "special case" of the GLM with the identity link, $LP = \mu$, and the variance function $= 1$. In other words, this is good old regression, or the linear model described in equation 6.1 – called the general (not generalized) linear model in some texts.

The steps involved in developing a generalized linear model for SDM or other types of data analysis are discussed in the following subsections and outlined in Fig. 6.1.

6.3.1 Transformations of the predictors

Non-linear responses can be achieved by including additional transformations of the predictors. As in the case of a linear model, flexibility in the specification of predictor variables in GLMs can be achieved by including polynomial terms, for example:

$$\hat{Y} = \hat{\beta}_0 + X_1\hat{\beta}_1 + X_1^2\hat{\beta}_2 + X_1^3\hat{\beta}_3 + \varepsilon.$$

In this way, a linear model can fit a curvilinear response (as an alternative to estimating a non-linear model, e.g., one that is non-linear in the parameters).

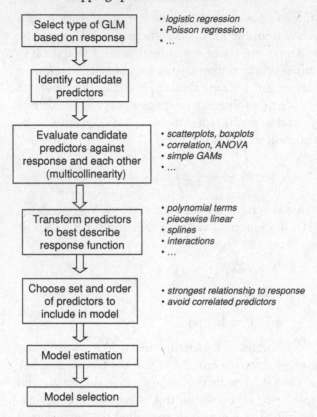

Fig. 6.1. Steps involved in developing a generalized linear model for SDM or other types of data analysis, described in Section 6.3.

Flexibility can also be achieved by specifying a piecewise linear function $(X > t)(X - t)$ where t is a threshold value of X, below which there is no effect of X on Y $(\beta = 0)$ and above which a linear coefficient $(X - t)$ is estimated. Parametric splines such as beta functions can be used to describe a non-linear response, which may be particularly appropriate for estimating skewed species responses (Austin *et al.*, 1994) expected from niche theory (Chapter 3).

Ordinal response variables (ordered factors) can be handled with generalized linear models (Guisan & Harrell, 2000). Binary or unordered factors (categorical variables) are treated as dummy variables, that is, for N categories, $N - 1$ new variables are created and cases are given values of 0 or 1 for each new variable. If a categorical variable has many classes, it requires a lot of data to parameterize – is uses many degrees of freedom in the GLM, and observations would need to be well distributed among all classes in order to estimate coefficients. Therefore, categorical variables should be simplified or aggregated into the smallest number of classes that are ecologically relevant and well represented by the data. This is especially relevant to SDM when these categorical variables come from a thematic GIS map (Chapter 5), the classification of which was developed for general purposes and may have many categories (e.g., soil type, land cover, vegetation class).

Another form of variable transformation often used in multiple regression is to include an interaction term between two or more predictors, that is, if multiplicative, not additive, effects of the predictors are expected. An interaction term is easily added by creating a new variable that is the product of the predictors. This is usually done on predictors where values have been centered on zero by subtracting their mean, to avoid collinearity of interaction terms with the lower-order terms (Quinn & Keough, 2002). Considering all possible pairwise and higher-order interactions is sometimes practiced in stepwise procedures (see below). However, with a large number predictors, the number of interaction terms can become very large. It has been our practice to only include interaction terms as candidate predictors in SDM if an interaction is hypothesized based on ecological theory. For example, interactions between temperature and precipitation variables have been hypothesized and found to be significant (Miller & Franklin, 2002).

6.3.2 Model estimation

Statistical model estimation or fitting involves defining a measure of goodness of fit between observed and fitted values, and estimating parameter values that minimize that measure, often the "deviance." Minimizing the deviance is done by maximizing the log-likelihood of the parameters for observed data (by setting their derivatives to zero), and is usually solved by an iteratively reweighted least squares (IRLS) algorithm. This is equivalent to minimizing the scaled deviance $D*(y; \mu)$ for observation y with mean $E(Y) = \mu$ and variance σ^2. Residual deviance corresponds to residual sum of squares in a linear model. The deviance between two models is the difference of their residual deviances (again, analogous to sum of squares).

In regression modeling it is typical to examine the patterns of the residuals of observations, in order to evaluate if the assumptions of the model are met (namely, linear models assume normally distributed errors) and to detect outliers, or observations that have a large influence on the estimated parameters. The graphical methods and tests used are often called regression diagnostics, and include plots of the residuals versus fitted values of Y, and tests of the influence of each observation on the fitted model. Further, the spatial pattern of the residuals, or lack thereof, may provide an important clue about whether the model has been mis-specified or has sufficiently addressed spatial dependence in the data (discussed in Section 6.7).

For GLMs, residuals can be described using different units or measurement scales. This is useful to keep in mind when using statistical software because the measurement scale used may not be the one the user was expecting. For example, the modeler may be most interested in residuals expressed in units of the response variable, whereas a particular program may display the residuals in units of the linear predictor. Caution is warranted when examining regression diagnostics so that they are interpreted correctly. There are four different measurement scales for residuals of GLMs (Chambers & Hastie, 1992, p. 205), whereas for plain old linear models (Gaussian families), all four types are identical. The four types of residuals are:

Response residuals are given in the units of the response variable, simply:

$$y - \hat{\mu}.$$

Deviance residuals are the signed square roots of the summands of the deviance,

$$r_i^D = \text{sign}(y - \hat{\mu})\sqrt{d_i},$$

where d_i is the contribution of the ith observation to the deviance, taking the same sign as $y - \hat{\mu}$. Deviance residuals are the units of the LP (linear predictor), so, for example, in logit regression the deviance residuals would be in units of the log odds ratio. By definition, the residual deviance of the model is:

$$D_M = \sum (r_i^D)^2.$$

Pearson residuals are normalized by the variance:

$$r_i^P = \frac{(y_i - \hat{\mu}_i)}{\sqrt{V(\hat{\mu}_i)}}.$$

Their sum of squares is the χ^2 statistic (sum from 1 to n):

$$\chi^2 = \sum \frac{(y_i - \hat{\mu}_i)^2}{\sqrt{V(\hat{\mu}_i)}}.$$

Working residuals are the difference between the "working" response and the linear predictor in the final iteration of the IRLS algorithm.

The residual deviance of a model is defined as twice the difference between the maximum achievable log likelihood and that attained under the fitted model (McCullagh & Nelder, 1989, p. 118). "The -2LL (log likelihood) statistic is the likelihood ratio. It is also called goodness-of-fit, scaled deviance, deviation chi-square, D_M, or L-square" (http://www2.chass.ncsu.edu/garson/PA765/logistic.htm). For the normal (Gaussian) distribution (summed from 1 to n observations), residual ("unexplained") deviance of a model D_M is simply the residual sum of squares (Chambers & Hastie, 1992, p. 243):

$$D_M = \sum_i (y_i - \hat{\mu}_i)^2$$

For the Poisson family (Poisson regression), residual deviance is calculated as (McCullagh & Nelder, 1989, p. 197):

$$D_M = 2 \sum \{y \log(y/\hat{\mu}) - (y - \hat{\mu})\}$$

For the binomial family (logistic regression) (McCullagh & Nelder, 1989, p. 118):

$$D_M = 2 \sum \left\{ y \log(y/\hat{\mu}) + (m - y) \log \left(\frac{m - y}{m - \hat{\mu}} \right) \right\}$$

where m is a weight for each observation.

D_m represents the unexplained (residual) deviance. With a slight adjustment, it can be used as a measure of model fit, equivalent to R^2. This adjusted measure, D^2, represents the "percent deviance explained by the model." Analogous to adjusted R^2 in a multiple linear regression, we can define adjusted D^2 as:

$$adj\_D^2 = 1 - \left[\frac{(n-1)}{(n-p)} \right] * [1 - D^2] \quad \text{where, } D^2$$

$$= \frac{(\text{Null deviance} - \text{residual deviance})}{\text{Null deviance}}$$

n is the number of observations and p is the number of parameters in the model (Guisan & Zimmermann, 2000). Although other measures than goodness-of-fit have been recommended for logistic regression (Hosmer & Lemeshow, 2000), D^2 is widely used. It is generally suggested that D^2 can be used to compare different models based on the same data, e.g., model selection.

6.3.3 Model selection and predictor collinearity

Whether the goal of modeling is to understand the role of the predictor variables in explaining the outcome, that is, inference or parameter estimation, or whether the goal is accurate prediction, searching for a parsimonious model that uses an appropriate subset of all available predictor variables, their non-linear transformations, and interaction terms, is important. This can be done manually, or it can be automated, by:

- Backwards elimination – removing insignificant terms from the model;
- Forward selection – adding significant terms to the model;
- "Stepwise" procedures (usually automated) using forward and backward strategy (sometimes called best subsets regression).

This is called variable selection or model selection because, essentially, different models are estimated using different subsets of predictors and then compared, by some criterion, in terms of their fit to the data.

In multiple regression, the presence of correlated explanatory variables, called multicollinearity, can render variables that may actually be causal/proximal to be insignificant (because a correlated variable is already in the model) or even reverse the sign of the coefficient. So, the order in which variables are considered (entered into the model) does matter. Before or during model selection, predictor variables can be examined for correlation or multicollinearity. Predictors that are strongly correlated with other predictors can be left out of the model prior to model selection. Logically, the predictors left out should be ones that are less strongly correlated with the response variable, and this can be determined in exploratory analyses of the relationships between single predictors and species response. However, another alternative is transforming one of the correlated predictors in relation to another, for example using normalization methods (Leathwick et al., 2005). This seems to be a very useful but relatively underutilized tool in SDM (Elith & Leathwick, 2009).

Recalling the statistical learning perspective, it is likely that, given the ever-present limitations of our data, a number of different models will be consistent with the data – they will be more or less equally plausible and it will be impossible to distinguish which among them is "best" or "correct" on the basis of the available data and existing theory. There are two alternative and somewhat philosophically different approaches to this problem, hypothesis testing and the information-theoretic approach (Burnham & Anderson, 1998). The latter is being used with increasing frequency in species distribution modeling and related ecological applications (e.g., Hoeting *et al.*, 2006).

In the hypothesis testing approach, an "analysis of deviance" (analogous to ANOVA) can be used to perform successive tests, adding variables in the order they are given (so, again, the order affects the results). The difference in the residual deviance between models as variables are added is approximately χ^2-distributed with the given degrees of freedom, and so a χ^2 test should be used to evaluate the significance of the change in fit (difference in residual deviance) between successive models. This tests the null hypothesis that there is no significant difference between two models at some specified significance level. The χ^2 test can be used as the criterion for model selection in stepwise procedures (however, the Akaike Information Criterion can also be used as the criterion; see below).

In the information-theoretic approach, a statistic that describes model parsimony, based on fit and number of parameters, is used as evidence to rank models and quantify the degree of difference between. It has been explicitly cautioned that this approach not be used with all possible subsets (automated stepwise) approaches to variable selection, because to do so is to ignore model selection uncertainty (Anderson & Burnham, 2002). This refers to the situation where several models have very similar levels of support but differ in the fitted terms and parameters. Stepwise variable selection seems to violate the spirit of this approach, where the modeler should propose a small number of alternative hypotheses and then determine to what degree the evidence is consistent with them. However, in practical terms, a stepwise approach may produce the same outcome because the few models selected *a priori* are nested within all possible subsets (of predictors).

The Akaike Information Criterion (AIC) (Venables & Ripley, 1994, p. 187), is a statistic that is commonly used in the information-theoretic approach to model selection, although there are others (Burnham & Anderson, 1998). The AIC is a measure of goodness of fit that takes into account the number of parameters:

$$AIC = D_M + 2p\hat{\varphi}$$

where D_M is the unexplained or residual deviance (as above), p is the number of parameters, and ϕ is scale factor. The scale factor is equal to the variance for the Gaussian family of distributions. The scale factor can be assumed to be equal to one if unknown, but must be held constant when comparing models. A smaller AIC (lower unexplained deviance) means a "better" model – that is, there is more deviance explained per number of explanatory variables (parameters estimated). Typically, hypothesis testing is not used to compare AIC values – rather, measures of the magnitude of the difference are examined. The AIC and related measures such as the Bayesian Information Criterion or BIC have been used for model selection in SDM (e.g., Li *et al.*, 2009).

Hypothesis testing and the information theoretic approach are not the only ways to select subsets candidate predictors (select models), and may not result in fundamentally different outcomes, in spite of their philosophical differences. Other "cutting edge" ways of selecting models are the so-called regularization procedures that combine variable selection with coefficient estimation such as shrinking the coefficients, lasso, and ridge regression (Hastie *et al.*, 2001; Reineking & Schröder, 2006; discussed in Elith & Leathwick, 2009).

6.3.4 Use of GLMs in species distribution modeling

GLM, in particular logistic regression, is one of the best established statistical frameworks for SDM (see Austin & Cunningham, 1981; Margules *et al.*, 1987). Case studies using GLM for predicting species distributions (and related biotic variables) were extensively reviewed in Franklin (1995), Guisan and Zimmermann (2000) and Guisan *et al.* (2002). A classic paper that was one of the best early expositions on the application of GLMs to species distribution modeling was by Nicholls (1989), and it is well worth reading. In that study, logistic regression models and spatial predictions are developed for several tree species (in the genus *Eucalyptus*) in southeastern Australia. The paper illustrates several of the components of statistical modeling discussed above including model fitting, variable or model selection, and examining residuals, as well as other model diagnostics, in a number of useful ways.

To cite just a few other examples, GLMs (primarily logistic regression) have been used to model species distributions of birds (Osborne

& Tigar, 1992; Tobalske, 2002), bats (Jaberg & Guisan, 2001), amphibians (Johnson et al., 2002), seabirds (Johnson & Krohn, 2002; Olivier & Wotherspoon, 2008), vegetation communities (Brown, 1994), fishes (Beger & Possingham, 2008) and multiple taxonomic groups (Pearce & Ferrier, 2000; Manel et al., 2001). A series of papers by G. S. Cumming (2000a,b, 2002) used logistic regression to model the distribution of ticks in Africa and addressed a number of interesting issues of sampling and model validation.

Another classic paper by Vincent and Harworth (1983) described Poisson regression models of species abundance, in contrast to the great majority of SDM studies that use logistic regression of species occurrence. There are also more recent examples of the use of Poisson regression in SDM (Jones et al., 2002; Vernier et al., 2002).

6.3.5 Summary

Generalized linear models are widely used in SDM because they are a generalization of the multiple regression model that uses the so-called link function to accommodate response variables that are distributed other than normally, namely the response distributions discussed above. Non-linear relationships between the predictor and response can be accommodated in linear models or GLMs using various transformations of the predictors, and interactions between predictors can also be specified (Fig. 6.1). This modeling framework is well suited to SDM because species distribution modeling almost always involves multiple predictors, non-linear response functions, and response variables that are binary, counts or ordinal. Because logistic regression is the form of GLM most commonly used in SDM to model species probability of occurrence, it is used as an example throughout this chapter.

The number of candidate predictor variables is frequently large in SDM and this makes a thoughtful approach to model selection particularly important. In a perfect world, ecological theory would suggest a small number of alternative candidate predictors and models that could be compared against the data. For example, should climate variables be entered into a model first (because order matters), followed by variables representing smaller-scale environmental variation (Chapter 5)? In reality, an exploratory approach might be required – the nature and strength of simple relationships between an individual environmental variable and species data can be explored using correlation, ANOVA, simple regression, graphical methods (scatterplots, boxplots), smoothing functions such as polynomials, splines, simple GAMs (see Section 6.3) or other methods such as decision trees (Chapter 7). These issues of data and data distributions also apply to each of the following methods discussed in the remainder of this chapter and throughout Part III.

6.4 Generalized additive models

Generalized additive models are a flexible and automated approach to identifying and describing non-linear relationships between predictors and response, and so from an applied perspective, they differ from GLMs in their ability to characterize the nature of the response function (Yee & Mitchell, 1991). An excellent recent resource on GAMs is by Wood (2006). GAMs are non-parametric extensions of GLMs (Guisan et al., 2002). Like GLMs, GAMs assume the Ys are independent and have a distribution belonging to the group of distributions discussed in Section 6.3 (Gaussian, Poisson, binomial, negative binomial and gamma). The formulation for the GAM can be written:

$$g\left(E(Y)\right) = LP = \hat{\beta}_0 + \sum_{j=1}^{p} X_j f_j + \varepsilon$$

where the coefficients of the GLM are replaced by some smoothing function, f. Typically this is a type of scatterplot smoother – a function that takes some kind of weighted local average of the data (Venables & Ripley, 1994, pp. 248–50). Semi-parametric models can also be estimated where some variables are fit parametrically (as in a GLM), others non-parametrically (using a smoother).

There are actually many approaches to scatterplot smoothing, a subtlety ignored by users like us (at our peril) who may tend to fit GAMs on "autopilot" (that is, accepting the default settings in the statistical software). It has also been said, however, that "what smoother to use is a matter of taste" (Chambers & Hastie, 1992, p. 255). Scatterplot smoothing methods include: taking the running mean or median, kernel methods (distance-weighted average of k-nearest neighbors, or neighbors within neighborhood width w), polynomial regression, and locally weighted polynomials ("Lowess"). Cubic smoothing splines are widely used. This involves dividing the ordered data at regular intervals of values or "knots," and in the interval between the knots the spline (smoother) is a cubic polynomial. It is because of the use of scatterplot smoothers that GAMs are considered non-parametric statistical models.

In addition to the type of smoother, the modeler may have to make some choices about the span (how large a neighborhood to take around a point when smoothing) and, for some smoothers, the number of degrees of freedom used in smoothing. The degrees of freedom of a smooth fit is usually a real number (rather than an integer), and it has been noted that "the theory for this area has not been fully developed" (Yee & Mitchell, 1991, p. 590).

Generally, the fit of a GAM is evaluated by testing the non-linearity of predictor. This is done by comparing a model with a linear fit for that predictor versus the non-parametric fit (scatterplot smoother). The difference in deviance between models is attributed to non-linearity (and its significance can be evaluated by a χ^2 test). Current best practices for choosing the degrees of freedom (d.f.) used in smoothing would also involve testing whether reducing the d.f. from a default setting of 4 to 3, or even 2, results in a significant difference in model fit.

Since a GAM does not produce coefficients or parameters that can be multiplied by grid maps of the predictors, how are spatial predictions made using a GAM? This can be done by importing new data (entire GIS datasets for all predictors – values for every pixel) into the statistical software package, predicting using the existing software tools, and exporting the results as a GIS map. In fact, this approach can be used with any modeling method. The ability to do so depends on the size of the dataset and software limits. An alternative approach has been to develop a "look-up table" to describe the response curves for each variable for a reduced number of values. Then the look-up tables are used in a GIS to reclassify predictors accordingly to their contribution to the model using logical operators or map algebra. For example: "If X1 is between values a and b, recode it to f(Xab); else if X1 is between values b and c, recode it to f(Xbc), etc. These new reclassified maps of f(X) are added together to produce the linear predictor (LP), which can be further transformed to the response variable scale. Specialized software tools have been developed for applying GAMs in SDM (see Table III.2), such as GRASP (generalized regression analysis and spatial prediction; Lehmann et al., 2002).

6.4.1 Use of GAMs in species distribution modeling

Yee and Mitchell's (1991) paper introducing GAMs to species distribution modeling asserted that, assuming the best-fitting response is non-linear, if a parametric curve fits the data, then for reasons of parsimony it is

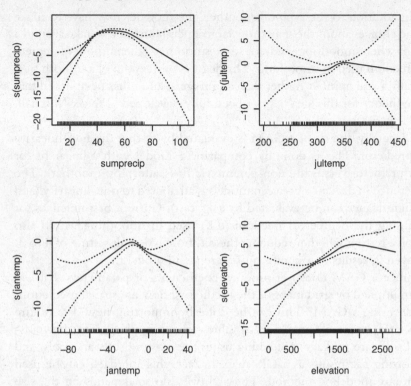

Fig. 6.2 Shape of response function (determined using a scatterplot smoother, a cubic spline) between the log-likelihood ratio of species presence (y-axis, labeled "s(X)") and predictors (x-axis) estimated by a GAM using the binomial distribution (logit link) for binary species presence/absence data. The plant species is Joshua Tree (*Yucca brevifolia*), using data from the Mojave Desert within California, USA (Miller & Franklin, 2002). The environmental predictors shown are average summer precipitation (sumprecip; mm), average July maximum temperature (jultemp; °C multiplied by 10), average January minimum temperature (jantemp; °C multiplied by 10), and elevation (elevation; m). The "rug" at the bottom of the graph shows the distribution of observed values of the predictors, showing that wide confidence intervals around the smoothed curve at extreme values are due to very few observations in that range. Note that response to jultemp does not seem to differ from zero. Plotted using the gam package in the R software (R Development Core Team, 2004).

preferred to a non-parametric curve (p. 591). They recommended using GAMs to suggest the shape of a parametric response curve (Gaussian, piecewise, curvilinear, symmetrical, skewed), that would then be parameterized as a GLM. So, for example, GAMs are very useful for characterizing non-linear species response curves in a graphical and exploratory way

(Fig. 6.2), perhaps suggesting a Gaussian or skewed unimodal response (Chapter 3). The way this is done is to plot X_j (the predictor) on the x-axis versus $f_j(X_j)$, the fitted function for that predictor, on the y-axis (Fig. 6.2). It is useful to plot the standard error of the function as well because broad confidence intervals in some portions of the range of values suggest great uncertainty (often due to few observations in that range), and might indicate the need for cautious interpretation or model simplification. However, GAMs cannot be used to calculate species response parameters such as optimum and tolerance, as some other methods can (Hirzel et al., 2002).

Another limitation of GAMs, either for exploration or prediction, is that they are additive, it is difficult to introduce interaction terms, and so it is not often done. Decision trees are particularly good for identifying interactions among predictor variables, as others have also recommended (Crawley, 2005, p. 193) (see Chapter 7). Further, scatterplot smoothers such as splines are highly data dependent and, as with any data–driven approach (high variance but low bias relative to GLMs), they could yield poor predictions with new data. However, in spite of these potential drawbacks, GAM is a very powerful and flexible modeling technique (Wood, 2006), and GAMs have become widely used for spatial prediction in SDM (Brown, 1994; Bio et al., 1998; Lehmann, 1998; Frescino et al., 2001; Guisan et al., 2002; Lehmann et al., 2002). In fact, in comparisons GAMs usually outperform GLMs in species distribution modeling (Austin, 2002; Ferrier et al., 2002b; Lehmann et al., 2002; Araújo et al., 2005a; Meynard & Quinn, 2007).

GAMs have been widely used operationally in SDM for conservation planning purposes (Platts et al., 2008; Lassalle et al., 2008 are just a few of many recent examples). They continue to be used as one of several, or as the sole, method of SDM for a wide range of questions, taxa and scales of analysis, for example, to model the distribution of shorebirds (Granadeiro et al., 2004), examine temporal habitat dynamics (Osborne & Suarez-Seoane, 2007) and regional variation in habitat preferences (Davis et al., 2007) of rare species, explore the effects of competitive interactions on congeneric beetle distributions (Jiménez-Valverde et al., 2007), spatially predict climate change impacts (Midgley et al., 2006; Trivedi et al., 2008), predict swordfish catch (Hazin & Erzini, 2008), and study the effect of forest composition and landscape structure on bird distributions (Venier & Pearce, 2007). GAMs have also been used to examine the effect of species traits (Poyry et al., 2008) and biotic interactions (Araújo & Luoto, 2007; Heikkinen et al., 2007) on SDM performance.

6.4.2 Summary

GAMs in species distribution modeling were originally suggested as a powerful, graphical method for detecting and describing non-linear response functions and then using this information to build a parametric model (GLM). It has been recommended that GAMs can be used this way generally in multiple regression modeling (Crawley, 2005, p. 193). GAMs are not designed to test or fit interaction terms. There are methodological challenges with prediction with new data using scatterplot smoothers, e.g., with spatial prediction using GIS, although this has been addressed with specialized software. Nonetheless, GAMs have been recommended as the method of choice for SDM and spatial prediction because they tend to have high prediction accuracy. They have been subjected to rigorous comparisons, especially with GLMs, and have proven to be informative and useful. It should be pointed out, however, that GLMs that include quadratic functions are often used as the basis for these comparisons. As noted previously, GLMs are not restricted to polynomial functions, and substituting an equivalent GLM function for the GAM smooth function, for example a piecewise linear function (Section 6.3.1), might serve as a better basis for comparison of GLMs and GAMs.

6.5 Multivariate adaptive regression splines

Although multivariate adaptive regression splines (MARS) have been around for a while (Friedman, 1991; Friedman & Roosen, 1995), and are closely related to the other methods discussed in this chapter and the next, they have not been used in SDM until fairly recently. There have been some applications in the fields of chemistry, data mining, engineering, medicine and epidemiology (see Kuhnert et al., 2000). As described by Hastie et al. (2001), MARS can be viewed as a generalization of stepwise linear regression, well suited to problems with large numbers of predictor variables, or as a modification of the regression tree approach (Chapter 7). Leathwick et al. (2006b) also compare MARS to regression trees (Hastie & Tibshirani, 1990) where the step functions of regression trees are replaced by piecewise linear functions. To our minds, they are also related to GAMs because they use piecewise splines, albeit specifically piecewise linear splines, as the functions relating the predictors to the response variable. These are referred to as "basis functions." While MARS are, like GAMs, able to model complex relationships between response and predictor variables, they are computationally fast, while GAMs can be slow or even impossible to fit for very large datasets (tens of thousand of observations, dozens of predictors).

As Leathwick *et al.* (2006b) describe, MARS is a non-linear regression method, where the so-called basis functions are defined in pairs, on either side of a knot. A knot, a value of a variable that defines an inflection point along a range of a predictor, is selected, and then a linear response is modeled for each section between knots. Coefficients are estimated that define the slope of the section (Fig. 6.3). Many potential knots are identified automatically (this is the "adaptive" part of MARS). In a forward stepwise fashion, knots and corresponding basis function pairs and their products are selected that give the greatest decrease in the residual sum of squares, evaluated using generalized cross-validation. A very large model is fit. Then, the model is pruned (reduced in complexity) by iteratively removing basis functions that contribute the least to model fit. This aspect of MARS is quite similar to decision trees (see Chapter 7), as is the user's ability to control the number of basis functions, or the model's parsimony.

MARS have several potential advantages for SDM that have been emphasized in the few studies that have thus far applied this novel method to the SDM problem. One advantage, already mentioned, is that MARS is computationally fast as compared to GAMs, widely applied in SDM, and therefore is practical for use with large datasets consisting of many observations and a great number of candidate predictors (Elith & Leathwick, 2007). These same authors noted another practical advantage, that while GAMs are difficult to use for spatial prediction (as we also discussed in Section 6.3), MARS models, consisting of piecewise linear basis functions, are much more straightforward to use for spatial prediction. For example, the basis functions corresponding to a simple model for *Yucca brevifolia* that uses the single predictor summer precipitation (sumpreci), shown in Fig. 6.3, are:

$$BF1 = \max(0, \text{sumpreci} - 60.000);$$
$$BF2 = \max(0, \text{sumpreci} - 31.000);$$

and the model equation is:

$$Y = -0.227 - 0.048*BF1 + 0.014*BF2.$$

Another unique advantage of MARS for SDM was emphasized in a study by Muñoz and Felicisimo (2004) (see also Leathwick *et al.*, 2006b).

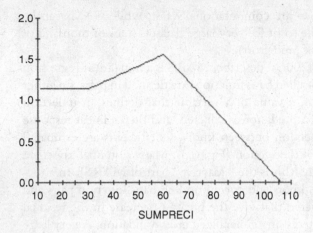

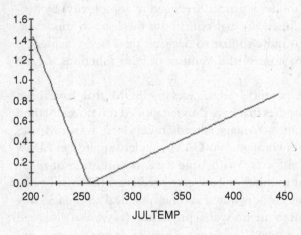

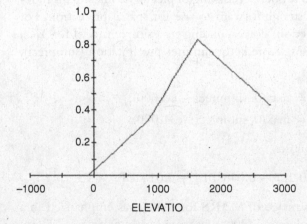

In contrast with global models, MARS considers interactions between variables, not globally (over the entire range of the predictors), but locally – between subregions of every basis function. For example, in their study, elevation and mean number of frost days in August only interacted to affect species occurrence below elevations of 3318 m, and then only when there were fewer than 21 frost days. Another hypothetical example might be that south-facing slope aspects (representing warmer, drier conditions in the northern hemisphere) are positively related to a species occurrence at high (cooler) elevations, unrelated at middle elevations, and negatively related at low (warmer) elevations in mountainous areas. Classification trees (Chapter 7) are also adept at describing these responses that act over subregions of the data space, but as discussed, MARS has the additional advantage of fitting linear functions (instead of constants) within those subregions of the data. However, other extensive tests of the MARS method in the species distribution modeling context did not report interaction results (Elith et al., 2006; Elith & Leathwick, 2007). Some actually found that using local interactions greatly overfit the data (Prasad et al., 2006).

A practical challenge of using MARS is that commonly used software implementations (for example, the mda library in R) rely on least squares fitting, and this is appropriate for data with normally distributed errors. This is not appropriate for binomial data (e.g., species presence-absence), and Leathwick et al. (2006b) overcame this limitation by fitting the MARS model, extracting the basis functions, and computing a GLM that used these basis functions as predictors of species presence/absence, as recommended by Friedman (1991). Other SDM studies using MARS with binary species presence/absence data using a different software

<hr>

fig. 6.3 Graphical example of basis functions from MARS showing the relationship between response variables, *Yucca brevifolia* presence/absence (see caption, Fig. 6.2), and environmental predictors: average summer precipitation (sumpreci; mm), average July temperature (jultemp; °C multiplied by 10), and elevation (elevation; m). Summer precipitation has a constant effect until ~31 mm, then precipitation has a positive effect on the likelihood of *Y. brevifolia* until ~60 mm, after which the relationship becomes inverse. *Y. brevifolia* appears to have a bimodal response July temperature – high likelihood at the lowest temperature, then decreased likelihood as temperature increases until the threshold of ~25.5 °C, after which the relationship becomes positive. This does not seem like an ecologically sensible response curve. The influence of elevation is more similar to the expected unimodal response, where *Y. brevifolia* likelihood increases with elevation until ~1600 m, after which the relationship is inverse.

implementation (MARS 2.0, Salford Systems, San Diego, CA) did not mention this limitation (Muñoz & Felicisimo, 2004), but it appears that this software has the same limitation. No doubt this is a technical obstacle that will soon be overcome by the development of new applications software.

Further, widely available software only allows for the use of continuous or quantitative predictor variables (not categorical). Some studies did not consider any categorical predictors. Elith and Leathwick (2007) excluded categorical predictors in their study, but also note in their paper that they have written code to fit them.

6.5.1 Use of MARS in species distribution modeling

MARS methods have been used in SDM in only a few studies, but these include some investigations that were quite broad in their scope. Because this is a relatively new method, these studies often involve comparisons of different modeling methods, and all used the binary response variable, species presence absence. Muñoz and Felicisimo (2004) showed that MARS and classification trees (CT) achieved higher prediction success when compared to logistic multiple regression in a landscape-scale case study, but not in another biogeographic-scale case study of plant species distributions. Spatial prediction was simpler using MARS because of the CT model complexity. Leathwick *et al.* (2006b) found that MARS performed comparably to GAMs but were much easier to use for spatial prediction, for the reasons discussed above. In an extensive comparison of methods applied to a large number of "presence only" datasets (Elith *et al.*, 2006), MARS performed comparably to GAMs and logistic multiple regression in terms of prediction accuracy; however, multiresponse MARS models (Hastie *et al.*, 1994), using multiple species as response variables, were among the highest performing types of models in this comparison. In contrast, Prasad *et al.* (2006) found that MARS performed poorly in comparison to ensemble decision tree methods (Chapter 7) when applied to plant distributions under current climate and future climate scenarios. Subject to further testing, this relatively new regression method, with a somewhat steep learning curve, may prove very useful for SDM and related forms of ecological data analysis and spatial prediction.

6.6 Multivariate statistical approaches to SDM

When a spatial prediction of community-level species composition is required, several variations on the SDM framework can be used (for review, see Ferrier & Guisan, 2006). Community types can be defined, *a priori*, based on species composition data, and then community type can serve as the categorical response variable in a predictive model (for example, Lees & Ritman, 1991; Brzeziecki *et al.*, 1993; Brown, 1994; Miller & Franklin, 2002). This is referred to as "classify, then predict." Alternatively, in the "predict, then classify" approach, individual species can be modeled and the resulting predictive maps combined in some way to define assemblages of species (e.g., Lenihan, 1993; e.g., Austin, 1998; Leathwick, 2001; Ferrier *et al.*, 2002a). Thirdly, using data on community composition, one can model various multivariate measures of overall species composition as the response variables, and even use community composition data to improve distribution models for individual species.

Multivariate statistical approaches are well developed in community ecology (Legendre & Legendre, 1998). Multivariate ordination methods are used to describe continuous variation in community composition and to relate that variation in composition to environmental gradients (in constrained ordination). An example of a widely used technique is Canonical Correspondence Analysis (ter Braak, 1987). In CCA, uni-modal symmetrical species response curves, of equal amplitude and maxima, are assumed, and the main axes of the correspondence analysis, defining the maximum dispersion of the species data, are constrained to be linear combinations of environmental variables. Although not widely used in SDM, predictive mapping of species or communities based on constrained ordination has been shown to be effective when the species data included information about multiple species (Ohmann & Spies, 1998; Guisan *et al.*, 1999; Dirnbock *et al.*, 2003). In a multivariate approach, species mapping is usually achieved by using a measure of the multivariate environmental distance from each unsurveyed grid cell to the centroid of a species calculated from the ordination analysis.

A variation on the ordination approach, referred to as generalized dissimilarity modeling (GDM), models the dissimilarity (multivariate distance) in community composition among all sites as a function of environmental gradients (Ferrier, 2002; Ferrier *et al.*, 2002a). As such, this approach models beta-diversity, or the differences in species composition between sites, using a non-linear extension of a matrix regression approach (Ferrier *et al.*, 2007). In contrast with an approach based on the raw (species abundance) data, such as CCA, GDM models the multivariate distances among sites, and is more appropriate for predicting variation in beta-diversity among those sites (Legendre *et al.*, 2005). The matrix regression approach generally correlates a matrix of compositional distance among sites with a matrix of environmental (or geographical) distance, and the non-linear extension by Simon Ferrier and colleagues addresses the fact that dissimilarity measures are asymptotic, and the rate of compositional turnover along environmental gradients is typically non-linear. The results of predictive GDM is a matrix of estimated pairwise compositional dissimilarities among all grid cells in a study area. In order to display and interpret these estimates the grid cells are then usually subjected to classification (clustering) or ordination so that a measure of species composition can be mapped (Ferrier *et al.*, 2007).

A number of single-response models discussed in this chapter and in Chapter 7 have also been extended to allow a multiresponse (multiple species) model to be estimated, including vector generalized linear or additive models (Yee & Mackenzie, 2002), multiresponse MARS (Leathwick *et al.*, 2005), neural networks (Olden *et al.*, 2006) and regression trees (De'ath, 2002), all reviewed by Ferrier and Guisan (2006). The basic premise of these approaches is that the important environmental correlates of overall community composition (as opposed to single species) can be identified, and can help to identify environmental predictors for the rarer species in the community. It might otherwise be impossible to select environmental predictors for rarer species because of the small number of observations.

In summary, a host of extensions of single response models, as well as adaptations of classical multivariate ordination and classification methods from community ecology, have been adapted for the SDM problem. They have proven to be useful for spatially predicting community type, continuous measures of community composition, and also individual species (based on "distance methods"). Multivariate methods are well known to community ecologists. However, these methods have seen limited application thus far in SDM. This may be because they are perceived to be complex or because analysts are unaware of them, or do not think they have the right kind of (multispecies survey) data. Obviously these methods can only be used with multispecies data, but many taxonomic groups are typically surveyed using multispecies observations (e.g., plants, invertebrates, fishes). These multivariate approaches seem to hold promise for future applications in spatial ecological prediction.

6.7 Bayesian approaches to SDM

Bayesian methods offer an alternative approach to statistical inference that differs from classical ("frequentist") statistical inference in fundamental ways. Namely, Bayesian inference estimates the probability that a hypothesis is true given the data, and defines that probability as the degree of belief in the likelihood of an event. Frequentist inference estimates the probability of the data given a hypothesis, and the probability is defined as the relative frequency of an observation (Wade, 2000; Ellison, 2004). Bayesian inference also treats model parameters as random variables, not fixed quantities whose values are estimated. Bayesian methods have been around for a long time and seem to be very useful. Consequently, it is unclear to us why they are not as widely taught and used as classical statistics. Perhaps it is because of the epistemological differences summarized above. Another distinguishing feature of Bayesian inference is that it incorporates prior knowledge about the hypothesis that is independent of the data being analyzed. It is this aspect of Bayesian analysis that was made use of, early on, in mapping and spatial prediction. Land cover mapping using remotely sensed data and, based on maximum likelihood classification, was improved using prior probabilities assigned to land cover classes based on their expected prevalence on the landscape (Strahler, 1980).

Applications in SDM soon followed as Bayesian models were developed for predictive mapping of species and communities (Skidmore, 1989; Fischer, 1990; Aspinall & Veitch, 1993; Brzeziecki et al., 1993). In these studies, prior probabilities of observing a species (based on the literature, a previous study, or an educated guess) were combined with their probabilities of occurrence conditional on the values of the environmental predictors (Franklin, 1995; Guisan & Zimmermann, 2000) using Bayes theorem:

$$P(H|Y) = \frac{F(Y|H)\pi(H)}{P(Y)}$$

Where $P(H|Y)$ is the probability of the hypothesis given the data (called the posterior probability), $F(Y|H)$ is the likelihood given the values of the environmental predictors (estimated from the sample data), and $\pi(H)$ is the prior probability. The denominator is the marginal probability of the data and is used as a normalizing constant (Ellison, 2004).

Bayesian methods continue to be used as an alternative approach to SDM (Tucker et al., 1997; Hooten et al., 2003; Thogmartin et al., 2004; Romero-Calcerrada & Luque, 2006; Termansen et al., 2006; Royle et al., 2007; Carroll & Johnson, 2008; Howell et al., 2008; La Morgia et al., 2008) and for other types of ecological analysis. Interestingly, Bayesian models have also been used recently for spatial prediction of risk in other disciplines, for example, to produce risk maps for wildfire, disease, landslides, and avalanches (Lee & Choi, 2004; Gret-Regamey & Straub, 2006; Beck-Worner *et al.*, 2007; Mathew *et al.*, 2007; Raul *et al.*, 2008).

Another useful application of Bayesian methods in SDM has been to combine separate SDMs that were based on any of the methods outlined in this book, by deriving prior probabilities from one model and revising those probabilities based on a second model (Pereira & Itami, 1991; Osborne *et al.*, 2001). Further, the use of Bayesian model averaging (Hoeting *et al.*, 2000) has specifically been described as an alternative to model (variable) selection in a multiple regression framework (Wintle *et al.*, 2003). Estimates or predictions from a number of plausible models can be combined, and the posterior probabilities can be weighted according to some measure of each model's fit (in Wintle *et al.*, they used naive or equal prior probabilities). This approach is one of several possible methods or tools that can be applied in ensemble model forecasting (Section 7.8).

6.8 Spatial autocorrelation and statistical models of species distributions

The statistical models described thus far in this chapter are not explicitly spatial – they make the standard assumption that observations of biotic distributions are independent. This assumption violates the "First law of geography" (Tobler, 1979) as well as ecology (Legendre & Fortin, 1989; Fortin & Dale, 2005) – that near things are similar (nearby locations have similar values) because they are likely to influence each other or be influenced by the same pattern generating processes (endogenous and exogenous, respectively; see below). Spatial autocorrelation (SAC) is defined as the covariation of properties with space so that values of a variable are related, either positively or negatively, as a function of proximity or distance (Cressie, 1991; Legendre, 1993; Anselin *et al.*, 2004).

Throughout this book, the spatial nature of species distribution data, environmental data, and the prediction maps has been emphasized. So, it may be somewhat puzzling that models that account for SAC in some way are not yet routinely used in species distribution modeling; in fact a recent survey found that < 20% of studies addressed SAC (Dormann, 2007b). We present the main concepts relevant to SDM. This has become a very active area of research, and we recommend that anyone interested in working in this area review the current literature before starting. Useful recent reviews are by Miller *et al.* (2007) and Dormann *et al.* (2007).

SAC in biotic distributions arises from one of two types of processes: endogenous or exogenous. Extending our modeling terminology, an endogenous process is generated by or associated directly with the response variable of interest (species distribution), for example, processes such as dispersal or competition. Exogenous processes occur independently of the variable of interest, for example, disturbance, historical barriers to dispersal or spatially structured environmental gradients, but can still result in SAC. Many different terms, often stemming from other disciplines that have been dealing with SAC longer than SDM, have been used to describe these processes and/or characteristics related to their manifested pattern (Table 6.1). What we refer to as an endogenous process is analogous to a second-order process that generally occurs at a finer scale than an exogenous (first-order) process (Haining, 1990).

Adding to the confusion, it may be impossible to distinguish between endogenous and exogenous processes based on empirical data because the same pattern can result from different processes and vice versa (Fortin & Dale, 2005). This has long been noted in spatial analysis in ecology (Ripley, 1981; Diggle, 1983). Even when the investigation focused on local scales in order to identify second-order patterns and processes, the same perceived pattern can result from different processes. A classic example is that trees in a forest can be clustered because of an endogenous process (limited ability to disperse away from a parent tree) or patchiness in tree density at the same scale could result from exogenous processes – fine-scale variation in soil conditions, or a disturbance process (Franklin, in press). While we often think of exogenous environmental factors as controlling species distributions at broad spatial scales, endogenous processes operating at more local scales, and noise or stochasticity occurring at fine scales, this is not always the case.

An issue, then, is whether SDM is usually conducted at a scale that is sufficient to detect or describe finer-scale spatial structure resulting from endogenous processes such as those listed above that affect biological populations and ecological processes.

6.8.1 Consequences of SAC data

What difference does all this make? There are several consequences of not accounting for spatial autocorrelation in the type of statistical models that are used for SDM and described in this chapter:

Table 6.1. *Different descriptions of the two sources of spatial dependence in data*

Process	Source	Example	Reference			
			(Haining, 1990)	(Anselin, 2002)	(Legendre & Legendre, 1998)	(Fortin & Dale, 2005)
Endogenous	Organism-specific process with spatial consequences	Dispersal limitations	Second order (fine-scale)	Substantive	False gradient	Inherent
Exogenous	Covarying spatially structured predictor	Soil type limitation	First order (broad-scale)	Nuisance	True gradient	Induced

(1) Precision of coefficients is decreased, resulting in increased likelihood of Type I errors or incorrectly rejecting null hypothesis of no effect;
(2) Variable selection may be predisposed towards more strongly auto-correlated predictors, termed the 'red-shift' (Lennon, 2002).
(3) Related to #2, broad-scale predictors are often selected over local or fine-scale predictors (Diniz-Filho *et al.*, 2003).
(4) Due to compromised residual variance structure, model selection based on AIC will tend toward models with more predictors (Ver Hoef *et al.*, 2001; Hoeting *et al.*, 2006; Latimer *et al.*, 2006; Diniz-Filho *et al.*, 2007a).

The primary aim of a modeling study, explanation or prediction, helps to determine the severity of the consequences described above. If the goal is to produce an accurate model for prediction, then variable selection that favors autocorrelated and/or broad-scale variables may be considered less problematic. However, if the goal is to explain the influence of environmental predictors, these consequences become much more problematic. All of the consequences listed above deal with some aspect of variable selection: either a variable is retained when it should not be, certain types of variables are more likely to be selected, or more variables in general will be selected. We know that the estimate of a coefficient will be less precise, but how does SAC affect the value of the coefficient itself, or restated in statistical parlance, "is the coefficient biased?"

The issue of whether estimation of coefficients is biased when non-spatial models are used with SAC data has recently been debated, and some of the conflicting conclusions stem in part from different uses of the word "bias." In a statistical context, an unbiased coefficient is one whose expected or average value is centered on the true value. In order to conclude that a coefficient is statistically biased, its average value must be either larger or smaller than the true known value. However, "bias" is often used in a more general sense, meaning "preference." For example, in describing the problem of ignoring SAC, Lennon (2002, p. 102) said that, "not only does ignoring or being unaware of this problem lead to serious difficulties in hypothesis testing, in the sense that significance levels are incorrect, much more importantly it also results in a systematic bias towards particular kinds of explanation for ecological patterns." This refers to a bias in the process of variable selection (consequences #2 and #3 above), rather than to biased coefficient estimates.

Some studies conclude that coefficients are biased when parameter estimates from non-spatial models differ significantly compared to those from spatial models (Dormann, 2007b; Kühn, 2007). However, this is more likely associated with the decreased precision of non-spatial model coefficient estimates, resulting in a wider distribution of potential coefficient values from which a single estimate is selected. Using simulated data (for which true coefficient values were known), Beale *et al.* (2007) showed mathematically that coefficients estimated by standard non-spatial and spatial models (OLS and GLS, respectively) are unbiased although the precision for OLS rapidly decreases as SAC increases. However, in another simulated data study that used binary data, Dormann (2007a) did find that the coefficients estimated with autologistic regression were biased (in a statistical sense) compared to logistic regression.

6.8.2 Solutions to SAC data

If ignoring spatial autocorrelation can lead to biased selection of variables or estimation of coefficients, what are the solutions? First, we describe ad hoc strategies to modify the data or specific elements of model selection in order to make spatially autocorrelated data suitable for non-spatial regression models (or vice versa).

One of the most obvious solutions is to (try to) remove the spatial structure from the data. With existing or compiled datasets, this is done by culling observations that are within some pre-determined distance of other observations (Hawkins *et al.*, 2007). At the data collection stage, this is done by manipulating the sampling strategy so that a minimum distance between observations is ensured (e.g., Legendre & Fortin, 1989; Davis & Goetz, 1990; Borcard *et al.*, 1992; Legendre, 1993). This minimum distance can be calculated from any number of statistical tools that plot similarity or dis-similarity (variance) of values between pairs as a function of distance between them, e.g., a variogram or a spatial correlogram (Fortin & Dale, 2005). These measures of SAC can be applied either to the response variable or to residuals from non-spatial regression. Residuals are checked in order to determine whether a potential SAC problem exists with respect to model formulation (Wagner & Fortin, 2005; Wintle & Bardos, 2006; Zhang *et al.*, 2008; Elith & Leathwick, 2009), presuming that any SAC in the response variable may be "taken care of" by a spatially structured predictor variable, i.e., in the case of exogenous SAC. Mapping the residuals (plotting their values on a map and examining visually) can be particularly useful for interpreting potential causes of spatially structured errors (Dormann *et al.*, 2007; Osborne *et al.*, 2007; Zhang *et al.*, 2008).

However, one perhaps equally obvious problem with this strategy is that the notion of throwing out what could be a substantial portion of data seems blasphemous to anyone familiar with fieldwork (and/or inferential statistics). In addition, selecting an appropriate minimum distance may be problematic, especially if there is periodicity or a scale-dependent pattern in the distribution of the response variable (Fortin & Dale, 2005). Finally, if the spatial autocorrelation is of the endogenous/inherent/substantive/second order variety (see Table 6.1), removing it may result in less informative or less accurate models.

The second solution addresses the first consequence described above, that coefficients will be accepted as significant when they might not be. As Lennon (2002) points out, in the equation to determine the significance of the ordinary (Pearson) correlation coefficient, r:

$$t = r\sqrt{\frac{n-2}{1-r^2}}$$

where n is the number of pairs of values and t has Student's t distribution with $n-2$ degrees of freedom; t is proportional to the square root of the degrees of freedom. When data are not independent, essentially the degrees of freedom are over-estimated, which increases t and can result in coefficients with erroneously higher levels of significance. Therefore, a very practical solution to this particular problem involves adjusting the effective degrees of freedom (Dutilleul, 1993; Thomson *et al.*, 1996; Rangel *et al.*, 2006), or even simply using a more stringent significance level (e.g., 0.01 instead of 0.05, Fortin & Dale, 2005).

The strategies above involve removing, decreasing, or at least neutralizing the effects of SAC in the data so that non-spatial models can be used. Recently, there has been a surge in the SDM literature that addresses explicitly spatial statistical models (Dormann *et al.*, 2007; Miller *et al.*, 2007). Generally speaking, these spatial models attempt to rectify the problem of SAC in the residuals, either by adding spatial structure to the model so that the resulting residuals will be independent, or by explicitly modeling the new, no longer independent, error structure (Table 6.2). Here we focus on the most commonly used family of spatial models, autoregression. This is not meant to be an exhaustive survey and readers are directed to the previous reviews for more details.

Autoregression
Autoregression models are a special type of regression, the most basic case of which has the response variable (y) regressed on a spatial lag of

Table 6.2. *Approaches to incorporating spatial dependence into statistical models of species distributions as described by Miller et al. (2007), who provided a comprehensive survey of more than 50 recent studies applying these methods*

Method	Data dist	Treatment of SAC	Negatives	Positives
Autoregressive (AR): Autologistic	B	Autocovariate	Data restoration, under-emphasis on environmental predictors; complicated to estimate AR parameters; forego GLM/OLS theory	Most appropriate for inherent SAC
Autoregressive: Conditional (CAR), Spatial (SAR)	G	Spatial variance-covariance (empirical)	Performance is very dependent upon model structure	Most appropriate for estimation and interpretation
Generalized Estimating Equations (GEE)	G, B, P	Divides data into clusters, parametric correlation function for SAC errors within clusters	Selection of appropriate correlation structure can be problematic; computationally intensive	Most appropriate for parameter estimation
Generalized Linear Mixed Models (GLMM)	G, B, P	Fits global fixed effects while SAC is treated as random effect and assumed only to occur within regions	Challenging to apply correctly; computationally intensive to fit	Most appropriate for parameter estimation
Geographically Weighted Regression (GWR)	G, B, P	Spatially varying regression coefficients as a function of nearby points	Does not explicitly deal with SAC; limited use for hypothesis testing	Can be used to explore scale of processes; Can be used to map coefficient values
Spatial filtering	G, B, P	Eigenvector of spatial factors removed	Computationally intensive	Uses GLM/OLS theory

GLM = Generalized Linear Model. OLS = Ordinary Least Squares. Data distributions: B = binary; G = Gaussian; P = Poisson.

itself. The idea behind this in a SDM context is that the distribution (occurrence, abundance) of a plant or animal species at a specific location is in some part influenced by the occurrence of the same species in the neighboring locations. While this description makes autoregression sound most appropriate for endogenous SAC, it can also be used for exogenous SAC, as the scale of the process (in the form of the specified location) can be broadened. The basic autoregression model is:

$$y = \alpha + \rho W y + \varepsilon,$$

where α is the constant term, ρ is the spatial autoregressive coefficient, Wy is the spatial lag for variable y, also called the autocovariate (W describes the neighborhood, usually based on distance), and ε is the error term. The basic autoregression model can be extended with the addition of other predictor variables, expressed as:

$$y = \alpha + \rho W y + X\beta + \varepsilon,$$

where $X\beta$ are the other predictor variables and coefficients.

The autoregression model for binary data is formally known as the autologistic model, originally described by Besag (1972, 1974) and subsequently modified to incorporate additional predictor covariates (Augustin et al., 1996; Gumpertz et al., 1997; Wu & Huffer, 1997):

$$\log\left(\frac{p_i}{1 - p_i}\right) = \alpha + \beta_1 \text{cov}_1 + \cdots + \beta_n \text{cov}_n + \beta_{n+1} \text{autocov}_i,$$

where an *autocovariate* at site i is defined as a weighted sum of observations in neighboring sites in a neighborhood defined by N_i (analogous to Wy above):

$$\text{autocov}_i = \sum_{j \in N_i} w_{ij} y_j.$$

If the coefficient for the autocovariate (β_{n+1}) is equal to zero, indicating that SAC has no effect, the model reduces to the ordinary logistic regression model. If the coefficients for all of the covariates ($\beta_1 \cdots \beta_n$) equal zero, the model follows Besag's (1974) autologistic model where only the autocovariate is used.

As you might guess, an important issue is how to estimate the autocovariate in a prediction context with "missing data," i.e., sample data. One way this has been addressed is by using non-spatial model predictions (based on environmental predictors only) to calculate neighborhood values (Dormann et al., 2007) or by using Gibbs sampler (Augustin *et al.*, 1996, 1998) or Markov chain Monte Carlo (MCMC) methods (Gumpertz *et al.*, 1997; Wu & Huffer, 1997). Very generally speaking, these techniques involve running iterations with simulations based on observed or predicted values until the output converges. However, in the presence of strong endogenous spatial dependence, the MCMC estimation methods can become numerically unstable (Wu & Huffer, 1997). Further, there are two issues that impede more extensive application of autoregression models in SDM: the procedures for fitting the full model are computationally intensive and software to do this is not readily available (but see Wintle & Bardos, 2006, for an application of Bayesian inference using WinBUGS), and selection of an appropriate neighborhood for calculating the autocovariate is highly subjective and very dependent upon the original sampling scheme. Autologistic models also underestimate the effects of the environmental predictors, and Dormann (2007a) found that Type II errors committed by the autologistic model were more serious than the Type I errors from the logistic model.

Along with binary data, autoregression models can also be used with normally distributed response data (auto-Gaussian, see below). Auto-Poisson models are more problematic in that only negative SAC can be accommodated in their original specification (Lichstein *et al.*, 2002), but recent alternative formulations have been suggested (Griffith, 2006).

Auto-Gaussian models are more formally specified as one of two types, mainly differentiated by how the spatially correlated error structure is defined (Haining, 1990): simultaneous autoregressive (SAR) or conditional autoregressive (CAR). The general formulas, followed by their respective error covariance matrices are (from Keitt *et al.*, 2002):

$$(\text{SAR})\, Y = X\beta + \rho W(Y - X\beta) + \varepsilon, \sigma^2[(I - \rho W)^T(I - \rho W)]^{-1}$$

$$(\text{CAR})\, Y = X\beta + \rho C(Y - X\beta) + \varepsilon, \sigma^2[(I - \rho C)]^{-1}$$

where W is a (possibly) asymmetric spatial matrix and C is a symmetric spatial matrix. A spatial model type closely related to SAR and CAR and also used with normally distributed data is generalized least squares (GLS). The main difference between SAR/CAR and GLS is that in GLS, the spatial structure in the variance–covariance matrix is a parametric function.

SAR models explain the relationship among the response values at all locations on the lattice simultaneously, while CAR models consider the distribution of a response variable at one location to be conditional on the values of its neighbors (Anselin, 2002). As can be seen from the formulas above, CAR models require a symmetric error matrix so should not be used when spatial patterns are a function of directional processes. Despite the similarities between the models (including sometimes producing nearly identical results, Lichstein et al., 2002), SAR have been used more often in SDM applications (see Kissling & Carl, 2008). Different SAR model types can be constructed based on which type of SAC process is indicated: spatial error SAR for exogenous process, spatial lag SAR for endogenous process, or a mixed model where both occur (Anselin et al., 2004; Fortin & Dale, 2005). Diagnostics (Lagrange multiplier tests) have been developed to test for the effects SAC on a non-spatial regression model, and to distinguish among error dependence, lag dependence, and their combined effects (Anselin & Rey, 1991).

We should also note that the concept of calculating an additional model term to describe the influence of neighboring observations, analogous to an autocovariate, can be used with many other statistical methods, such as decision trees (Miller & Franklin, 2002) and GAMs (Leathwick, 1998).

Applications of autoregression methods in SDM

Augustin et al. (1996) provide one of the first SDM applications of the autologistic model. They modeled the distribution of red deer in Scotland and found that autologistic models resulted in the lowest misclassification rate while logistic models were more appropriate for estimating global characteristics such as total occupied area or overall abundance. More recent studies have suggested several general trends when (non-spatial) regression models are compared to autoregression models using the same data: when model fit is used as a comparison measure, spatial models often have improved fit over non-spatial models (Tognelli & Kelt, 2004; Ferrer-Castan & Vetaas, 2005; Dormann, 2007b; McPherson & Jetz, 2007; Platts et al., 2008). Further, the relative importance of predictor variables shifts (Lennon, 2002; Diniz-Filho et al., 2003; Tognelli & Kelt, 2004; Diniz-Filho & Bini, 2005; Segurado et al., 2006; Kühn, 2007). In reviewing more than 20 studies that compared regression to autoregression models, Dormann (2007b) found that the average increase (autoregression – regression models) in adjusted model R^2 was 0.060 (min $= -0.10$, max $= 0.20$). He also found that, where coefficients could be compared directly, the values were always different (he used the term "biased", p. 133), but he also cautions that the errors on the coefficient estimates could be large enough to render the differences statistically insignificant (Dormann, 2007b).

In a different study using artificial data and focusing only on binary response data, Dormann (2007a) found that, not only were the coefficient estimates different for the logistic and autologistic models, the logistic model coefficients were, in fact, closer to the "real" values. He also found that the Type I errors committed by the autologistic model (rejecting a truly important variable) were more egregious than the Type II errors committed by the logistic model (retaining a truly unimportant variable). In this study, SAC was introduced to the data only in the form of spatial errors, and Dormann notes that more realistic distribution patterns could be used to investigate these issues further. However, this study illustrates potential problems associated with using autologistic regression for SDM.

Kissling and Carl (2008) compared OLS regression to three different types of SAR models (spatial error, spatial lag, and mixed) using artificial data and varying neighborhood distance, coding of the spatial neighborhood, W, and the kind of SAC that was present. They concluded that SAR model parameter estimates were not always more precise than the OLS estimates, but that the spatial error SAR was the most robust of the SAR models.

Generalized estimating equations and generalized linear mixed models

Generalized estimating equations (GEE) are an extension of GLMs that can be used with data that are measured repeatedly through time or space (Cressie, 1991). GEE takes these potential correlations into account by dividing the data into groups, within which the correlation matrix is defined, and between which correlation is expected to be zero. GEE has some application in SDM studies (Albert & McShane, 1995; Gotway & Stroup, 1997; Gumpertz et al., 2000; Mugglestone et al., 2002; Dormann et al., 2007), but it is considered to be most appropriate for parameter estimation, rather than as a method that explicitly includes SAC for spatial prediction (Table 6.2).

Generalized linear mixed models (GLMM) combine the properties of GLMs (able to handle non-normally distributed response variables, such as binary species presence–absence) and linear mixed models which incorporate both random and fixed effects (Bolker et al., 2009). Random effects traditionally comprise the blocking in experimental treatments, but in ecology and evolution can also encompass variation among individual, genotypes, species, or geographical regions. When applied to spatial data coming from distinct regions, GLMMs can be used to fit global fixed effects while spatial autocorrelation is only assumed to occur within regions (Dormann *et al.*, 2007). GLMMs have been used to a limited extent in SDM and spatial prediction (e.g., Das *et al.*, 2002; Stephenson *et al.*, 2006; Dormann *et al.*, 2007). For example, a recent study using GLMM identified significant geographical variation among regions in species–environment relationships. Key predictors of koala (*Phascolarctos cinereus*) distributions, such as habitat patch size, were found to from region to region owing to the interactions of edaphic factors, landscape history, and contemporary land use (McAlpine *et al.*, 2008).

It has been cautioned that it is challenging to apply GLMMs correctly for ecological applications (Bolker *et al.*, 2009). Both GEE and GLMM are most useful if the spatial regions comprising groups of observations have been defined *a priori*. If the relationship between the response and predictors is expected to vary over geographical space, but the spatial structure of that variation is unknown, then geographically weighted regression, discussed in the next section, may be useful.

Geographically weighted regression
The next model represents a bit of a shift from the rest of the models described here. Although the other spatial models, in part, attempt to describe fine-scale endogenous SAC (i.e., by including an autocovariate that is a spatial lag of the response variable), the results are ultimately equations with global parameter estimates. The relationships they describe, e.g., between temperature and species distribution, are consistent throughout the region of interest. However, in reality there may be other processes and patterns that make the species–environment relationship vary spatially. For example, precipitation may have a different effect (either magnitude or direction) on species distribution, given the temperature regime in different parts of the region.

When global relationships are not adequate (i.e., regression coefficients vary across space), the process is considered to be non-stationary, which understandably is a violation of global model assumptions. When non-stationarity is evident, data partitioning has been used to divide the data into sections in which the processes are stationary (Osborne & Suarez-Seoane, 2002).

The spatially varying coefficients can be modeled using geographically weighted regression (GWR) (see Fotheringham *et al.*, 2002). Each observation can potentially have a different coefficient for a variable, and while all observations are used to fit the regression parameters, the observations that are closest to a point are weighted more heavily than observations farther away. GWR extends a global regression model such as:

$$y_i = \beta_0 + \sum_k \beta_k x_{ik} + \varepsilon_i$$

by allowing parameter estimates to vary locally:

$$y_i = \beta_0(u_i, v_i) + \sum_k \beta_k(u_i v_i) x_{ik} + \varepsilon_i$$

where (u_i, v_i) are the coordinates of the ith point in space.

A spatial kernel defines how the influence of other observations will be measured, and the user determines its geometric shape (circle, square), bandwidth (distance within which observations will have influence), and functional form (ex. Gaussian, exponential). When data are relatively sparse, an adaptive spatial kernel can be used in which the bandwidth varies based upon how many points are included rather than a specific distance (Fotheringham *et al.*, 2002).

The fact that GWR can result in n (number of observations) different regression equations makes it a truly local method; however, its approach to SAC is much more implicit than the other methods described. SAC is incorporated in the model only through the way in which it affects the relationship between predictor variable(s) and response variable. Another consequence of the n different variable coefficients estimated is that GWR is not appropriate for more traditional hypothesis testing. Following Table 6.1, GWR would seem to be a more appropriate method for SAC resulting from exogenous processes.

While GWR and its predecessor, the spatial expansion method, have been primarily used in human geography applications (Jones & Cassetti, 1992), there have been a few recent applications of it to SDM or related problems (Foody, 2004; Bickford & Laffan, 2006; Shi *et al.*, 2006; Kupfer & Farris, 2007; Osborne *et al.*, 2007).

Bickford and Laffan (2006) found that the strength and nature of the relationship between pteridophyte species richness and water availability across Australia varied spatially and was scale dependent. While they did not directly compare GWR to (global) regression methods, they concluded that the complex and regional patterns detected by GWR may be important to understanding broad-scale diversity relationships. In a study that modeled the relationship between tree basal area and environmental factors in Arizona, Kupfer and Farris (2007) found that not only did GWR produce a better model fit than OLS, it also helped to elucidate fine-scale patterns, such as the influence of aspect, that were overlooked in the global model. Osborne *et al.* (2007) compared GWR to GLM and GAM in models relating bird distribution to environmental factors. In addition to fitting better models, GWR was also used to identify the spatial scale at which the relationship between an environmental variable and the response became stationary (Table 6.2).

GWR has some limitations with respect to its use in SDM applications. The majority of SDM studies use presence–absence data and current GWR software (GRW 3.0, Fotheringham *et al.*, 2002,) has been less extensively tested with binary data to date, and is unable to predict on independent data. Austin (2007) argued that, when GWR is compared to a linear global regression model, what appears to be non-stationarity (and subsequently results in a better fit by GWR) could result from incorrectly specified (linear) models. Unless all subregions have the same range of predictor values, non-stationarity may be impossible to distinguish from changing terrain. Many authors suggest that GWR should be used as a complement to global regression, rather than as a substitute (Jetz *et al.*, 2005; Osborne *et al.*, 2007).

Spatial filtering methods
One of the newest spatially explicit methods on the scene has roots in earlier techniques that used partial regression or trend surfaces of geographic coordinates (Pereira & Itami, 1991; Borcard *et al.*, 1992; Legendre & Legendre, 1998; Lobo *et al.*, 2002; 2004; Nogues-Bravo *et al.*, 2008). However, unless they were specified to be extraordinarily complex, the polynomial equations defined in the trend surface analysis were capable only of describing broad-scale SAC, such as that related to exogenous processes. The selection of polynomial order was also arbitrary and, unless they were orthogonal, lack of independence could affect model selection (Griffith & Peres-Neto, 2006).

Spatial filtering methods (also similar to spatial eigenvector mapping) have been proposed as a way to remove SAC by explicitly including spatial information that has been decomposed from the predictor variables (Diniz-Filho & Bini, 2005; Dray *et al.*, 2006; Griffith & Peres-Neto, 2006). Essentially, a distance or connectivity matrix based on the geographic coordinates of the dataset is used to derive a series of orthogonal "spatial filters" that explain different amounts of variance in the response variable, at different scales. Once these spatial filters have been extracted, a subset is selected (based on how much residual SAC their inclusion decreases) to become new predictors along with environmental predictors. The residuals should now be independent, ensuring the appropriate use of standard GLMs.

The biggest advantage of this method is that it is the only one described here that deals with SAC explicitly while still allowing classical GLM implementation (Table 6.2). The spatial filters (eigenvectors) can also be mapped, in order to examine the different patterns of SAC (Diniz-Filho & Bini, 2005). However, extracting eigenvectors from large datasets (>7000 observations) can be problematic (Dormann et al., 2007).

6.8.3 Summary

We can summarize this discussion of spatial autocorrelation by outlining several reasons that the spatial models described above should not be considered uniformly superior to non-spatial models in SDM applications:

- As a corollary to #3 in the consequences of ignoring spatial dependence, spatial models may underestimate the influence of broad-scale predictors (Diniz-Filho et al., 2007a). The severity of this depends on the scale of the study (if it is a relatively broad-scale study, spatial models may not be appropriate). Similarly, spatial models, specifically those with autocovariates, may underestimate effects of all environmental predictors (Miller & Franklin, 2002; Segurado et al., 2006; Dormann et al., 2007). This results in models that will not be appropriate for goals of extrapolation.
- Spatial models overall are much more complicated to fit, and involve more subjective considerations, such as neighborhood shape/size and spatial covariance matrix definition. Many methods are computationally intensive at best and intractable at worst for generating model predictions (but see Augustin et al., 1998).
- Naïve use of spatial models (i.e. failure to test different model types, neighborhood distances) may result in less reliable results than non-spatial models (Kissling & Carl, 2008). It has not been well studied how spatial methods act under different SAC structures, which makes this decision difficult.
- Spatial models may not be robust where there is non-stationarity in the data (Haining, 1990), and it is very likely that there will be non-stationarity in biological survey data. Of course, non-spatial models have similar problems in the presence of non-stationarity. Geographically weighted regression is particularly appropriate for non-stationary data.

- The utility of autoregression can be limited when sample density is low (Wintle & Bardos, 2006), that is, in data-sparse situations that are also typical of species surveys. This may be addressed by using different sampling strategies that are designed specifically for use with spatial models.
- A spatial pattern in model residuals, in addition to suggesting spatial autocorrelation in the response data, may also result from model mis-specification, and changing a linear model term to a polynomial in a non-spatial model may produce a much more efficient model, with potentially a more ecologically realistic explanation (Austin, 2002, 2007).

It is also important to remember that SAC rarely occurs in isolation of other related problems that affect non-spatial models. Multicollinearity, non-stationarity, and scale-dependent patterns also wreak havoc on non-spatial models, and often are to blame for differences that are observed when spatial models are compared to non-spatial models.

With that said, the evidence overwhelmingly supports the use of some type of spatial model in SDM, if only to compare with non-spatial model results. When explanation or clarification of the relationships between environmental drivers and species distribution is the goal, precise coefficient estimates are extremely important. Environmental predictors should be selected for the model based on their own effect, not just for some spatial structure that is coincident with the species pattern. When the goal is prediction, being able to include in the model some term that describes processes that are otherwise not available as a predictor, or simply not measurable or observable, is potentially very valuable.

7 · *Machine learning methods*

7.1 Introduction

As discussed in the overview of Part III, species distribution modeling can be treated as a supervised learning problem – observations of a response, such as species presence or absence, and associated environmental predictors, are used to develop rules that can be used to classify new observations where the values of the predictors, but not the response, are known. Statistical or machine learning approaches can be used to solve a supervised learning problem. In Chapter 6 it was noted that the linear (regression) model can be thought of as a model-driven or parametric approach to statistical learning, in which certain assumptions are made about the form of the model, and also a "global" method, meaning that all of the data (observations) are used to estimate the parameters. In other words, the problem in supervised learning is to construct a function that "maps" inputs X to outcome Y. In statistical inference the distributional form is chosen by the analyst and its parameters are estimated from the data. Machine learning methods, in contrast, are various kinds of algorithms that are used to learn the mapping function or classification rules inductively, directly from the training data (Breiman, 2001a; Gahegan, 2003).

As we also saw in Chapter 6, GAMs and MARs have been described as "non-parametric" extensions of GLMs because they are not global, but use a local subset of the data (in predictor measurement space) to estimate the response by assuming a particular structured form to that response (e.g., using piecewise linear functions, smoothing splines or polynomials). So, as mentioned in the overview of Part III, it would have been just as valid to put GAMs and MARSs in this chapter as in Chapter 6, and some would disagree with the way I have organized these topics (but I trust the readers to use the table of contents to find what they are looking for). Hastie *et al.* (2001), for example, grouped decision tree methods with GAMs and MARSs because each assumes a different but nonetheless structured form for the unknown regression relationship. Decision trees, for example, partition feature space into a set of rectangles, and fit a constant to each one. In this chapter decision trees are discussed along with other machine learning methods.

The methods that will be described in this chapter have all been applied in SDM, have been called inductive (supervised) machine learning and have been developed in the fields of artificial intelligence and statistics. They include decision tree-based methods, artificial neural networks, genetic algorithms, maximum entropy and support vector machines. I will begin by focusing on decision trees because they have been widely used in SDM, and then will give an overview of the others. Ensemble forecasting methods will also be discussed in this chapter. I recommend, in addition to Hastie *et al.* (2001), the well-written discussion of statistical learning, machine learning, data mining and inference applied to geographical data by Gahegan (2003), as background reading. Also, an edited book and a recent review paper provide an introduction to and overview of machine learning methods for ecological applications (Fielding, 1999; Olden *et al.*, 2008).

7.2 Decision tree-based methods

7.2.1 How decision trees work

Tree-based methods, referred to as classification and regression trees, or, collectively, decision trees (DT), are the first methods I ever used for SDM and I have learned a lot about seeking patterns in data from using them. I found several book chapters to be useful general introductions to decision tree modeling, and although they are associated with particular software implementations, I still consult them (Clark & Pregibon, 1992; Venables & Ripley, 1994 or the newer edition). A nice description of decision trees is also given in Elith *et al.* (2008). In this book, I have focused on the special case of classification trees with binary outcomes (presence versus absence). However, in the field in which I was first trained, remote sensing, the goal is often to map many categories of land cover, for example, using remotely sensed imagery and other variables. Classification trees are a particularly useful method of supervised classification when the response is a categorical variable with many (more than two) categories and when the predictors include both categorical and continuous variables. Regression examples are relatively uncommon in machine learning, but we expect them to be particularly useful in ecology.

Decision trees are divisive, monothetic, supervised classifiers. What does that mean? The goal in decision tree modeling is to partition (divide) the data into subgroups that are homogeneous, that is, where the response variables have similar values or are members of the same class, based on ranges of values of predictor variables. This takes place in three stages, tree building or growing, tree stopping, and tree pruning or optimal tree selection (Olden *et al.*, 2008). In tree building, first the multivariate data are sorted according to the values of each predictor variable, one at a time, and then every possible threshold value of each predictor is examined. For unranked categorical predictors (nominal variables), every possible grouping of classes is examined. Each potential threshold or grouping is referred to as a "candidate split" that can possibly be used to divide the data.

The two subsets of the data that result from any one of these candidate splits are described in terms of their homogeneity in the response variable. This is easier to show in an example than to describe in words (Table 7.1). The subsets may be of any size. For a continuous response variable, reduction in variance or deviance (sum of squares) is a measure of homogeneity, and the decision tree is called a regression tree. For categorical responses, usually some measure of the homogeneity or "purity" of class membership in the resulting subsets is used – an information or entropy statistic – and the resulting tree is referred to as a classification tree.

The single candidate split – the threshold value (or categorical grouping) of the single predictor – that gives the greatest increase in homogeneity or purity (reduction in deviance or entropy) of the subsets is used to divide the data into those subsets. The single split, resulting in a tree with nested binary decision thresholds (like a dichotomous key), is why the method is called monothetic. Then, in an iterative and nested fashion, the same procedure is applied to the two subsets of the data, again using all predictor variables to search for candidate splits. The subsets of the data in the tree are often referred to as nodes, or sometimes as leaves.

A number of purity measures have been used to select among candidate splits for a classification tree. These include the misclassification error (the proportion of observations misclassified) of the individual split, cross-entropy, the Gini index and the deviance. The deviance or likelihood ratio was described in Chapter 6. If there are 1 through K classes, and p_{mk} is the proportion of class k observations at node m, then the so-called information statistic, entropy statistic, or cross-entropy is:

$$-\sum_{k=1}^{K} \hat{p}_{mk} \log \hat{p}_{mk}$$

Table 7.1. *A tiny example of recursive partitioning to develop a classification tree*

Y (P/A)	X1 (soil)	X2 (elevation)
0	A	930
0	A	1100
1	A	760
0	A	880
1	B	545
1	B	650
0	B	750
1	B	700
1	B	590

The dependent variable Y is the presence or absence of a species coded 1 or 0. There are two explanatory variables, soil type ($X1$) which is categorical (A or B), and elevation ($X2$), which is continuous. First, nine observations are sorted according to values of the first predictor, $X1$. In this simple example, there is only one possible threshold value or grouping for this categorical variable, shown by the dashed line. The purity of class membership (the two possible classes of Y) of the resulting groups is measured by calculating the reduction in entropy or deviance (see text) – the change in value from the unsplit dataset to the two groups. Then the process is repeated by sorting the observations by $X2$, and examining all nine possible splits in that case. Whichever candidate split of whichever variable ($X1$ or $X2$) results in the most pure grouping in the response is used to split the data. The process then is repeated for each subgroup.

The Gini index is:

$$1 - \sum_{k=1}^{K} \hat{p}_{mk}(1 - \hat{p}_{mk})$$

Deviance has been preferred in some software implementations (Venables & Ripley, 1994) and the Gini index in others (Breiman *et al.*, 1984), although the Gini index and entropy have similar properties (Hastie *et al.*, 2001, p. 271).

To summarize thus far, in tree building the data are divided using recursive binary partitions (Hastie *et al.*, 2001). The result is a "decision tree" – a set of nested, binary decision rules that can be used to classify observations into subgroups (nodes) based on threshold values of the predictors. The decision tree can be presented as text (Table 7.2) or

Table 7.2. *Example of a classification tree. Decision rules are organized like a dichotomous key*

(1) root 1470 360.100 0 (0.973469 0.026531)	
(2) jan < 6.565 1383 137.800 0 (0.991323 0.008677)	
(4) jan < 4.525 889 0.000 0 (1.000000 0.000000) *	
(5) jan > 4.525 494 112.900 0 (0.975709 0.024291) *	
(3) jan > 6.565 87 107.800 0 (0.689655 0.310345)	
(6) precip < 265.5 22 8.136 0 (0.954545 0.045455) *	
(7) precip > 265.5 65 87.490 0 (0.600000 0.400000)	
(14) precip < 289 49 67.910 1 (0.489796 0.510204) *	
(15) precip > 289 16 7.481 0 (0.937500 0.062500) *	

The dependent variable is the presence/absence (1/0) of *Ceanothus verrucosus*, a rare shrub in coastal southern California, USA. In this dataset, there are 1470 observations, and *C. verrucosus* is present in 39 of them. There are two predictors in this model, average minimum January temperature (jan), and average annual precipitation (precip). The elements in each line are the node number, the splitting rule (variable and threshold), the number of observations at that node, the deviance of the node, the predicted value (0, 1) and the proportions of classes 0 and 1 in that subgroup of observations. For example, *C. verrucosus* is classified as present only at terminal node 14 (∗ denotes a terminal node; see Fig. 7.1) based on a probability threshold of 0.5. At that node there are 49 observations, the deviance is 67.91, the predicted *Y* value is "1" (*C. verrucosus* present) and the proportions of 0s and 1s are roughly 0.49 and 0.51. A quick calculation shows that 51% of 49 observations at node #14 equals 25 (64% of the 39 total) occurrences of this species correctly classified by this model, or about a 36% omission error rate. This does not account for varying the classification thresholds or for errors of commission (see Chapter 9).

graphically (Figs. 7.1, 7.2). The splitting rules define the branching at the "internal nodes" of the tree, and the composition of the "terminal nodes" or leaves of the tree defines the predicted value. The predicted value is the average value of the training data in that node in the case of regression trees, or the majority class in the case of classification trees.

When do you stop partitioning the data or growing the tree? Taken to its logical extreme, rules could be derived to classify every observation in the training data into a one-member terminal node. These rules would classify the data perfectly. However, the resulting decision rules would probably not be very good at correctly predicting new observations. This is a classic example of "over fitting" the training data. Using the terminology introduced in Chapter 6 we would say that a tree model carried to this extreme has low bias but high variance (it makes very few assumptions and can make accurate predictions for the training

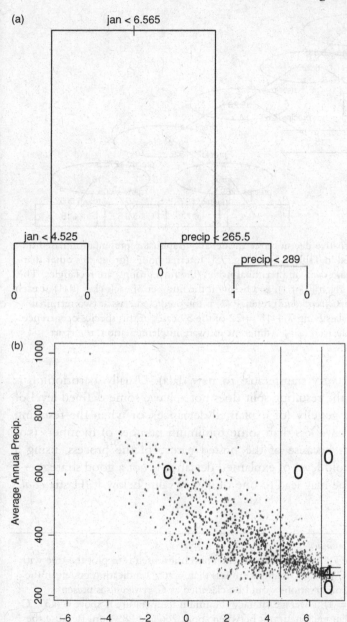

Fig. 7.1. Graphical representations of the decision tree presented as text in Table 7.1: (a) the tree diagram is useful because the length of the branch is proportional to the decrease in impurity or the deviance explained by the split.

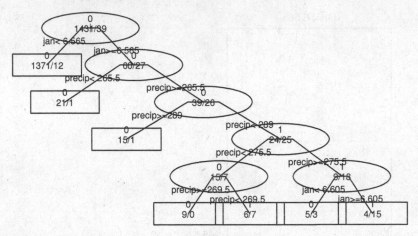

Fig. 7.2. An alternative decision tree model and graphical representation based on the same data used in Table 7.2 and Fig. 7.1. Internal nodes (groupings other than the final groups) are ovals and terminal nodes (the final groups) are rectangles. The predicted class is given (0 or 1), and below it the number of each class (0/1) at each node. In this case *C. verrucosus* presence (1) is the predicted class at two terminal nodes, correctly classifying 7 + 15 = 22 of the 39 cases of this species occurrence. This model was derived using a different software implementation, the rpart package in R.

data but is not very transferable to new data). Usually, partitioning is stopped when the resulting split does not achieve some defined level of increased homogeneity (or explained deviance), or when the resulting subsets would have less than some minimum number of members (say, five). However, because of the nested nature of the process, using a rigorous threshold level of explained deviance is not a good strategy – a split of low value may lead to one of higher value below it (Hastie *et al.*, 2001).

←———

Caption Fig. 7.1 continued.
Comparing Table 7.2 with this figure, the convention used is to plot the tree with the decision rule ("January temperature less than 6.565") indicating go left in the decision tree. So, observations would be classified as *C. verrucosus* is present (predicted value = 1) if average January minimum temperature is above ~6.57 °C and average annual precipitation is between about 266 and 289 mm. Because the model only has two predictors, it can also be depicted as shown in the bottom graph, (b) showing the decision thresholds for the two environmental predictors shown on a scatterplot of the data. This tree model was developed using the tree package (Ripley, 1996) in the R software (R Development Core Team, 2004).

Tree-based methods have been developed over the past 25 years (Breiman *et al.*, 1984), and the approach that has proven most effective is to "grow" a large decision tree using fairly liberal stopping criteria, and then to "prune" the tree, which means to remove the splits (that is, collapse the internal nodes) that add the least to overall subgroup homogeneity (according to those same criteria of explained deviance or purity used to define the splits). Often, the change in the misclassification error rate is used as the criterion for pruning nodes from classification trees. The goal is to prune the decision tree to a size (number of final groups or terminal nodes) that is likely to provide robust predictions for new data. That size is usually determined through some form of k-fold cross-validation procedure – dividing the training data into k subsets, developing a tree model with $(k-1)/k$ of the data, testing it with $1/k$ of the data by calculating error on that $1/k$, and repeating k times (Fig. 7.3). Then the error rates for the k trees are calculated for all possible sizes as the trees are successively pruned from their terminal to root nodes by successively collapsing nodes that produce the smallest per-node decrease in a purity measure (or increase in error rate). The best tree size is the smallest one that produces an estimated error rate within one standard error of the minimum (Breiman *et al.*, 1984; De'ath & Fabricius, 2000), reflecting a balance between size and error. This procedure is also called cost-complexity pruning (Hastie *et al.*, 2009).

It should be noted that, when a classification tree is used for prediction, the majority (plurality) category can be predicted at a node, or alternatively the proportion of training observations in the majority category at that node (Table 7.2, Fig. 7.2) can be predicted. This is particularly useful for spatial prediction in SDM because that proportion has been interpreted as the probability or "suitability" (Pontius & Schneider, 2001) of the category or event, e.g., occurrence of a species. This makes the output of prediction from a CT analogous to the probability of an event predicted by logistic regression (GLM), or its GAM analog. The other machine learning methods described below also yield probabilistic predictions, e.g., on some kind of continuous scale.

7.2.2 When are decision trees useful?

Why are classification and regression trees useful as an alternative modeling method in SDM? What are their advantages and disadvantages? Decision trees have been used in many fields, including medical diagnosis, and have been developed for "mining" large, messy datasets for patterns and information. SDM is just one of these applications. They

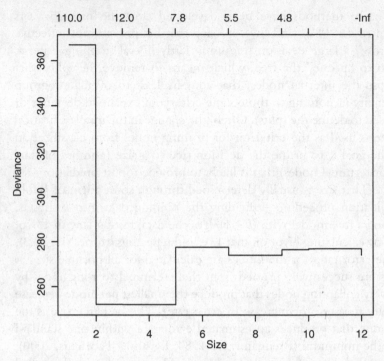

Fig. 7.3. Cross validation to determine the optimal tree size – that is, a tree model that makes robust predictions for unseen data (not used to train), and is therefore not overfitted to the training data. This shows the average deviance explained (*y*-axis), based on ten-fold cross validation, for nested trees of various sizes (number of terminal nodes, *x*-axis), calculated by pruning each tree from its maximum size (12 nodes) to the root (1 node). The minimum unexplained deviance at 2–5 nodes indicated that, for this example, a tree pruned to 2–5 terminal nodes will make robust predictions for unseen data and not be overfitted to the training data. The scale on the upper axis shows the cost-complexity parameter (a measure of the change in fit per number of terminal nodes) associated with the trees of each size (Breiman *et al.*, 1984; Venables & Ripley, 1994). This plot is based on the data for *Ceanothus verrucosus*, described in Table 7.2, and the resulting tree model shown in Table 7.2 and Fig. 7.1 is the best five-node (pruned) tree, not the full tree. Although a two-node tree would be the most parsimonious based on this corss-validation, a five-node tree is shown for illustrative purposes.

are noted to be particularly good at handling some kinds of data and problems:

(1) *Categorical predictors*: Decision trees handle both ordered and categorical predictors "in a simple and natural way" (Breiman *et al.*, 1984, p. 56). In SDM, categorical predictors such as vegetation type, soil

type or land cover can be difficult to parameterize and interpret in a linear model if they have many categories. While it is possible to treat each category as a dummy variable (0/1) in linear models, the results are often difficult to interpret, and it uses up a lot of degrees of freedom. It is best to aggregate categorical variables to as few categories as possible, based on ecological criteria, when using them in GLMs and GAMs (Chapter 6). In contrast, in classification trees, categorical predictors are easily associated with categorical responses in an approach that is analogous to contingency table analysis. This is especially useful with a categorical response such as species occurrence.

(2) *Hierarchical interactions*: Tree-based methods characterize interactions between variables, or "... [make] powerful use of conditional information" (Breiman *et al.*, 1984, p. 56). They identify and model non-linear and non-additive relationships between predictors and a response in relatively simple ways. In addition, hierarchical responses (when the response to a predictor is conditional on the values of another predictor) are very naturally described by decision trees. For example, a certain level of soil moisture may be adequate to support a species only if the maximum temperatures in the warmest season are below some threshold level. The species may be able to survive at higher temperatures given more moisture. These kinds of interactions would have to be specified at the outset in a linear model (Michaelsen *et al.*, 1994).

(3) *Threshold responses*: Decision trees characterize threshold effects of predictor variables on response. Multivariate species response functions are likely to be complex, or appear to be because of data limitations (Barry & Elith, 2006), and may be characterized by threshold effects, rather than linear or smooth responses (Chapter 3). The method of recursive partitioning is very effective at characterizing a threshold effect of an environmental variable on species response given that the splitting rules are based on threshold values of ordered response variables.

(4) *Informative output*: Trees are effective at exploring, graphically and quantitatively, complex relationships in multivariate data (De'ath & Fabricius, 2000). For many people, the structure of a decision tree, as it is displayed graphically, is an intuitive and informative way to describe patterns in data ("if elevation is above X, and slope aspect is north-facing, and soil type is A, then the species is present"). On the other hand, very large trees can be difficult to interpret.

(5) *Missing data*: Decision trees handle missing values and outliers in a very robust way. These observations get isolated in the tree without influencing the estimation of other model "parameters," that is, decision rules (Clark & Pregibon, 1992; De'ath & Fabricius, 2000).

(6) *Classifying new data*: Trees were developed to be used for prediction (Breiman *et al.*, 1984; De'ath & Fabricius, 2000). Although applying the classification rules in a GIS for spatial prediction can be cumbersome for large decision trees, the process is conceptually straightforward. Some software applications generate a simplified set of classification rules (C5 or "See5," Quinlan, 1993) and spatial prediction from decision trees has been automated in some software systems (see Table III.1).

There are some drawbacks to classification and regression trees. When a species response is linear or smooth, it is not well characterized by trees because they use rules based on thresholds, and would have to approximate a linear response with a step function (De'ath & Fabricius, 2000). Further, because later splits are based on fewer and fewer observations, trees may require large samples to detect patterns (Vayssiéres *et al.*, 2000), especially when there are many predictors. With species presence/absence data, the limiting factor may not be the overall sample size but the number of observations characterizing a rare class (such as presence of a rare species).

Finally, trees can be very unstable (Benito Garzón *et al.*, 2006; Prasad *et al.*, 2006; Hastie *et al.*, 2009). This means that varying the inputs, either by sampling from a set of observations, or varying the set of explanatory variables used, can result in very different models (predictors, decision rules), although the error rate or predictive ability may be very similar (for example see Scull *et al.*, 2005). This illustrates several points. One is that this method is very data-driven or non-parametric, and a slightly different set of training data can yield a different selection of threshold values or variables. Another is that, given a large set of potential predictors, applying the criterion of maximizing the increase in node purity does not necessarily distinguish well among them. In other words, two variables may give very similar results and the tree model arbitrarily selects the one that gives an ever-so-slightly greater increase in purity. So, paradoxically, although trees are touted as effective and efficient at variable selection, it can be difficult to interpret variable importance from these models. The instability of decision trees is being addressed recently through the improved tree-based methods described in Section 7.3.

7.2.3 A note about multivariate decision trees

Multivariate decision trees (Brodley & Utgoff, 1995), that is, trees with multiple response variables, have also been proposed as a method of predicting species composition based on environmental data (De'ath, 2002). I have not yet seen this approach applied in SDM, that is, for mapping multiple species, although decision trees have been used to map multi-species assemblages or communities (e.g., Lees & Ritman, 1991; Franklin et al., 2000; Ferrier et al., 2002a). Frequently ecological surveys collect information about multiple species (Chapter 4), and given the demonstrated strength of multivariate response modeling of multiple species using MARS (Leathwick et al., 2006b), measures of compositional dissimilarity (Ferrier et al., 2002a) and other approaches (see Section 6.5), I anticipate that these multivariate approaches will become more widely used.

7.2.4 Application of decision trees in species distribution modeling

An excellent "introduction" to classification and regression trees for ecologists was published in 2000 (De'ath & Fabricius, 2000). However, there were applications of decision tree (DT) models in ecological analysis, biospatial prediction and landscape stratification published prior to that (for review, see Franklin, 1995; Guisan & Zimmermann, 2000). Table 7.3 summarizes some of the recent and not so recent studies specifically related to SDM. In these and other studies, DTs have been applied to the SDM problem and used for spatial prediction because of the advantages listed above, or have been compared to other modeling methods. As was discussed, an emerging generality about the efficacy of decision trees for SDM is that they are somewhat unstable and have lower classification (prediction) accuracy than other methods discussed in Chapters 6 and 7. This has led to the development of new decision tree methods described in the next section.

7.3 Ensemble methods applied to decision trees – bagging, boosting, and random forests

New, computationally intensive methods have been developed that address some of the shortcomings of classification and regression trees. Because these all involve estimating a large number of tree models based on subsets of the data and then averaging the results, they are considered

Table 7.3. *Examples of studies using decision trees (DT) and random forests (RF) for species distribution modeling and related applications*

Citation	Response variable	Application	Comments
(Michaelsen et al., 1987)	Oak seedling survival	Ecological data analysis of hierarchical responses	Early application of DTs to ecological data analysis.
(Walker, 1990)	Kangaroo species occurrence	Potential habitat in relation to climate change	DTs can identify local interactions.
(Moore et al., 1991a)	Vegetation class	Vegetation mapping	Early application of DTs for predictive mapping of vegetation classes.
(Lees & Ritman, 1991)	Vegetation class	Vegetation mapping from remotely sensed and GIS data	Combined gradient modeling of species composition with imagery to distinguish land cover using DT.
(Michaelsen et al., 1994)	Spectral vegetation index	Ecological land classification	Noted usefulness of DTs with hierarchical interactions among predictors.
(Lynn et al., 1995)	Vegetation types	Forest mapping	Focus on accuracy assessment.
(Bell, 1996)	Bird species	Potential habitat	Demonstration of DT method.
(O'Connor et al., 1996)	Bird species richness	Predict large-scale biodiversity patterns	DTs can identify hierarchical relationships among predictors.
(Franklin, 1998)	Plant species occurrence	Compare modeling methods	DTs outperformed GAMs and GLMs but on training data.
(Vayssiéres et al., 2000)	Plant species occurrence	Demonstrate method and compare to GLM	DTs outperformed GLMs in 2/3 of cases, selected similar predictors, were more interpretable and better at detecting interactions.
(Meentemeyer et al., 2001)	Plant species abundance	Ecological inference	Used method because of ability to describe hierarchical interactions among predictors.

Reference	Target	Application	Comments
(Franklin, 2002)	Plant species occurrence	Potential species distribution maps required for landscape simulation model	Refining a regional vegetation map with distribution of dominant species predicted using DTs.
(Miller & Franklin, 2002)	Vegetation class	Compare modeling methods with and without spatial dependence	DTs lower accuracy than logistic regression (illustrating the overfitting problem).
(Flesch & Hahn, 2005)	Invasive bird species (brood parasite)	Compare native and colonizing range of invasive species	Used DT because predictors not expected to have a common effect across the entire sample, and hierarchical interactions expected.
(Fertig & Reiners, 2002)	Plant species	Species range maps for conservation planning and management	DT and logistic regression selected different predictors, and while they had similar commission error rates, DTs had lower omission error.
(Thuiller et al., 2003)	Plant species	Compare modeling methods at multiple scales	DT performance slightly lower than GLMs, GAMs, especially at finer scale.
(Accad & Neil, 2006)	Vegetation class	Map potential distribution of vegetation classes from remnant patches	DTs modeled presence of 28 vegetation types and combine results a posteriori for a more realistic representation of vegetation patterns.
(Rehfeldt et al., 2006)	Plant communities and plant species	Evaluate potential climate change impacts	RF used because of high accuracy in preliminary analysis.
(Prasad et al., 2006)	Tree species	Evaluate potential climate change impacts	RF and bagging trees yielded better interpolations and projections than DT and MARS.
(Poulos et al., 2007)	Vegetation types and fuel classes	Predictive mapping for forest and fire management	DTs used because they were able to deal with outliers and categorical predictors, and yield interpretable classification rules.
(Benito Garzón et al., 2008)	Tree species	Predict impacts of climate change	RF used because of better performance than DT and ANN.

(By no means a comprehensive list – see also De'ath & Fabricius, 2000; Olden et al., 2008). The selected studies emphasized spatial prediction.

forms of model averaging or "ensemble modeling" (see Section 7.8) and are beginning to be used in SDM and related ecological applications with great success.

In order to avoid developing a tree model that is over-fit to the training data, bootstrap aggregation or "bagging" (Breiman, 1996) works by repeatedly (say, 30–80 times) sampling the data with replacement (bootstrapping) and developing trees for each dataset using some stopping rule but without pruning. Typically about 1/3 of the data are held out of each sample ("out-of-bag") and used to evaluate the model, while other data are replicated to bring the "in-bag" sample to full size (Prasad *et al.*, 2006). Then the predictions based on all of the trees are averaged (using a plurality voting rule in the case of a categorical response such as species presence/absence). In other words, each of the many models is used to make a prediction for each observation in a new dataset, and these predictions are averaged.

Another variation called "boosting" (Freund & Schapire, 1996; Ridgeway, 1999) is somewhat similar to bagging except that each observation, instead of having an equal probability of being selected in subsequent samples, is weighted to have a higher probability of selection if it is a "problem" observation (tended to be misclassified by previous models). Therefore, among the methods available for developing and then combining the results from many tree models, boosting is unique because it is sequential.

A form of boosting called stochastic gradient boosting (SGB) (Friedman, 2002) builds many small tree models sequentially from the residuals (referred to as stepping down the gradient of the loss function) of the previous tree. The loss function is defined as some measure such as deviance (Chapter 6) that quantifies the lack of fit of a sub-optimal model. Again, at each step, a model is developed with a random subsample of the data. Hundreds to thousands of tree models are developed in this way (Elith *et al.*, 2008).

Boosting, particularly SGB, has been used for SDM and spatial prediction of other ecological variables and has been shown to perform better that ordinary decision trees (Elith *et al.*, 2006; Leathwick *et al.*, 2006a; Moisen *et al.*, 2006; De'ath, 2007; Elith *et al.*, 2008). These recent papers give a useful introduction to boosted tree methods and practical guidance for their application to SDM and similar ecological questions. For example, measures of the relative importance of predictor variables are available for Boosted Regression Trees (based on how many times the variable was used, in how many trees, weighted by the deviance it explained), and partial dependence plots show the effect of the value of the predictor on the response, yielding a response curve (Elith *et al.*, 2008). With these diagnostic tools, this ensemble modeling method does not have to be a "black box" with regard to ecological interpretation of the model.

Random forests (Breiman, 2001b) is a form of bagging that builds a large number of de-correlated trees (Hastie *et al.*, 2009) and averages them. As in bagging, many trees are developed with subsets of the data, but in addition, each split in each tree model is also developed with a random subset of candidate predictor variables. A large number (500–2000) of trees are "grown" to a maximum size (without pruning) and then the resulting predictions are averaged. The "out-of-bag" (test) sample, the set of observations held back, is used to estimate model error and variable selection or importance.

Variable importance is estimated in two ways for random forests (RF). For each decision tree there is a misclassification error rate (in the case of classification of a categorical outcome) or mean squared error (regression) calculated from the out-of-bag sample. The difference between this error rate and the error rate calculated by randomly assigning the values of a predictor variable, and then passing the test data down the tree to get new predictions, is a measure of the importance of that predictor. This decrease in accuracy if the variable were randomly permuted, the difference between the error rates or mean squares errors, divided by the standard error, is calculated as one measure of variable importance (Cutler *et al.*, 2007). Another measure of variable importance, based on the training (in-the-bag) data, is the reduction in sum of squares (deviance) achieved by all splits in the tree that use that variable, averaged across all the trees (Prasad *et al.*, 2006).

With all of these ensemble tree methods, the tendency to over-fit the data is overcome by averaging the predictions from a large number of models based on subsets of the data. Random Forest modeling is beginning to be used in SDM (e.g., Benito Garzón *et al.*, 2008; Lawler *et al.*, 2009), and, like SGB, has been shown to have higher prediction accuracy than ordinary decision trees in SDM and other applications (Prasad *et al.*, 2006; Cutler *et al.*, 2007). While the methods for determining variable importance are a great improvement over interpreting variable importance from single trees (because of the instability problem), the ease of interpretation of the rule sets for individual decision trees is lost in the ensemble methods (bagging and boosting). That is, these methods do not allow one to determine if "the likelihood of Species A occurring declines drastically if annual precipitation is less than 200 mm," as from a single classification tree. However, fitted functions can be visualized for all ensemble tree methods using partial dependence plots, as noted above (Elith *et al.*, 2008). Although these graphs do not perfectly represent the effect of each predictor, especially if there is strong multicollinearity (Friedman, 2001), they are extremely useful for interpreting species response functions, and analogous to graphical tools used for GLMs or GAMs.

7.4 Artificial neural networks

Artificial neural networks, ANNs, (Ripley, 1996; Lek & Guégan, 2000), so-called because they were first developed as models for the human brain, represent another machine learning approach that has been widely using in remote sensing image classification (Benediktsson *et al.*, 1993; Civco, 1993; German & Gahegan, 1996) and related applications. As Gahegan (2003) described it, the familiar distributional forms used in statistical modeling do not perform well on high-dimensional problems (e.g., with many predictor variables), This is due, in part, to the "curse of dimensionality" (Hastie *et al.*, 2001) – most of the environmental hyperspace is empty, making it very difficult to fit parametric response functions in high-dimensional space. Neural networks, decision trees, and related approaches that hierarchically partition the data into subareas, focusing on the informative portions and ignoring the empty parts, work particularly well in this setting.

This class of machine learning method was developed in parallel in separate fields, statistics and artificial intelligence (machine learning), each using slightly different terminology (Hastie *et al.*, 2001). The basic step is to derive features (new, composite variables) that are linear combinations of the predictors (inputs) and then model the output (response, target) as a non-linear function of those features. As Hastie *et al.* (2001) point out, in spite of the daunting terminology (jargon) associated with neural networks, they are basically non-linear statistical models– they liken them to projection pursuit regression.

Neural network now refers to a large group of models, but as an example I will describe the single hidden layer back-propagation (or single layer perceptron) networks. These are the most commonly used ANNs in ecology (Olden *et al.*, 2008). This is a two-stage classification

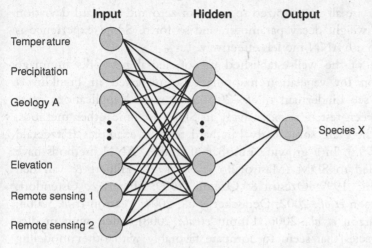

Fig. 7.4. Schematic diagram of a single hidden layer back-propagation neural network, a classification model based on non-linear statistics. For K-class classification there are K units (classes) in the output layer, and the m derived features in the hidden middle layer, Zm, are created from linear combination of the p inputs (predictors). Then the output classes are modeled as a function of linear combination of the Zm. This is an example of a one-class classification predicting the occurrence (probability of presence) of Species X (modified from Olden *et al.*, 2006).

or regression model, and for classifying K classes, there are K units in the output layer. For a single class (species occurrence) or continuous response variable there is typically one unit in the output layer. The neural network has a hidden layer comprising derived features that are linear combinations of the predictor (input) variables, scaled by an "activation function" that is non-linear (often logistic or sigmoidal). The response variable, or "output" in neural network terminology, is a weighted combination of the derived features in the hidden layer (Fig. 7.4). The coefficients are estimated through an iterative optimization method.

Overfitting the training data using a large network is avoided by limiting the number of iterations of the estimation procedure using cross-validation (Moisen and Frescino, 2002). However, as Hastie *et al.* (2001, p. 355) note, there is quite an art to estimating these models or "training" neural networks. The analyst must decide the starting values for the weights, the number of hidden units (and layers), the scaling of inputs (often they are all standardized to mean of zero and standard deviation of one), a weight decay parameter, and so forth. Some experience is required to use ANN models effectively.

In spite of the well-established use of neural networks in image classification for vegetation mapping (also reviewed in Franklin *et al.*, 2003; see Linderman *et al.*, 2004) and related applications, they have not been tested as extensively in SDM as some other methods, although there are some early but hard to find examples (Fitzgerald & Lees, 1992). In a growing number of studies, ANN methods have been applied to SDM (Mastrorillo *et al.*, 1997; Hilbert & van den Muyzenberg, 1999; Ozesmi & Ozesmi, 1999; Hilbert & Ostendorf, 2001; Pearson *et al.*, 2002; Dedecker *et al.*, 2004; Pearson *et al.*, 2004; Benito Garzón *et al.*, 2006; Lippitt *et al.*, 2008), often with prediction accuracies that seem to compare favorably with other modeling methods.

The details of the classifier developed using neural networks are not obvious or interpretable, e.g., it is considered a "black box" approach to classification, and so the main advantage of this method is that it sometimes achieves much higher classification accuracy than other methods in high-dimensional, complex classification problems (Carpenter *et al.*, 1999). Recently, tools have been developed that allow one to examine the contributions of the predictor variables in an ANN model, that is both the magnitude and direction (sign) of the weights, analogous to coefficients in regression (as shown in Olden *et al.*, 2008). However, when ANNs have been compared to other modeling methods in a SDM context, they have sometimes performed worse, or no better, than methods such as GLMs, GAMs, DTs, and random forests (Thuiller *et al.*, 2003; Benito Garzón *et al.*, 2006; Dormann *et al.*, 2008). From a practical perspective ANNs can be more difficult to use operationally than other modeling methods (Moisen & Frescino, 2002). This "steep learning curve" for implementing ANNs may be the key to the seemingly contradictory findings when ANNs have been applied to the species distribution modeling problem. ANNs take a fair amount of skill and experience to apply them effectively to machine learning problems. In studies by experienced ANN modelers acceptably high predictive performance was achieved, while in comparison studies with other modeling methods, ANN did not necessarily perform better than other statistical and machine learning methods.

7.5 Genetic algorithms

Genetic algorithms are another machine learning approach to supervised classification that was first developed in the 1970s. It is so named because a population of classification rules is generated and then the rules "evolve" by a process analogous to natural selection (random mutation and selection based on fitness) until an optimal solution is reached. Genetic algorithms search among a population of potential sets of classification rules, and use probabilistic transition rules to develop new solutions for iterative testing (Jeffers, 1999). Like neural networks this approach is useful when there is a large search space for a solution and where there are complex relationships between variables. Rules are developed by searching for appropriate conditional probabilities (Stockwell, 1999), and the resulting model is expressed in terms of conditional decision rules. For example: "Species A is present if annual rainfall > 20 cm and average July temperature <14 °C." In this sense, GAs are similar to classification tree models, except that classification tree rules are derived from recursive partitioning of the data.

Recently, a genetic algorithm approach to SDM was developed that has the advantage of imposing minimum requirements for each environmental factor such that if one factor drops below a critical threshold, habitat suitability drops sharply (Termansen et al., 2006). This sounds similar in practice to Habitat Suitability Indices (Chapter 8) that are based on the geometric (rather than arithmetic) mean of several habitat factors. In contrast, additive models such as GLMs and GAMs essentially treat environmental factors as if one can substitute for another (their effects are additive), which may not always be ecologically realistic (Chapter 3). However, in that study, their approach performed slightly worse than GAMs and GLMs, but slightly better than DTs based on predictive accuracy (Termansen et al., 2006).

GAs per se are not as widely used in environmental data analysis as the other machine learning methods we have discussed (Olden et al., 2008). A notable exception to this is that genetic algorithms have been extensively used in SDM through the application of the genetic algorithm for rule-set production (GARP) software (Stockwell & Peters, 1999). The GARP software implements an ensemble method that generates a population of prediction rules based on several different types of models, and then uses a genetic algorithm to select among them and develop a final rule set to make predictions. Further, because the output from GARP is stochastic, it is typical to run the model multiple times, and then average a subset of the best models. The application of GARP to the SDM problem will be reviewed in Chapter 8.

7.6 Maximum entropy

Maximum entropy or "Maxent" is a general-purpose machine learning method that has been developed in statistical mechanics and is being applied in fields as disparate as finance and astronomy. Maximum entropy is a principle from statistical mechanics and information theory that states that a probability distribution with maximum entropy (the most spread out, closest to uniform), subject to known constraints, is the best approximation of an unknown distribution because it agrees with everything that is known but avoids assuming anything that is not known (Phillips *et al.*, 2006). When applied to the species distribution modeling problem, for example, the distribution being estimated is the multivariate distribution of suitable habitat conditions (associated with species occurrences) in environmental feature-space. The unconstrained distribution is that of all factors in the study area, and the constraint is that the expected value is approximated by an empirical set of observations of species presence. A software application designed specifically for SDM has recently been developed (Phillips *et al.*, 2006; Phillips & Dudík, 2008). This method has generated great interest because in comparisons it has shown higher predictive accuracy than many other methods when applied to "presence-only" species occurrence data (Elith *et al.*, 2006). Because it was specifically developed for use with presence-only data, and to overcome problems of small undesigned samples, applications of Maxent in SDM will be discussed further in Chapter 8.

7.7 Support vector machines

Another recently developed, statistically-based machine learning algorithm that has been applied to SDM is called support vector machines (SVM). These algorithms are typically designed for a two-class problem where the SVM seeks to define a hyperplane in predictor space that separates two classes, such as species presence/absence (Guo *et al.*, 2005). In this sense it is analogous to discriminant analysis.

A one-class version of SVM has been developed that estimates the "support" of the statistical distribution of predictor variables where the class (species in the case of SDM) is found to occur. This means that, instead of estimating the probability density function, the range of conditions (set of points) where the probability is greater than zero is modeled by estimating a kernel. In other words, hyperplanes are fit around the data points. The one-class SVM, like Maxent and GARP, has been touted as being particularly effective for presence-only species observations (Drake *et al.*, 2006). It was suggested by Drake *et al.* that, because one-class SVM estimates the support of a distribution based on presence-only observations, its interpretation is consistent with the concept of the species niche as a portion of environmental hyperspace. In contrast, other "presence-only" methods that discriminate occupied versus available habitat using background data (ENFA, Maxent) would be more consistent with the resource selection function concept (discussed in Chapter 3).

However, one-class and two-class SVMs were compared in an SDM application where only presence observations were available to predict the distribution of an invasive forest pathogen (Guo *et al.*, 2005). In the two-class case, pseudo-absence (background) data were generated (Chapter 4). This study found that one-class SVMs tended to predict a much larger area of suitable habitat or potential distribution when compared to other kinds of models. In applications where minimizing omission error is important, e.g., estimating risk of forest disease, this could be an advantage. But the tendency to overestimate species distributions is a characteristic that one-class SVM shares with some other presence-only methods (Chapter 8).

SVM has been applied to the presence-only specie distribution modeling problem only to a limited extent, but the one-class model has been shown to work well with as few as 40 observations (Drake *et al.*, 2006). In another study focused on predicting forest pathogen invasion, two-class SVM was compared with four other methods: expert rule-based (Chapter 8), logistic regression, GARP, and decision trees. SVM yielded similar spatial predictions to the other models and had the lowest omission error, although validation data were very limited (Guo *et al.*, 2007). This machine learning method seems to hold some promise for SDM and similar applications, although it is difficult to tell if it is possible to interpret its classification rules from these models in order to verify its ecological meaning or validity (in terms of the importance of predictors and the form of their relationship with the response). Tools to interpret and visualize the classification rules would be important to develop if SVM is to be more widely and effectively used in SDM.

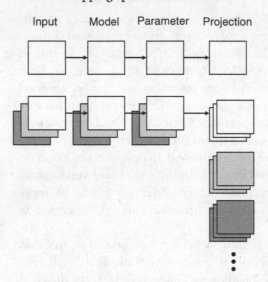

Input   Model   Parameter   Projection

Fig. 7.5. Schematic representation of ensemble model forecasting in SDM (modified from Araújo & New, 2007; Beaumont, 2008). Ensembles of forecasts (projections) can result from varying the input (initial conditions), for example, the species occurrence data and predictors; the model type used in SDM; and the parameters (may vary with model selection and input predictors). New input predictors may result from environmental change (e.g., climate change) scenarios that are derived from multiple realizations of models and inputs.

7.8 Ensemble forecasting and consensus methods

Ensemble forecasting (Fig. 7.5) can be defined as making multiple simulations (projections) across some range of initial conditions (data inputs), model types, parameters (coefficients), and/or boundary conditions (new predictors) (described for SDM by Araújo & New, 2007). An ensemble of predictions can result from different realizations of the same class of model, either because the model has a stochastic element, or because the inputs or other modeling decisions are varied, or an ensemble of predictions can result from different classes of models. In the second case, more than one model is developed and they are used together in some way either to explore the range of predictions, better understand model uncertainty, or get a better projection by using the predictions together. Averaging predictions from different classes of models has also been called model fusion (Elder, 2003).

Several machine learning methods described in this chapter are referred to ensemble modeling methods because many models are estimated, often using subsets of the data, and then their predictions are averaged or composited in some way. For example, bagging, boosting and Random Forests use this approach where the predictions from many decision tree models are averaged. Operationally, GARP and ANN models are often run multiple times, and then a subset of the best models is averaged.

When thinking about how to combine predictions or forecasts from different models we might automatically think of taking their average. "Consensus method" refers to a way of finding the majority view or agreement among different model outputs, and this can involve more than just simple averaging. For example, five consensus methods were compared, in the context of SDM, and these included taking the average, median, weighted average, and median PCA (principal components) of the predictions, as well as using the single best model (Marmion *et al.*, 2009). That study found that average and weighted average predictions were more robust than other consensus methods or single model predictions. Other studies have found that other consensus methods worked equally well or better, and that consensus predictions are not always an improvement over single models. So, while a variety of consensus methods exist, the best approach will vary with the input data and according to whether extrapolations are being made over space or time.

BIOMOD is software dedicated to SDM that typically uses an ensemble modeling approach based on multiple classes of models (e.g., model fusion). Different kinds of models are used to estimate SDM (specifically GLM, GAM, DT and ANN), and the resulting projections are combined using principal components analysis (Thuiller *et al.*, 2003; 2006). Similarly, BIOMOD was used in one case to develop four models per species in a study projecting climate change impacts on species distributions. However, instead of combining the four model outputs using PCA, the entire ensemble of forecasts, using four models, two post-processing rules, and five climate change scenarios, was combined using clustering to detect groups of forecasts making similar projections (Araújo *et al.*, 2006).

Recent SDM studies using an ensemble of forecasts to improve reliability and characterize uncertainty have emphasized temporal extrapolation of species distributions under climate change scenarios (for example,

Thuiller, 2004; Araújo *et al.*, 2005a). These studies are particularly concerned with characterizing SDM variability because, in addition to uncertainty caused by variations in the SDM modeling methods, inputs, and parameters, coupling SDMs with climate change scenarios involves using multiple realizations of different climate models (Araújo & New, 2007; Beaumont, 2008) to establish new inputs (boundary conditions, Fig. 7.5).

However, whether it was explicitly called "ensemble forecasting" or not, a number of recent SDM studies have also fused or averaged predictions across different types of models, for example, to examine the risk of invasive species establishment (Kelly *et al.*, 2007; Crossman & Bass, 2008). Consensus approaches are particularly useful in cases where there are not adequate data to evaluate which model makes the most accurate predictions, as is the case with species invading new areas, projecting future distributions under environmental change scenarios, and in very data-poor regions (e.g., Hernandez *et al.*, 2008).

7.9 Summary

Supervised machine learning methods have been used in SDM and related ecological modeling applications for several decades (Skidmore, 1989; Lees & Ritman, 1991; Stockwell & Noble, 1992, and see Table 7.3). However, the application of newer methods to the SDM problem has greatly increased recently. A pattern is emerging whereby, when they are compared, "classic" single decision trees tend to perform somewhat worse than both the statistical methods discussed in Chapter 6, and the newer machine learning methods discussed in this chapter. Single decision trees can be very useful for exploratory analysis, especially with categorical predictors, because of the interpretable decision rules presented in graphical form. Boosted regression trees and random forests are ensemble decision tree methods that show a great deal of promise for SDM and other complex classification problems. Other machine learning approaches that can be implemented as iterative or ensemble methods, including artificial neural networks and genetic algorithms, tend to perform well given complex classification problems, but can be difficult to use – there can be a steep learning curve for their effective application. Further, interpreting the modeled relationships between predictors and response is not always straightforward in some machine learning implementations. GARP and Maxent are machine learning-based approaches with dedicated software developed specifically for SDM using presence-only data. They are discussed in detail in Chapter 8. Maxent, in particular, is emerging as a tool for SDM that is able to make robust predictions with very few presence-only species observations, e.g., in remote and poorly studied regions.

8 · *Classification, similarity, and other methods for presence-only data*

8.1 Introduction

Many of modeling methods described in Chapters 6 and 7 require observations of species presence and absence, preferably a lot of them (in order to characterize complex response functions), well distributed in space and along environmental gradients. These data are required to estimate model parameters or to derive decision rules for supervised classification. If presence and absence data are available, the modeling approaches that are designed for binary response variables, discussed in the previous chapters, generally give more accurate predictions than models based on presence-only data (e.g., Brotons *et al.*, 2004), but not always (Hirzel *et al.*, 2001). But what if only observations of species presence (but not absence) are available, or what if there are no georeferenced observations at all, but simply some expert knowledge on species habitat requirements? The methods described in this chapter can be, and have been, applied to SDM in these situations. It is actually a very common predicament, to have species presence data, or no species location data at all, and the reasons for that are reviewed here.

If a species has been recorded as being present in a location, we can be fairly certain that it occurs there (except for taxonomic misidentifications). We then make the assumption the occurrence of an organism indicating habitat *use*, *occupancy* or *suitability* for the purpose of modeling. This assumption may perhaps be more easily made with some kinds of data (abundance, observations of foraging, nesting, maps of home ranges) than others (point counts, trapping, species lists). In some types of surveys, if observations are made at a location that do not include a species, e.g., if it is recorded as "absent" (implying non-use or non-suitability), we are less certain about this absence because of the species detectability issues discussed in Chapter 4. Wide-ranging or highly mobile organisms may potentially use or occupy the location, but not at the time of the survey, cryptic organisms may require a more extensive search, and so forth.

Other types of data provide no information about absence but only include locations where a species was observed. As Boyce *et al.* (2002, p. 282) stated, "in some sampling situations we cannot estimate a sample of unused sites." This includes georeferenced natural history collection records (Graham *et al.*, 2004), opportunistic "sightings" that are recorded, and radiotelemetry data. Even these three examples include widely varying types of data with regard to their ability to identify suitable habitat. In contrast with specimen records and informal sightings, radiotelemetry data are generally collected densely on a small portion of the landscape with the expressed purpose of determining activity, movement or habitat use by an individual.

This chapter describes three general approaches to modeling habitat suitability when only data on species presence are available:

(a) For unvisited (new) sites, calculate some measure of similarity to suitable habitat as described by the values of environmental variables for observed species locations (Section 8.2).
(b) Model presence (use) versus availability of habitat, characterizing the available habitat by a sample of other locations in the landscape, or by complete census of all locations, made possible through GIS (Section 8.3).
(c) Develop a habitat suitability model based on expert opinion to assign weights to, and defined transfer (mapping) functions, for environmental predictors (Section 8.4).

These approaches will be discussed and compared in this chapter. An important concept to keep in mind is that models fit to presence-only data predict the relative likelihood of species presence at a site, or the relative habitat suitability (as articulated by Elith & Leathwick, 2009). This is because presence-only data lack information about species prevalence (Chapter 4). Presence/absence data, on the other hand, can be used to predict the probability of species presence (or of encountering the species if sampling with the same method used to derive the training data) using the methods described in Chapters 6 and 7.

8.2 Envelope models and similarity measures

A number of approaches, including some of the first ones developed in SDM, quantify, in some way, the environmental conditions associated with species presence. These have been called "profile methods" (Pearce & Boyce, 2006), and they are based on the ranges, means, or some other

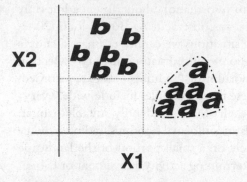

Fig. 8.1. Graphical representation of environmental envelopes shown for two environmental variables, X1 and X2. The b's represent observations of Species B, enclosed by a box whose boundaries are defined by the maximum and minimum values of X1 and X2 where Species B is observed. This is the boxcar or parallelepiped classifier used in BIOCLIM (typically the box is defined that encompasses 95% of the observations; see text). In contrast, the a's, representing observations of Species A, are encompassed in a minimum bounding box or convex hull, a simple example of the approach used in HABITAT (see text).

description of the values of environmental variables for locations where a species has been recorded.

8.2.1 Environmental envelope methods

One of the first species distribution modeling software systems, BIO-CLIM (Busby, 1986, 1991), developed by researchers in Australia, used a simple "hyper-box" classifier, also called boxcar or parallelepiped, to define species potential range as the multi-dimensional environmental space bounded by the minimum and maximum values for all presences (or 95% of them, or other similar variations). In other words, this is a minimum rectilinear envelope (Fig. 8.1, Species B) that gives a binary classification of suitable and unsuitable habitat. Others have since developed their own implementations of a hyper-box classifier applied to bioclimatic predictors for SDM, and it is generally referred to as the "Bioclim" method (e.g., as implemented in the DIVA-GIS software; www.diva-gis.org). The original BIOCLIM software was developed hand in hand with a data and information system containing appropriate spatial predictors (e.g., climate, geology) for large-scale species range modeling. That system emphasized the derivation and mapping of bioclimatic variables most relevant to species distributions (Chapter 5). Improvements on the simple box classifier it used were soon to follow. HABITAT (Walker & Cocks, 1991) refined the environmental "envelope" by enclosing presence points within a convex hull (Fig. 8.1, Species A), and using the relative density of presence points within subareas as a measure of the degree of membership of that subregion in the potential range of the species.

These early modeling methods were very soon supplemented with now more familiar approaches, such as decision trees and logistic regression (Walker & Moore, 1988; Walker, 1990), applied to presence/absence or presence/available data (see Section 8.3 below). However, the Bioclim algorithm (as well as the distance-based DOMAIN; see Section 8.2.2) continue to be used for large-scale species range modeling today and are packaged with global climate datasets, e.g., in free software developed to support the analysis of species distribution data (Hijmans et al., 2001, 2005). Simple envelope approaches have been and continue to be used to predict the impacts of biological invasions (Robertson et al., 2004), plan species reintroductions (Pearce & Lindenmayer, 1998) and to determine the potential distribution of a widespread species (Zhao et al., 2006). Also, recent improvements to the Bioclim envelope method have been developed using a fuzzy classification approach (Robertson et al., 2004), not unlike HABITAT.

Simple envelope models using large scale climatic variables as predictors also continue to be widely used in phylogeographical research (for example, Diniz-Filho et al., 2007b) and studies of impacts of climate change on species distribution (Erasmus et al., 2002; Midgley et al., 2002). Further, the reader should be aware that there is some confusion about the terminology used in the recent literature. Sometimes species distribution modeling is referred to as "(bio)climatic envelope modeling" if it is primarily aimed at modeling species–climate relationships and uses large-scale climatic variables as predictors, even if it does not use one of the environmental envelope classification methods just described (Thomas et al., 2004; Heikkinen et al., 2006).

8.2.2 Environmental distance methods

Ecologists may already be familiar with multivariate coefficients of similarity that can be applied to community data (multiple species and sites; see also Section 6.5). For example, Legendre and Legendre (1998, p. 299) list more than 20 metrics. Similarly, there are a number of measures of association among objects (observations) based on multivariate descriptors of those objects (Legendre & Legendre, 1998, p. 300). For example, the Euclidean multivariate distance (dissimilarity) between two objects can be calculated based on several descriptor variables.

DOMAIN was one of the first implementations of distance-based method for SDM (Carpenter *et al.*, 1993). DOMAIN used "a point-to-point similarity metric to assign a classification value to a candidate site based on the proximity in environmental space of the most similar record" (p. 670–671). A Gower metric is a multivariate measure of similarity (Legendre & Legendre, 1998), and is used in DOMAIN to map predictions of this continuously-varying similarity measure.

The Gower coefficient of similarity (Legendre & Legendre, 1998, p. 259) between two observations, x_1 and x_2, is:

$$G(x_1, x_2) = \frac{1}{p} \sum_{j=1}^{p} s_{12j}$$

or the average, over the p descriptors, of the similarities, s, calculated for each descriptor. For quantitative descriptors, similarity is calculated as the complement of a normalized distance, where R_j is largest distance found across all sites or in a reference population:

$$s_{12j} = 1 - [\gamma_{1j} - \gamma_{2j}]/R_j$$

Another measure of multivariate association is Mahalanobis generalized distance. This measure takes into account the correlations among descriptors by scaling the difference in the mean vectors describing two groups by their covariance. Thus, this distance measure is independent of the scales of the various predictors (say, elevation and mean annual temperature). This is a very useful quality in a measure of multivariate environmental similarity between unsurveyed locations and locations that a species occupies when the descriptors typically show multicollinearity.

Although designed to compare groups of observations or sites (Legendre & Legendre, 1998) (p. 280), when applied to habitat suitability modeling, Mahalanobis distance is usually used to compare a single observation to a group of sites (Farber & Kadmon, 2003). The Mahalanobis distance statistic is a measure of dissimilarity and can be written, in matrix notation:

$$D^2 = (x - \hat{\mu})^T V^{-1} (x - \hat{\mu}),$$

where, as it is applied in SDM, $\hat{\mu}$ is the mean vector of habitat characteristics based on some number of observations of species occurrence, x is the vector of habitat characteristics associated with a new observation or point, such as a grid cell in a map, T indicates the transpose of the vector, and V^{-1} is the covariance matrix associated with $\hat{\mu}$. While Mahalanobis distance has long been used as a measure of multivariate similarity in ecology (see Greig-Smith, 1983), I believe that its first application to habitat modeling was by Clark et al. (1993). From the very first, this dissimilarity measure was applied directly to environmental maps in a GIS – the distance statistic was calculated from each grid cell in the map (the stack of mapped environmental variables) to the mean vector calculated from a sampling of locations where the species was recorded, scaled by the covariance matrix for species presences. The raw Mahalanobis distance values can be depicted in a map, or can be scaled in some way (Fig. 8.2). For example, assuming multivariate normality, the statistic has a χ^2 distribution, and the P-values can be mapped (Clark et al., 1993).

Since it was first presented in the SDM context and used for spatial prediction, Mahalanobis distance has been used as a habitat similarity index (Knick & Rotenberry, 1998) to predict the distribution of suitable habitat in a number of GIS-based studies. Interestingly, this approach seems to be most frequently applied in landscape-scale wildlife management applications of SDM (Knick & Dyer, 1997; Watrous et al., 2006; Hellgren et al., 2007), including species restoration, recovery and reintroductions (Corsi et al., 1999; Thatcher et al., 2006; Thompson et al., 2006; Telesco et al., 2007).

A variation improving on Mahalanobis distance as a measure of habitat similarity, based on partitioning, has been described (Rotenberry et al., 2002, 2006). This approach, instead of using dissimilarity to optimal habitat, as D^2 conceptually measures, identifies the minimum set of basic habitat requirements by partitioning Mahalanobis distance using principal-components analysis (PCA) applied to the multivariate environmental data associated with species occurrence. Although it may seem counterintuitive to those who have used PCA for other kinds of data-reduction applications, it is actually the components (or partitions) with the smallest eigenvalues (smallest variation) that are of interest in this case – these represent the variables that maintain a constant value where the species occurs (Rotenberry et al., 2006). The rationale is that environmental variables whose values vary widely among species locations are not very informative, while those that vary very little represent factors that limit a species distribution. This approach is beginning to be applied in wildlife management (Browning et al., 2005) and global change studies (Preston et al., 2008).

Distance-based approaches have limitations – they work best when organisms are using optimal habitat, are well-sampled in environmental

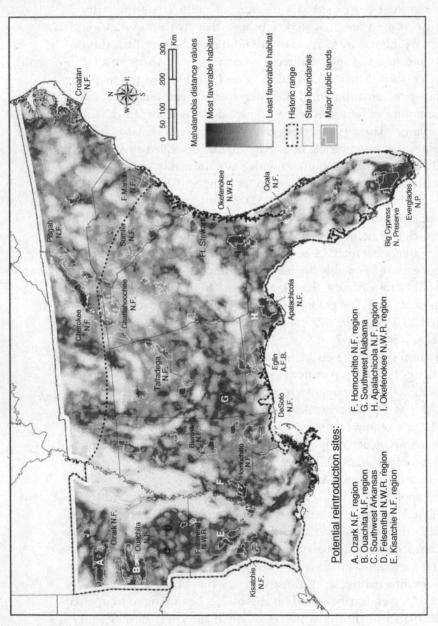

Potential reintroduction sites:

A. Ozark N.F. region
B. Ouachita N.F. region
C. Southwest Arkansas
D. Felsenthal N.W.R. region
E. Kisatchie N.F. region

F. Homochitto N.F. region
G. Southwest Alabama
H. Apalachicola N.F. region
I. Okefenokee N.W.R. region

Fig. 8.2. An example of Mahalanobis distance used as a measure of habitat suitability, estimated for an endangered species, the Florida panther (*Puma concolor coryi*), and mapped for the southeastern USA (from Thatcher *et al.*, 2006). Mahalanobis distance is shown on a relative scale (darker tones indicate greater suitability). This analysis was used to identify potential reintroduction sites (indicated by letters A–I). Copyright © The Wildlife Society; Used with permission.

space, and when habitat variables are not dynamic (Knick & Rotenberry, 1998). Mahalanobis distance is more restrictive than other SDM methods in that it assumes that predictors are equally weighted, have normal error distributions and only considers linear relationships of species to environmental predictors. The metric is based only on continuous, quantitative, predictor variables. To overcome this last limitation, categorical predictors that are often associated with wildlife habitat suitability, such as land cover or vegetation type, are typically expressed numerically on a continuous measurement scale as an area or proportion of the area within some radius of a species location (e.g., Knick & Rotenberry, 1998; Johnson & Gillingham, 2005). However, I did find one application that appeared to use categorical predictors including vegetation type (Thompson *et al.*, 2006), perhaps by expressing each category as a 0/1 "dummy variable."

How does Mahalanobis distance compare to other SDM methods applied to presence-only data? Farber and Kadmon (2003) showed that D^2 described the bioclimatic species envelope more accurately than the BIOCLIM hyper-box classifier, while Tsoar *et al.* (2007) found that D^2 was among the most accurate of six presence-only methods compared. In another comparison, D^2 was evaluated using independent observations of caribou occurrence (Johnson & Gillingham, 2005) and performed comparably to a variety of methods. The other methods tested included logistic regression, used to estimate a resource selection function (habitat use versus availability), an HSI expert model, and GARP – these methods are discussed below. Notably, in that study, although prediction accuracies were comparable, the spatial distribution of predicted suitable habitat varied widely among methods. Disadvantages of Mahalanobis distance noted in that study were that it does not provide a measure of the influence of individual environmental variables on the distribution of the species (statistical inference, parameter estimation), nor is it amenable to statistical tests or information theoretic approaches to variable selection (as logistic regression is). Another disadvantage is that predictors are weighted equally. The distance based methods described here do not appear to offer any particular advantages over currently available methods described in Section 8.3. More comprehensive comparisons of presence-only SDM methods are discussed in Section 8.5.

8.3 Species presence versus habitat availability

Another class of models contrasts habitat use, as characterized by observations of species presence, with habitat availability. Habitat availability can be characterized by the entire range of environmental conditions found in a region, essentially all grid cells in the map, or by a sample of those conditions or cells. When a sample is used, it can be conceptualized either as approximately describing all available habitat, sometimes called background, or describing habitat that is likely to be unsuitable or unused, sometimes called pseudo-absences. Each of these ways of dealing conceptually and operationally with a sample of presence-only observations is discussed in the following subsections. The first subsection (Section 8.3.1) describes how the statistical and machine learning models that discriminate between classes of events (species presence and absence), such as logistic regression and related methods described in Chapters 6 and 7, can be applied to presence and availability data. The remaining subsections describe methods designed specifically to model species distributions relative to "background" environment. These include ENFA, GARP, and Maxent.

8.3.1 Resource selection functions using discriminative models

Any of the discriminative (sensu Phillips & Dudík, 2008) modeling approaches that have been considered for binary or categorical data (Chapters 6, 7), such as logistic regression, GAMs or decision trees, can be applied to presence-only data if a sample of available locations is generated appropriately. Developments in the wildlife biology literature are beginning to converge with advances in SDM and provide a framework for this. A resource selection function (RSF) is defined as any function that is proportional to the probability of the use of a resource unit by an organism (Manly et al., 2002). A RSF is a quantitative, statistical form of a Habitat Suitability Index (Section 8.4). Probability of habitat use can be estimated from presence/absence data in exactly the same way that a species distribution can be estimated, for example, using logistic regression (Chapter 6). However, when observations of unused sites are not available, e.g., if only presence data exist, then the appropriate conceptual framework is to contrast used and available sites, where "available" can be based on a complete census of available habitats, or a representative random sample of landscape locations. When used in this way with presence and availability data, it has been recommended that the statistical model is more appropriately used as an estimating function, describing the relative likelihood of habitat use, and not for statistical inference (Boyce et al., 2002).

There has been some debate about whether it is appropriate to use logistic regression to estimate RSFs from use/availability data, and statistical procedures are available that correct for the problem of some utilized sites (presences) occurring in the "available" data (Keating & Cherry, 2004). However, it has been shown that logistic regression is robust even when a fairly high proportion (20%) of the available sites are actually used ("false absences") (Johnson *et al.*, 2006). But, a distinction can be drawn between models of used/unused sites versus used/available sites (Pearce & Boyce, 2006). Pearce and Boyce describe the first instance as contrasting occupied sites with those where the species has not been recorded, and the SDM predicts the "relative probability of habitat occupancy" (species occurrence). In the second case, all resource units within the sample area are considered to be available for use, but some are used more frequently than others, and the model estimates the "likelihood of use." To illustrate the second case, the authors give the example of radiotelemetry data for large territorial animals such as bears that might potentially be recorded at any location within their home range, but use some locations more frequently than others. Available locations, in this case, might be drawn randomly from among unused locations (no use recorded by telemetry) within the delineated home range.

Pearce and Boyce suggest four approaches to modeling use/availability: ENFA, case-control logistic regression, logistic regression using an exponential model (Manly *et al.*, 2002; for example, Alexander *et al.*, 2006), and logistic regression to estimate a logistic discrimination model. Modeling the relative probability of habitat occupancy, as already noted, can be done using logistic regression and related methods (for a recent example using GAMs see Zielinski *et al.*, 2006) as an approximation. In this case, presences versus available environment are discriminated, acknowledging that available sites may actually include some occupied sites (false absences). Or, as noted above, logistic regression can be corrected for the effect of false absences in the "available" site data (Keating & Cherry, 2004). Ward (2007) recently proposed an expectation-maximization (EM) algorithm to estimate the underlying presence-absence logistic model using presence/background data.

Data on available environments are often called "pseudo-absences" in the literature, and this term can be confusing because it suggests that they represent absences (that to some degree may be contaminated by false absences), when in fact they can simply represent the range of available environmental conditions. In order to characterize available sites, a random sample is often selected from among the grid cells in a GIS throughout a study area that may be somewhat arbitrarily defined (for example, Gibson *et al.*, 2007). Recent studies have shown that approaches other than random selection, for example environmentally stratified random samples (Zaniewski *et al.*, 2002), improve the resulting SDMs derived from logistic regression, GAMs and related methods. Excluding locations with historical species presence data, based on natural history collections, from the background sample (Lutolf *et al.*, 2006), and accounting for spatial clustering of the species (Olivier & Wotherspoon, 2006), has also improved the accuracy of SDMs using discriminative methods with presence-only data.

It has been proposed that the results of a preliminary habitat model, such as from ENFA (see next section), can be used to stratify the sample by inversely weighting or using a threshold value of the SDM to select pseudo-absences in areas of low likelihood of species occurrence (Engler *et al.*, 2004; Chefaoui & Lobo, 2008; Le Maitre *et al.*, 2008). However, Elith and Leathwick (2009) and Phillips *et al.* (2009) strongly caution that background data should be located in areas that could reasonably have been sampled for species presence/absence, not in the most unlikely areas.

No matter what method is used to select background data or pseudo-absences from within the study area, it is important to keep in mind that determining habitat use versus availability depends critically on defining the correct predictors, those variables that actually limit distributions, and upon which the organism depends (Johnson, 1980). This caveat applies to all SDM methods. Further, the estimation and interpretation of use versus availability (habitat selection) will vary depending on how the extent of the study area is defined (Ciarniello *et al.*, 2007; Phillips *et al.*, 2009). Extent will determine what range and frequency distribution of environmental conditions are assumed to be available. It has also been emphasized that validation methods for presence/absence models may not be appropriate for evaluating use/availability models, and other evaluation methods have been proposed (Boyce *et al.*, 2002), and are discussed in Chapter 9.

A recent comparison study found that SDMs based on presence-only data were improved if they were developed using background data that were sampled with the same bias as the species occurrence data (Phillips *et al.*, 2009). In other words, when location records for a target group of species in a region (say, all museum collection records for birds in Ontario), rather than a random sample of locations, were used as background data to model the occurrence of an individual species within that target group, this yielded a more accurate SDM. This strategy worked because the target-group background (as they called it) effectively estimated the distribution of survey effort in the region and therefore had the same bias (in geographical and environmental space) as the occurrence data for any species in the target group. The improvement in model performance was the same order of magnitude as the performance differences among model types. This is a promising and useful approach to improving SDMs based on presence–only data derived from natural history collections (Chapter 4) and related sources, and one that is easy to implement (see Phillips et al., 2009, for details).

The following subsections describe SDM approaches developed for presence–only data that explicitly model species distributions (habitat suitability) relative to the available environment. The available environment is characterized by background data describing the range of conditions found in the modeled region (Phillips et al., 2009). A random sample, all non–presence locations, or all locations (grid cells) can be used to represent background. If all locations are used, this necessarily includes some presence locations, but this may only add some noise to a model.

8.3.2 Ecological niche factor analysis

Ecological niche factor analysis (ENFA), proposed by Hirzel et al. (2002), is, like Mahalanobis distance, based on a multivariate description of species occurrence locations. Mahalanobis distance relates habitat suitability to the unbounded distance from the average conditions where a species is found (the species mean). In contrast, ENFA estimates species niche more explicitly based on the magnitude of the difference between this species mean and the entire range of environmental conditions observed (background data). For example, the Marginality of a species distribution in one environmental dimension is defined as the difference between the global mean and the species mean, normalized by dividing by 1.96 standard deviations of the global distribution (Fig. 8.3). Specialization of a species is calculated as the ratio of the standard deviation of the global distribution to the species distribution, and describes how wide or narrow the species range of tolerance is. As these authors emphasize, these measures are highly dependent on the global set of data used for reference, that is, the extent of the defined study area.

Hirzel et al. extended this to the multivariate case and related their approach, conceptually, to calculating the Hutchinsonian niche hypervolume (Chapter 3). Environmental variables are normalized and standardized. Marginality and Specialization are extracted as factors from the multivariate data using factor analysis procedures analogous to Principal

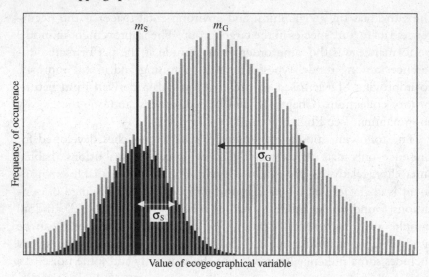

Fig. 8.3. The frequency distribution of values where the focal species is present (black bars) contrasted with the distribution for all cells in the study area shown for a single environmental ("ecogeographical") variable (Fig. 1 from Hirzel *et al.*, 2002). In Ecological Niche Factor Analysis (ENFA), the difference between the means, m_S and m_G, is defined as the Marginality of the species. The difference in the magnitude of the standard deviations (σ_S and σ_G) is defined as the Specialization (or its inverse, Tolerance). In contrast, Mahalanobis distance (for example) would be measured as the distance from any data point to m_S, normalized by the variance. Copyright © The Ecological Society of America; used with permission.

Components Analysis (Fig. 8.4). Factors are extracted successively by computing the translation and rotation that maximizes the variance of the global distribution relative to the species distribution in multivariate space (Hirzel *et al.*, 2001). Marginality is described by the first factor while Specialization can be described by several subsequent orthogonal factors. The factors then can be scaled and used to derive a species distribution or habitat suitability map using procedures described elsewhere (Hirzel *et al.*, 2002; Hirzel & Arlettaz, 2003; Hirzel *et al.*, 2006). For example, habitat suitability can be scaled as the Manhattan distance in transformed space. Its developers have provided specialized software to perform ENFA called Biomapper (http://www2.unil.ch/biomapper/).

ENFA has been used in conservation biology and wildlife management to predict suitable habitat for a reintroduced species, the bearded vulture (Hirzel *et al.*, 2004), for cryptic bat species (Sattler *et al.*, 2007), and

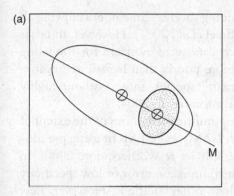

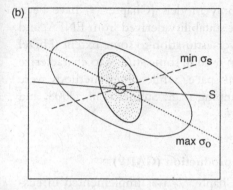

Fig. 8.4. Marginality and Specialization shown graphically in two environmental dimensions (Fig. 2 from Hirzel *et al.*, 2004). The white ellipse shows the distribution of all grid cells and the gray ellipse shows the distribution of cells where the focal species is present. The crossed-circles mark the centroids of each distribution. In Ecological Niche Factor Analysis (ENFA), the marginality factor is the line passing through both centroids (a). The specialization factor is the axis maximizing the ratio of σ_G to σ_S, and is intermediate between the lines of minimum species variance (dashed) and maximum global variance (dotted). Copyright © Wiley-Blackwell; Used with permission.

for large grazing ungulates (Traill & Bigalke, 2006). A habitat suitability map from an ENFA model was used to estimate the population size of an endangered shorebird (Long *et al.*, 2008). A recently proposed modification to ENFA allows skewed species–environmental responses in marginal habitats to be more easily calculated, which may be useful for modeling suitable habitat for certain conservation-relevant species (Braunisch *et al.*, 2008).

ENFA was compared to logistic regression using simulated data (a "virtual species") and shown to produce robust predictions of suitable habitat even for a spreading or expanding species – one not occupying all suitable habitat on the landscape (Hirzel *et al.*, 2001). However, in other studies where presence and absence data were available for modeling and evaluation, ENFA performed more poorly than logistic regression, especially for wide-ranging "generalist" species, those without highly restricted ecological requirements (Brotons *et al.*, 2004).

ENFA tended to produce overly optimistic predictions of the extent of suitable habitat in a comparison with GAMs and GLMs fit using pseudo-absence data (Zaniewski *et al.*, 2002; Olivier & Wotherspoon, 2006). In other words, ENFA is prone to high commission error or low specificity (Chapter 9) in comparison with these other methods. However, new methods for evaluating presence-only models (Chapter 9) have been used to scale the measure of habitat suitability derived from ENFA, and this has addressed the problem of over-estimation to some extent (Hirzel *et al.*, 2006). Finally, ENFA has been used in some studies to characterize the species niche in environmental space, while other methods were used for predictive distribution mapping (e.g., Steiner *et al.*, 2008, and see Table 10.2).

8.3.3 Genetic algorithms for rule production (GARP)

A genetic algorithm framework (Chapter 7) was implemented in customized algorithms by David Stockwell who developed the "genetic algorithm for rule-set production" or GARP software (Stockwell & Noble, 1992; Stockwell, 1999; Stockwell & Peters, 1999; Stockwell *et al.*, 2000). The currently available software has been widely applied to species distribution modeling, especially using large-extent, low resolution "presence only" datasets such as are found in natural history collections. Free software is available for non-commercial use (http://www.nhm.ku.edu/desktopgarp/).

GARP is a "super-algorithm" for binary (two-class) classification that first generates a population of rules for classifying a species present or absent based on four different methods (Stockwell & Peters, 1999; Stockwell *et al.*, 2006) that are now somewhat familiar to the reader: (a) atomic rules use single values of environmental variables, for example, "if soil type is A then species is present;" (b) "Bioclim-type" rules use ranges of bioclimatic or other variables as in BIOCLIM (Busby, 1986), e.g., boxcar classifiers; (c) range rules are generalizations of Bioclim-type rules (see Stockwell *et al.*, 2006, for details); and, (d) logit rules are an adaptation of logistic regression where coefficients are estimated to weight some predictor variables in order to predict probability of presence (logit rules predict presence if the probability is > 0.75). The rules are generated by sampling, with replacement, an equal number (1250) of presence and background (not recorded as present) observations. Observations are divided into training and testing subsets, and rule sets evolve by developing a population of rules using a training sample and randomly selected methods, searching them heuristically (by random walk) using the genetic algorithm, testing them, retaining or rejecting rules based on how accurately they classify the test data, modifying rules (using crossovers, truncation, point changes), and repeating until 1000 iterations or convergence is reached (Peterson & Kluza, 2003).

Because of the stochasticity in the GARP algorithm, and because different rule sets can be equally "good" when evaluated with sample test data, different GARP runs applied to the same data will yield different models (Anderson et al., 2003). Therefore, it has been recommended that a number of models be retained (say 10 or 100), each rule set be used to predict presence or absence, and then the proportion of models predicting presence for an observation (pixel) "can be interpreted as probability of occurrence" (Stockwell et al., 2006, p. 143). This is typically how GARP is implemented, and then the same evaluation statistics used to evaluate other models that predict probability of presence (Chapter 9) can also be applied to GARP models (Peterson et al., 2007).

GARP treats absence and background data as equivalent concepts. All non-presence points are considered to be absence data (though referred to as pseudo-absences), in the sense that the algorithm develops rules that predict presence or absence. Although, in principle, actual absence observations could be used to train the model (Stockwell & Peters, 1999), it is not possible to do so in the current implementation of GARP (Anderson et al., 2003).

GARP has been used extensively with presence-only data for many of the applications of SDM outlined in Chapter 1, including large scale conservation assessment (Anderson et al., 2003; Peterson & Kluza, 2003; Ortega-Huerta & Peterson, 2004; Peterson et al., 2006), predicting the impacts of climate change on species distributions (Peterson et al., 2002; Martinez-Meyer et al., 2004; Thomas et al., 2004), and phylogeographical studies (Peterson et al., 1999; Martinez-Meyer & Peterson, 2006; Kambhampati & Peterson, 2007). In particular, GARP has recently been used to predict the geographical range of species invasions (Peterson & Vieglais, 2001; Peterson, 2003) for Barred Owls (Peterson & Robins, 2003), Argentine ants (Roura-Pascual et al., 2006), plants (Peterson et al.,

2003; Underwood *et al.*, 2004; Fonseca *et al.*, 2006; Mau-Crimmins *et al.*, 2006), parasitic plants (Mohamed *et al.*, 2006), mudsnails (Loo *et al.*, 2007), and Chinese mitten crabs (Herborg *et al.*, 2007). GARP is also increasingly being used to model the potential distribution of disease vectors (Costa *et al.*, 2002; Adjemian *et al.*, 2006; Benedict *et al.*, 2007; Monteiro de Barros *et al.*, 2007). Given that GARP is a generic machine learning method for spatial prediction (Stockwell & Peters, 1999), it is interesting to note that it has overwhelmingly been applied for species distribution modeling.

How well does GARP perform in comparison with other SDM techniques? This will be addressed in the following subsection because many studies involve direct comparison of GARP with Maxent.

8.3.4 Maximum entropy

Maximum entropy has been called both a general machine learning method and a statistical method, so I think of it as a statistical learning method. As noted in Chapter 7, it has been developed and used in other fields (for example, physics). However, because a software application, Maxent, was specifically developed to develop SDMs with "presence-only" species occurrence data (Dudik *et al.*, 2007), it is discussed in more detail here. Recall that, in maximum entropy, the multivariate distribution of suitable habitat conditions in environmental feature-space is estimated according to the principle of maximum entropy that states that the best approximation of an unknown distribution is the one with maximum entropy (the most spread out) subject to known constraints. The constraints are defined by the expected value of the distribution, which is estimated from a set of species presence observations.

Maxent has been described as a generative modeling approach (as opposed to discriminative; Phillips & Dudík, 2008) that models the species distribution directly by estimating the density of environmental covariates conditional on species presence. The raw output of Maxent is an exponential function that assigns a probability to each site. Raw values are dependent on the number of background and occurrence sites used because they must sum to one, and so they are converted to a cumulative probability (Phillips *et al.*, 2006). In the cumulative format values from 0–100 represent the range of probabilities predicted by the model. The cumulative format, while scale independent, is not necessarily proportional to the probability of presence, for example, if probability values are similar across the entire region, e.g., for a generalist species. Therefore, newer versions of Maxent have introduced a logistic output format that estimates the probability of presence (Phillips & Dudík, 2008).

An important distinction between Maxent and logistic regression-type models is that in Maxent locations without species occurrence records are not interpreted as absences, but rather as representing the background environment. When the regression models are applied to presence/background data, interpretation of the resulting prediction is not clear cut because, instead of modeling probability of occurrence, the prediction is one of relative suitability (Phillips et al., 2006), or the relative likelihood of habitat use (Boyce et al., 2002). However, Ward (2007) have recently described an expectation-maximization (EM) algorithm to estimate the underlying presence-absence logistic model using presence/background data.

Phillips et al. (2006) outlined some advantages and disadvantages of Maxent for SDM compared to other methods. Maxent only requires presence data plus environmental information for the whole study area. The results are amenable to interpretation of the form of the environmental response functions (Fig. 8.5). Maxent has properties that make it very robust to limited amounts of training data (small samples), namely, that it is a density estimation method, not a regression method, and it is well-regularized (Phillips & Dudík, 2008). Because it uses an exponential model for probabilities, it can give very large predicted values for conditions that are outside the range of those found in the data used to develop the model. However, the new logistic output format addresses this problem (Phillips & Dudík, 2008). Extrapolation outside of the range of values used to develop an SDM should be done very cautiously no matter what modeling method is used (Elith & Graham, 2009). Its developers provide freely available Maxent software customized for SDM (http://www.cs.princeton.edu/~schapire/maxent/).

Maxent has been used in studies of species richness (Graham & Hijmans, 2006; Pineda & Lobo, 2009), invasive species (Ficetola et al., 2007; Ward, 2007), climate change effects on species distributions (Hijmans & Graham, 2006; Fitzpatrick et al., 2008), endemism hotspots (Murray-Smith et al., 2009) and to estimate the extent of occurrence (Pearson et al., 2007; Sergio et al., 2007) and quality of protection (DeMatteo & Loiselle, 2008; Thorn et al., 2009) of rare species. Maxent has also been used to investigate the degree to which climate constrains distributions of species (Wollan et al., 2008; Echarri et al., 2009; Cordellier & Pfenninger, 2009), including pathogens and their hosts. Maxent has also been employed as a tool to address ecological questions

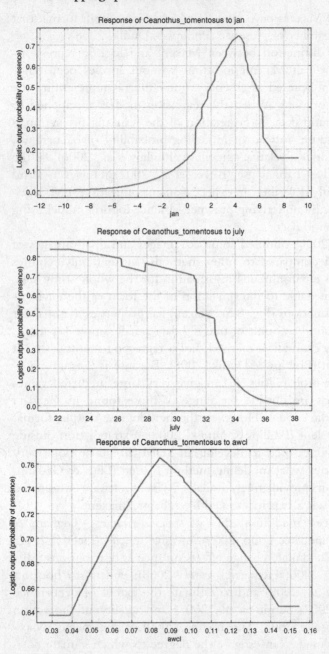

Fig. 8.5. Example of response curves as estimated from Maxent showing the log response of the focal species (*Ceanothus tomentosus*), a woody shrub (data described in Fig. 4.1), to three environmental predictors in a southern California, USA, study area (Fig. 5.1): average January minimum temperature (jan), average July maximum temperature (july) and soil water-holding capacity (awcl, derived from STATSGO data; see Section 5.2.3).

concerning seasonal changes in habitat use (Suárez-Seoane et al., 2008) and the relative importance of abiotic conditions and biotic interactions in determining species distributions (Cunningham et al., 2009).

In several of these examples, Maxent was used in conjunction with, or was compared to, other presence-only modeling methods, primarily GARP, ENFA, BIOCLIM and DOMAIN. In these comparisons, Maxent outperformed GARP in terms of several measures of prediction accuracy (Phillips et al., 2006; Elith & Graham, 2009), especially with small sample sizes (Hernandez et al., 2006; Pearson et al., 2007). Further, in a comprehensive study of a large number of SDM methods applied to presence-only data, Maxent was usually among the top-performing methods in terms of prediction accuracy (Elith et al., 2006). Other novel methods such as MARS, boosted decision trees, and multivariate models (generalized dissimilarity modeling) also fell into the top performing group of models in that study. Maxent performed somewhat better than GLMs and GAMs (applied to presence/background data), and substantially better than envelope methods (BIOCLIM) and GARP. In another comparison, Maxent again performed marginally better than a GLM, and both models suggested similar environmental response curves (Gibson et al., 2007). Elith and Graham (2009) compared the ability of several models to capture known response curves using simulated species data, and found that boosted regression trees (Chapter 7), Maxent and random forests (Chapter 7) performed best in this regard, slightly better than GLMs, and much better than GARP. In particular, GARP was not able to model responses to categorical predictors. GARP models tend to overpredict the extent of species distributions, that is, to have higher commission errors (Hernandez et al., 2006; Elith & Graham, 2009; Phillips, 2008).

A study that addressed the ability of Maxent versus GARP to predict habitat suitability under novel conditions (geographical transferability or extrapolation) claimed that GARP models were better able to estimate potential distributions under novel conditions, that is, in unsampled regions (Peterson et al., 2007). However, that study did not actually adequately address extrapolation because background data from the entire study area were used, and only three species models were examined qualitatively (Phillips, 2008). Rather, it addressed sample bias, also an important issue when using presence-only data.

In one comparison with BIOCLIM and DOMAIN, as well as mechanistic models, both Maxent and GAMs were better able to model species range shifts under future climate change (Hijmans & Graham,

2006). Using SDMs of any kind to extrapolate to novel environmental conditions and locations is a very important and challenging application (Chapter 1) and a topic that needs further investigation (Chapter 10).

8.4 Habitat suitability indices and other expert models

Habitat suitability indices (HSIs) and cartographic overlay models are models that identify combinations of environmental conditions that are similar to known suitable habitat conditions. In this case the suitability index values, or variable ranges, can be calibrated from presence/absence data, presence-only data, or assigned using "expert knowledge." Sources of expert knowledge include qualitative or quantitative descriptions interpreted from publications or field guides, or expert opinion solicited directly (verbally or in writing) from the experts, using informal or formal methods.

HSI modeling was first developed in wildlife management as a tractable quantitative approach to describing wildlife–habitat relationships based on expert knowledge or limited data (Verner *et al.*, 1986; Morrison *et al.*, 1992). In the narrow sense, "HSI" modeling was specifically designed to support the habitat evaluate procedures (HEP) developed and used by the US Fish and Wildlife Service (US Fish and Wildlife Service, 1981). The purpose of an HSI is to document the quality and quantity of available habitat for a wildlife species. Methods for HSI modeling were first developed several decades ago before powerful desktop computers and GIS data were ubiquitous and so relied on simplicity. However, it was not long before HSIs and related modeling approaches were being applied to GIS data in order to make spatial predictions of suitable habitat (Schulz & Joyce, 1992; Stoms *et al.*, 1992; Duncan *et al.*, 1995).

HSI development steps include (a) defining a conceptual model of what factors define suitable habitat for the species (Figure 8.6), (b) selecting the most important habitat variables, (c) quantifying their importance by applying a transfer function to the habitat variable, transforming its range of values to a suitability index (SI) value scaled 0–1, and (d) combining them into a composite index, the HSI, also scaled 0–1, whose value is proportional to habitat quality. Originally, HSIs as developed by the US Fish and Wildlife Service were interpreted to be carry-capacity indices, where an HSI value of 1 indicated some maximum carrying capacity and 0 indicated unsuitable habitat (Cooperrider *et al.*, 1986). However, HSIs can be more generally interpreted as indices of relative

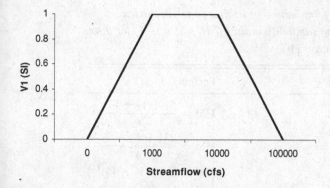

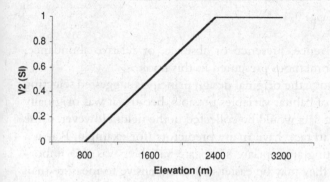

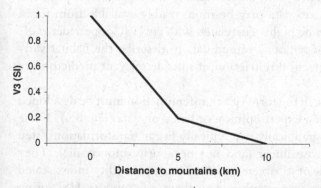

Fig. 8.6. Example of the graphical transfer functions used to generate Suitability Index (SI) values from three environmental variables, stream flow (V1), elevation (V2) and distance to mountains (V3), from a hypothetical conceptual habitat model for the American Dipper (*Cinclus mexicanus*), a bird. The verbal model corresponding to this is, species is found in "fast flowing streams near mountains." A quantitative model is "Streams > 1000 cubic feet per second flow, above 1600 m elevation, within 10 km of mountains." For a published example concerning a butterfly species, see Bojorquez-Tapia *et al.* (2003).

Table 8.1. *Different formulations of habitat suitability index (HSI) models based on suitability indices (SI), scaled 0–1, for three hypothetical habitat factors (V)*

Type of model	Formula
Simple additive model	$HSI = \dfrac{V_1 + V_2 + V_3}{3}$
Weighted additive model	$HSI = \dfrac{2V_1 + V_2 + V_3}{4}$
Multiplicative model	$HSI = (V_1 \times V_2 \times V_3)^{1/3}$
Limiting factor model	$HSI = Min(V_1, V_2, V_3)$

The HSI also has the range 0–1.

likelihood of occurrence, presence or absence, or relative abundance, much like the other methods presented in this book.

On a historical note, the original design principles suggested selecting the fewest number of habitat variables possible, because it was originally assumed that habitat data would be collected in the field. However, some older HSI models, in fact, have many predictors (for example, Raleigh et al., 1986). Selecting appropriate surrogate variables was also important, again because they may be easier or less expensive to measure than the causal factors (see Chapters 3 and 5). For example, forest canopy cover may be correlated with habitat quality for a particular species of bird, but tree basal area data may be more readily available from forest inventories, and may be highly correlated with cover (Cooperrider et al., 1986). Using GIS or remotely sensed data to describe the habitat variables removes this practical restriction, if suitable relevant predictors are available.

A suitability index in the form of a transfer function must be developed for each variable using expert opinion or based on data (Fig. 8.6). These SIs, often depicted graphically, are typically linear transformations, step functions, or piecewise linear (and not necessarily monotonic). They transform the values of a habitat variable into a suitability index scaled from 0–1. These are then combined to form a composite HSI, using one of several approaches (Cooperrider et al., 1986): additive, weighted additive, multiplicative or limiting factor (Table 8.1).

Although, to some extent, the HSI approach has been supplanted or replaced by some of the other methods described in this book (see Scott et al., 2002), it still remains a useful method, especially when species location data are not available to develop an empirical (inductive sensu Stoms et al., 1992) model. Spatial predictions of habitat from HSIs or very similar approaches have been implemented using GIS (for example, Duncan et al., 1995; Wu & Smeins, 2000; Gerrard et al., 2001; Fornes, 2004; Inglis et al., 2006; Rubio & Sanchez-Palomares, 2006; Barnes et al., 2007; Eisen et al., 2007; Boitani et al., 2008). In fact, scaling mapped environmental variables to suitability indices, and combining them into a summary index of suitability, is conceptually very similar to site suitability modeling in urban planning. Site suitability modeling was one of the original and core applications of cartographic overlay modeling, laying the foundations for the early development of GIS (McHarg, 1969; Tomlin, 1990).

When species location data are available, empirical models based on data (such as those described throughout this book) tend to outperform those based on expert knowledge (Clevenger et al., 2002; Bojorquez-Tapia et al., 2003), although sometimes only marginally so (Pereira & Duckstein, 1993; Inglis et al., 2006). One advance that has been made in linking expert knowledge to habitat models is the application of multicriteria decision methods such as the analytical hierarchy process, and other objective methods of soliciting and formalizing expert opinion derived from the field of decision theory (Anselin et al., 1989; Pereira & Duckstein, 1993; Store & Kangas, 2001; Bojorquez-Tapia et al., 2003; Doswald et al., 2007). Another approach is to use expert systems formalisms, implemented in GIS, to develop habitat criteria and models from expert knowledge (Yang et al., 2006). Fuzzy HSIs have also been developed using the formalism of fuzzy sets theory (Rüger et al., 2005). These approaches hold a lot of promise to improve the usefulness of HSIs when the data or the application suggests that HSI modeling is the best approach to SDM. Carter et al. (2006) made a strong case for using existing HSIs as a basis for developing more quantitative models of species–habitat relationships for wildlife species, because of the significance of HSIs in management decisions (within certain agencies), and their usefulness in developing hypotheses.

8.5 Summary

Modeling species distributions with presence-only data is such an important practical problem that a number of comparisons among the various approaches have been made, and some of those have already been mentioned in the chapter. It seems that different methods developed over time because information about species distributions was required for different kinds of applications, at different scales, and using various kinds of "presence-only" data. In general, "profile" methods, ENFA, and later the more complex machine learning approaches, GARP and Maxent, were developed to infer species distributions from georeferenced museum records (natural history collections). They are tailored to small numbers of species occurrence records and, when predictions encompass entire species ranges they are frequently used with coarse-scale environmental predictors (Chapter 5) for spatial prediction.

Mahalanobis distance and related distance methods, as well as HSI methods, have a strong tradition in wildlife ecology and management, have frequently been developed for vertebrates, and tend to be applied at ecological or landscape scales. In these applications, environmental predictors are proximally or distally related to the availability of food and shelter and include factors such as vegetation characteristics (type, cover), soil, topographic and hydrological features (landform, proximity to streams). While distance-based methods are derived quantitatively using occurrence data, HSI methods are useful when georeferenced observations are lacking but expert knowledge of species habitat requirements exists. The straightforward HSI approach is now complemented by powerful new methods for eliciting and formalizing expert opinion as decision rules.

There are some clear distinctions among the approaches presented in this chapter. HSI, or more generally, models based on decision rules derived from expert opinion, do not require any georeferenced species occurrence data, but it is also possible for decision rules to be estimated from data in this framework. However, HSI models tend to perform more poorly than quantitatively estimated models when training data are available (but see Pereira & Duckstein, 1993).

Distance-based methods are not well-suited for categorical predictors, they equally weight predictors, and assume linear relationships between environmental variables and habitat suitability. An even more significant drawback is that distance-based methods do not provide information about, or estimates of, the influence of environmental predictors on species occurrences (the ecological response functions). ENFA presumably is best able to discriminate species niche from background when both are normally distributed so that means are the best measures of central tendency (Figs. 8.3 and 8.4), a condition that is rarely met in nature.

In spite of the potential theoretical advantages of the expert, envelope and distance-based methods that only use presence data to describe the species niche, in model comparisons, methods that use presence and background data, including those discussed in Chapters 6 and 7, tend to show greater predictive performance. Specifically, they have less of a tendency to overpredict, and are able to make more discriminating predictions. Comparisons suggest that the statistical learning methods such as Maxent and ensemble tree methods are particularly robust when only a small and biased sample of observations is available.

Part IV
Model evaluation and implementation

Essentially, all models are wrong, but some are useful.
George E. P. Box

This section outlines methods for the evaluation of species distribution models (Chapter 9) and presents a summary and framework for their implementation (Chapter 10). Evaluation of species distribution models (SDM) has tended to focus on predictive performance as the most important measure of model validity. But predictive performance is really only one aspect of model evaluation. Ecological realism and acceptability to the user community (model credibility) are also important evaluation criteria. Very broadly defined, a valid model is one that meets performance requirements that have been specified. Performance requirements for SDMs may be difficult to specify or quantify in some cases, and all models simplify reality and have prediction errors. SDMs are used to make spatial predictions of species distributions and therefore the spatial nature of the predictions and errors should be explicitly considered when the models are subsequently used to address a question.

In the model evaluation step, many criteria could be used for validating the output of a model of species–habitat relations (Chapter 10 in Morrison *et al.*, 1998). Evaluation is distinct from calibration when the model is used to make predictions based on new or different data. If a strictly independent dataset with suitable attributes is not available, it is common to divide the dataset into "training" and "testing" data prior to modeling, or to use some kind of resampling method (such as bootstrapping) to estimate, from the training data, what the prediction accuracy of the model would be if it were applied to new data. This is the subject of Chapter 9. I discuss model uncertainty and validation or verification in this section, separate from the modeling chapters of Part III, because most of the validation concepts and approaches can be applied to all types of models.

Chapter 10 synthesizes the findings of recent SDM research into a framework for implementation of SDM for various applications. Attributes of species, occurrence data, environmental data, and modeling methods, affect the applicability of SDM for different purposes. Finally, a step-by-step summary of the decisions involved in SDM implementation is given.

9 · Model evaluation

9.1 Introduction

In the previous chapters on statistical methods used to develop species distribution models, it was noted that an important aspect of model building is model (variable) selection based on measures of model fit such as D^2 (explained deviance) or information theoretic measures such as Akaike Information Criterion (AIC; see Chapter 6). This chapter addresses the important step of model evaluation. In species distribution modeling, evaluating habitat suitability models and the resulting predictive maps has focused on quantifying prediction accuracy as a measure of model performance or validity (Table 9.1; criterion 7), as described in Section 9.3. But predictive performance is really only one aspect of model validity. In this introduction, I will outline, more broadly, the many faces of error or uncertainty in SDM.

One broad and useful definition that has been given for model validity is: validation means that a model is acceptable for its intended use because it meets specified performance requirements (Rykiel, 1996). Performance can be measured by a number of criteria (Morrison et al., 1998). These criteria can be applied at different stages of model development described in the introduction to Part III – conceptual formulation, statistical formulation and model calibration – as well as in the subsequent model evaluation steps (Table 9.1). To reiterate, in SDM, model evaluation has tended to focus on *predictive* performance, but other criteria, such as ecological realism, spatial pattern of error, and model credibility (acceptability to the user community) are also important.

All models have prediction errors because models are simplifications of reality. However, because SDMs, as described in this book, are used to make spatial predictions, those predictions, in the form of a digital map, are subsequently used in a variety of applications without necessarily giving sufficient attention to the nature and degree of those errors (Barry & Elith, 2006). It is useful to consider what is meant by

Table 9.1. *Criteria for evaluating species distribution models that address different kinds of uncertainty arising during model formulation and calibration*

Modeling step	Criterion	Description	Reference
Conceptual formulation	Precision	Ability to replicate system parameters	(Morrison *et al.*, 1998)
	Specification	Does the model address the problem? Does it describe the true relationship?	(Barry & Elith, 2006)
	Ecological realism	Is conceptual formulation consistent with ecological theory?	(Austin, 1980, 2002; Austin *et al.*, 2006; Barry & Elith, 2006)
Statistical formulation	Realism	Account for relevant variables and relationships?	(Morrison *et al.*, 1998)
	Verification	Is the model logic correct?	(Rykiel, 1996)
Model calibration	Calibration	Parameter estimation or model fitting and selection	(Rykiel, 1996; Chatfield, 1995)
Model evaluation	Validity, performance	Capability to produce empirically correct predictions to a degree of accuracy that is acceptable given the intended application of the model	(Morrison *et al.*, 1998; Rykiel, 1996; Barry & Elith, 2006)
	Appeal, credibility	Accepted by users, matches user intuition, sufficient degree of belief to justify use for intended application	(Morrison *et al.*, 1998; Rykiel, 1996)

"error" – error includes not only "mistakes" but also variability in the statistical sense (Goodchild, 1994). Although spatially explicit environmental modeling using GIS is well established (Goodchild *et al.*, 1993, 1996), it is still challenging for many people to think of maps that depict real places as "realizations" or outcomes of models (Franklin, 2001), and to perceive map error as statistical or analytical uncertainty rather than a cartographic blunder (Goodchild, 1988; Goodchild & Gopal, 1989).

To emphasize the concept of statistical variation, the terms error and uncertainty are sometimes used interchangeably (Barry & Elith, 2006). However, uncertainty in species distribution modeling extends beyond the data and model estimation (see Table 9.1). In complex modeling tasks such as SDM, uncertainty arises not only about determinate facts, that is, epistemic uncertainty (measurement error, natural variability, model uncertainty), but also from vague or ambiguous meanings of concepts (linguistic uncertainty; Elith et al., 2002; Regan et al., 2002).

Prediction errors in species distribution models can result from data errors or errors in model specification (Barry & Elith, 2006). Most discussion of model uncertainty and evaluation criteria tends to focus on errors arising from model specification (identifying the correct predictor variables, estimating the parameters), but in SDM, uncertainty resulting from data errors can be substantial and deserve attention (see Chapters 4 and 5). An important source of data error is missing predictors, owing to lack of complete knowledge of what environmental factors determine species distributions, or more frequently, to lack of spatial data that represent those factors (Chapters 3 and 5).

In the remainder of this chapter, Sections 9.2 and 9.3 describe data and methods for evaluating predictive performance of SDMs. Specifically, these portions of the chapter focus on performance evaluation of models that spatially predict probability of species occurrence or habitat suitability, and that are evaluated using observations of species presence-absence, the type of data that have been emphasized throughout this book. Section 9.4 provides a summary framework for model evaluation in SDM by considering uncertainty at all stages of model development, implementation and application.

9.2 Data for model evaluation

SDMs (and forecasting models in general) are best validated using new or independent data – data not used to estimate the parameters or fit the model (Fielding & Bell, 1997; Barry & Elith, 2006). Using the same data to calibrate and evaluate SDMs (e.g., Franklin, 1998), called "resubstitution" (Fielding & Bell, 1997), tends to overestimate the predictive performance of the model for new observations (Edwards et al., 2006). Some studies have independent data available for validation (for example, Franklin, 2002; Elith et al., 2006), or they collect new data for validation, which is referred to as prospective sampling (Fielding & Bell, 1997).

However, often it is not feasible to collect new data. In this situation, one approach very commonly used to validate SDMs (for review, see Guisan & Zimmerman, 2000) is to divide or partition the data into one portion used to calibrate the model, called the training data, and one portion used to validate the predictions, called the testing data (Smith, 1994; Miller & Franklin, 2002). Fielding and Bell (1997) discussed data partitioning for species distribution modeling and, as they noted, a range of partitioning methods can be used (Verbyla & Litaitis, 1989). They suggested a rule of thumb described by Huberty (1994) for determining the optimum partitioning of training and testing data for presence/absence models. The proportion of testing data should be $1/(1 + \sqrt{p - 1})$, where p is the number of predictors. So, if there are two predictors, the train:test ratio should be 50:50; if there are five, it should be 67:33, and if there are many (>10), it should be 75:25.

Partitioning the data into one train and test group for model evaluations is an example of two-fold cross-validation. In k-fold cross-validation, the data can be split k times, yielding k estimates of accuracy that can be averaged (as in Dormann *et al.*, 2008). For example, the data can be split into tenths, with 9/10 of the observations used to train the model and the remaining 1/10 used to estimate performance; this is repeated ten times and the estimated performance measures are averaged (Fielding & Bell, 1997).

Often, the available species distribution data are so sparse and hard-won (Chapter 4), and the potential number of explanatory variables so great (e.g., Chapter 6, the curse of dimensionality), that all of the available observations are needed for training (estimating) the model. Therefore, another approach often used (Guisan & Zimmerman, 2000) in partitioning data for SDM validation is bootstrap sampling (sampling with replacement) to estimate prediction accuracy (for example, Wintle *et al.*, 2005; Leathwick *et al.*, 2006b). Fielding and Bell (1997) suggested that some sort of partitioning should be used to estimate prediction accuracy, but that after validation, all available data should be used to estimate the final model parameters if the aim is to develop a robust predictive model.

Whether the evaluation data are derived from partitioning or an independent source, the same guidelines that apply to sampling for model development are relevant to model evaluation. A SDM can be most thoroughly evaluated using a sample that is representative of the environments that are being predicted. This is generally accomplished using a sample that is geographically and environmentally representative of the study area (Chapter 4).

9.3 Measures of prediction errors

What do we do with these testing data? Models that predict the value of a continuous response variable, be they statistical, simulation, or mathematical models, are conventionally evaluated using all kinds of measures of the correlations or differences between observed and predicted values. These measures include R^2 (proportion of variance explained), root mean square error (RMSE), mean (absolute) (percent) error (MPE, MAE or MAPE), and so forth. These measures of fit are applied to many types of models in fields as disparate as economics and hydrology (Pindyck & Rubenfeld, 1981; Armstrong & Collopy, 1992; Legates & McCabe, 1998). However, this chapter will focus on those measures of accuracy that can be applied to predictions that are categorical (suitable/unsuitable habitat), ordinal (high, medium, low habitat suitability) or probabilistic (probability of species occurrence), and that are validated using categorical data (species presence/absence). These are the types of responses and validation data frequently used in SDM.

Most modeling methods used in SDM to assign a categorical class membership (e.g., "classifiers"), actually predict a probability (or some output on a continuous scale related monotonically to likelihood) of class membership. These include logistic regression, maximum likelihood classification, classification and regression trees, artificial neural networks, and maximum entropy, as discussed in Chapters 6–8. Conventionally, a probabilistic prediction is converted to a categorical one by using a threshold probability value (see next section) to distinguish predictions of an event from a non-event (e.g., species presence versus absence). Alternatively, for each observation the class with the greatest predicted probability is assigned (as in maximum likelihood classification). Subsequently, so-called "threshold-dependent" measures of categorical prediction accuracy are used to evaluate SDMs as well as other similar applications such as thematic map accuracy assessment (Card, 1982; Stehman & Czaplewski, 1998). Measures of categorical prediction accuracy are discussed in Section 9.3.1. Alternatively, the probabilistic predictions can be evaluated and used without imposing a threshold by using methods that compare the distributions of probabilities predicted

Table 9.2. *An error matrix or confusion matrix for a two-class problem, such as species presence–absence, showing the cross-tabulation of observed and predicted values as a two-by-two contingency table; also showing the marginal sums*

		Observed		
		Present	Absent	Sum
Predicted	Present	TP (true positive)	FP (false positive)	Total predicted present
	Absent	FN (false negative)	TN (true negative)	Total predicted absent
	Sum	Total observed present	Total observed absent	Total number observations

for different categories (presence, absence) with reference to a set of validation data. Threshold-independent accuracy assessment will be discussed in Sections 9.3.2 and 9.3.3.

9.3.1 Threshold-dependent measures of accuracy

If the observed and predicted values of the response variable are categorical, then the results of model validation, which is a cross-tabulation of observed and predicted (fitted) values, can be arranged in an error matrix, also called a confusion matrix (Table 9.2). Specifically, for a binary outcome such as species presence/absence, an error matrix is a two-by-two contingency table cross-tabulating true and false positives (presences) and true and false negatives (absences). This deceptively simple summary can be interpreted in many more ways than there are cells in the table! While percent correct classification (PCC) – sum of diagonals (true positives and negatives) divided by the number of observations (Table 9.3) – seems like a logical summary measure of prediction accuracy, there are two problems with it. First, it does not distinguish between errors of omission (false negative error rate) and errors of commission (false positives). Second, it does not account for the prevalence (proportion or frequency) of observations in each class. As a trivial example, if the test data consist of 95% absences and 5% presences, and this reflects the true prevalence of the species on the landscape (a realistic scenario for a rare species), a "null" model that predicted absence in all cases (e.g., no model at all) would be correct 95% of the time – 95% accuracy sounds pretty good!

Table 9.3. *Threshold-dependent accuracy measures for species presence–absence models based on the error matrix, where n is the total number of observations used for validation;* $n = TP + TN + FP + FN$ *(Table 9.2). These accuracy measures can be calculated for any probability threshold used to define categorical predictions, except for the true skill statistic which, by definition, is based on the probability threshold for which the sum of sensitivity and specificity is maximized*

Measure	Calculation
Sensitivity	$TP/(TP + FN)$
False negative rate	$1 -$ Sensitivity
Specificity	$TN/(TN + FP)$
False positive rate	$1 -$ Specificity
Percent correct classification	$(TP + TN)/n$
Positive predictive power	$TP/(TP + FP)$
Odds ratio	$(TP \times TN)/(FP \times FN)$
Kappa	$\dfrac{[(TP+TN)-(((TP+FN)(TP+FP)+(FP+TN)(FN+TN))/n)]}{[n-(((TP+FN)(TP+FP)+(FP+TN)(FN+TN))/n)]}$
True skill statistic	$1 -$ maximum (Sensitivity $+$ Specificity)

Therefore, models of binary outcomes such as species presence/absence are often evaluated using the measures Sensitivity (proportion of actual presences that are accurately predicted) and Specificity (proportion of actual absences that are accurately predicted; Table 9.3). It is particularly useful to examine both of these measures if the cost of a commission error is different from the cost of an omission error. For example, omission errors (predicting a species absent where it is present, or predicting unsuitable habitat where there is suitable habitat) might be more serious than commission errors if the SDM is used to identify sites for species reintroduction or sites at risk of species invasion. In these cases, validation data may not include species occurrences in all suitable habitat and so sites recorded as commission "errors" might actually represent suitable but unoccupied habitat. Commission errors (predicting presence where the species is absent) might be more serious if designating critical habitat for a species of concern. It could be very costly in biological and economic terms to set aside habitat that is not, in fact, suitable for the target species.

Kappa is a measure of categorical agreement that describes the difference between the observed agreement and chance agreement. That is to say, a random map with class proportions equal to the marginal sums

(Table 9.2) would have a certain amount of agreement, by chance, to observed data whose classes occur in certain proportions. Kappa has been widely used to assess categorical predictions of land cover from remote-sensing based mapping (Congalton, 1991; Stehman & Czaplewski, 1998), e.g., of more than two categories. The formula for the two-class case is shown in Table 9.3. While kappa, a well-developed measure of improvement of chance agreement for categorical data, has been used in SDM (Segurado & Araújo, 2004; Elith *et al.*, 2006), it is sensitive to the prevalence of an event in the sample (McPherson *et al.*, 2004; Vaughan & Ormerod, 2005). Although it has been suggested that Kappa may be insensitive to prevalence under certain conditions (Manel *et al.*, 2001), strictly speaking, threshold-dependent measures of prediction accuracy should not be compared among models with different species prevalence. This effect of prevalence on accuracy measures has provided motivation for the use of threshold-independent accuracy measures (see Section 9.3.3).

Another metric that has been used to measure the extent to which a model prediction is an improvement of random assignment of class labels is the normalized mutual information statistic (NMI). However, the NMI is based on logarithms of the cells in the error table and cannot be computed for tables that have zeroes in some cells (unless a small constant value is added). It has not been widely used in SDM (but see Manel *et al.*, 2001).

Finally, the true skill statistic (TSS) is defined as $\{1 -$ maximum (sensitivity + specificity)$\}$ where sensitivity and specificity are calculated based on the probability threshold for which their sum is maximized (see Section 9.3.2). This has been suggested as an alternative to kappa (Allouche *et al.*, 2006) when a threshold-dependent measure of performance in needed. TSS responds to species prevalence differently than kappa, but nonetheless, the statistic has been shown to be negatively related to prevalence (Allouche *et al.*, 2006).

9.3.2 Choosing a threshold for classification

The threshold probability, or criterion for classification of cases, is the probability value above which a case is predicted to be positive (e.g., species present). By convention, this threshold is often set at 0.5, for example in logistic regression implemented in statistical software packages. However, selecting an optimum threshold for use with any of the threshold-dependent accuracy measures discussed in the previous subsection depends on the cost of the different types or misclassification, which, in turn, depends on the intended use of model (Liu *et al.*, 2005). The effect of the threshold on omission and commission error rates also depends on the prevalence of positives in the sample. The threshold probability at which false positive rates and false negative rates are equal to each other tends to equal the prevalence of the events (e.g., species presence) in the sample for many types of models (Fielding & Bell, 1997; Franklin, 1998; Manel *et al.*, 2001). In other words, if the prevalence of positives in the sample is about 10%, then the threshold probability at which sensitivity equals specificity is roughly 0.1. By examining sensitivity, specificity, and other error measures based on varying the probability thresholds, a threshold can be selected for a given application based on trade-offs between these omission and commission errors (Fig. 9.1).

For many of the applications of SDM discussed in Chapter 1, it may be necessary to categorize predictions into a binary map (species presence vs. absence or suitable vs. unsuitable habitat) rather than using a continuous representation of probability of species presence. For example, a reserve design algorithm that selects areas to include in a reserve network with the goal of maximizing the number of species protected might require binary species presence/absence maps as input (Wilson *et al.*, 2005). Landscape simulation models of ecological dynamics may similarly require baseline information on species distributions in the form of discrete maps (Mladenoff & He, 1999). Calculating the extent of available habitat within a management area is often based on a discrete representation of suitable versus unsuitable habitat (for example, Fernandez *et al.*, 2006). Area of occupancy (AOO) for a species of conservation concern, required in order to determine its threatened status, might also be based on binary maps (Rondinini *et al.*, 2005; Elith & Leathwick, 2009).

Two recent studies have evaluated the use of different criteria or methods for selecting a probability threshold (Table 9.4) based on SDM prediction accuracy and the overall pattern of errors, and given guidelines on choosing a threshold criterion (Liu *et al.*, 2005; Freeman & Moisen, 2008a). Freeman and Moisen (2008a) found that, for species with high predictive performance (measured by AUC; see Section 9.3.3), and prevalence near 50%, the optimal thresholds based on all criteria tend to converge. When prevalence is low, different criteria can result in quite different thresholds. In some applications of SDMs, such as estimating extent (species range) or area of species occupancy, correctly predicting prevalence is important (see also Section 9.3.3 on Calibration).

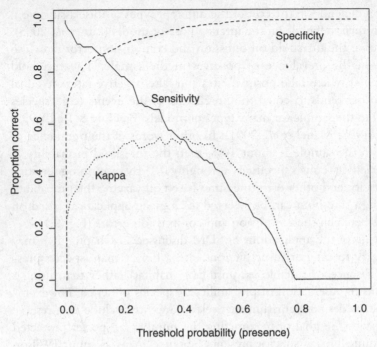

Fig. 9.1. Example of sensitivity (solid line), the proportion of presences correctly predicted, versus specificity (dashed line), the proportion of absences correctly predicted, and Kappa (dotted line; see Table 9.3), as the lower threshold probability of an event (e.g., probability of species presence) predicted from a model is varied from zero (always predict present) to 1 (always predict absent). Shown for the data in Table 9.5. The threshold probability where sensitivity = specificity is 0.13; and the threshold where Sensitivity + Specificity is maximum 0.09. These thresholds tend to be similar to the prevalence of the events (presence) in the sample, which is 10% (see Table 9.5). In this example, the threshold where Kappa is maximum is 0.41 and the threshold where the predicted proportion of presences equals observed is 0.34.

Setting the probability threshold so that predicted prevalence equals observed prevalence (if it is known) obviously results in the lowest bias in predicted prevalence, but so did using maximum Kappa (the probability threshold at which Kappa is greatest; see Fig. 9.1) as a threshold criterion in Freeman and Moisen's (2008a) study. Liu *et al.* (2005), on the other hand, suggested that using (a) observed prevalence, (b) average predicted probability, (c) the sum of Sensitivity and Specificity, (d) Sensitivity = Specificity, or (e) the point on the ROC plot nearest the upper left corner, as the threshold all gave roughly the same results. They

Table 9.4. *Optimization criteria for selecting probability thresholds used to convert probability maps into binary maps*

Criterion	Reference
Fixed threshold: traditional default of 0.5 is often used in logistic regression and related modeling methods	(Manel *et al.*, 1999b)
Sensitivity = Specificity: Probability threshold where the false positive error rate equals the false negative rate	(Fielding & Bell, 1997)
Maximum (Sensitivity + Specificity); Threshold that gives the highest total value of Sensitivity and Specificity; equivalent to finding the point on the ROC curve whose tangent is 1.	(Manel *et al.*, 2001)
Maximum Kappa: Probability threshold where Kappa is highest	(Beerling *et al.*, 1995; Guisan *et al.*, 1998; Moisen *et al.*, 2006)
Maximum percent correct classification (PCC): Probability threshold where PCC is highest	–
Threshold where predicted prevalence is equal to observed prevalence	(Cramer, 2003)
Threshold equal to observed prevalence	(Cramer, 2003; Laurent *et al.*, 2005)
Threshold equal to mean or median predicted probability	(Fielding & Haworth, 1995; Cramer, 2003)
Threshold that minimizes the distance between the ROC curve and the upper left corner of the ROC plot; minimize $(1 - \text{Sensitivity})^2 + (\text{Specificity} - 1)^2$.	(Cantor *et al.*, 1999)
Weighting omission and commission errors by their costs relative to a given application of the SDM	(Wilson *et al.*, 2005; Elith & Leathwick, 2009)
Maximum Sensitivity (Specificity) that also achieves user-defined minimum Specificity (Sensitivity)	(Freeman & Moisen, 2008a)

(From Liu *et al.*, 2005; Freeman & Moisen, 2008a).

Table 9.5. *Portion of a table showing the number of true- and false-positive (TP, FP), and true- and false-negative (TN, FN) predictions from a species distribution model for each threshold minimum probability value, prob_Present*

prob_Present	Sensitivity	Specificity	TP	TN	FP	FN
0	1.000	0.000	87	0	0	873
0	0.989	0.630	86	550	323	1
0.01	0.966	0.705	84	615	258	3
0.02	0.954	0.741	83	647	226	4
0.03	0.931	0.764	81	667	206	6
0.04	0.908	0.781	79	682	191	8
.	.	.	.	.	.	.
.	.	.	.	.	.	.
.	.	.	.	.	.	.
0.75	0.058	0.999	5	872	1	82
0.76	0.046	0.999	4	872	1	83
0.77	0.023	0.999	2	872	1	85
0.78	0.000	0.999	0	872	1	87
0.85	0.000	1.000	0	873	0	87
1	0.000	1.000	0	873	0	87

Observations with predicted probabilities greater than or equal to prob_Present are predicted present and those less than prob_Present are predicted absent in a binary classification. Sensitivity and Specificity are calculated at each probability level and used to calculate the AUC for the ROC plot shown in Fig. 9.2. The data for this example are from a Generalized Additive Model (for example, see Fig. 6.2) for a plant species (*Yucca brevifolia*, Joshua Tree) in the Mojave Desert, California USA (see Miller, 2005), that had high prediction accuracy (AUC = 0.934; Fig. 9.2). The test data consist of 87 presences and 873 absences, so the species has a prevalence of 10%.

did not recommend using maximum Kappa in their study, but nor did they examine correct estimation of prevalence as a performance criterion.

Figure 9.1 illustrates an example of several criteria that could be applied to the predictions from an SDM (data shown in Table 9.5). In this example, the model's predictive performance is high (AUC = 0.934; Fig. 9.2 and see the following section). The species prevalence is 0.10, the threshold probability where Sensitivity = Specificity is 0.13, and the threshold where Sensitivity + Specificity is maximum is 0.09. In other words, these three criteria give roughly the same threshold. On the other hand, the threshold where Kappa is maximized is much higher, 0.41, and the threshold where the predicted proportion of presences equals observed is 0.34. This example illustrates the findings in Freeman and

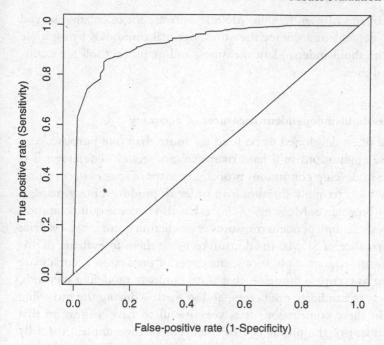

Fig. 9.2. A receiver operating characteristic (ROC) plot for the accuracy data shown, in part, in Table 9.5. The Area Under the Curve (AUC) of this ROC plot is 0.934 ± 0.012 (1 s.e.).

Moisen's (2008a) study. This species is relatively uncommon (10% prevalence in the survey data), and so thresholds near the species prevalence tend to balance omission and commission errors – that is, most known occurrences are correctly predicted, while not too great a number of known absences are predicted to be in suitable habitat. A threshold in this range (0.09–0.13) should be used in cases where predicting too much unsuitable habitat to be suitable is costly. An even lower threshold might be chosen if the model is expected to correctly predict all known occurrences, perhaps because recorded absences are suspected to include suitable but unoccupied habitat. An example would be when identifying suitable habitat for a species reintroduction. On the other hand, if a higher threshold based on maximum Kappa were chosen, this would yield a result very similar to using setting the predicted proportion of presences equal to observed, so a threshold in this range (0.34–0.41) would be more appropriate for estimating AOO, for example. Using a lower threshold would predict suitable habitat for most known occurrences, but does so at the expense of overestimating the AOO.

Freeman and Moisen (2008b) provide software for calculating several threshold-dependent accuracy measures, optimal thresholds by multiple criteria, threshold-independent measures, and graphical tools for examining all of these.

9.3.3 Threshold-independent measures of accuracy

SDMs are often developed to be used for more than one purpose, e.g., for diverse applications in a land management agency. Therefore, it is preferable to develop continuous probability maps of species occurrence, and allow users to apply thresholds in order to produce binary maps, if necessary (Freeman & Moisen, 2008a). It has also become quite common to use threshold-independent measures of prediction accuracy to evaluate the performance of SDMs. In addition to using them to evaluate SDMs for use in subsequent applications, measures of prediction accuracy are often used to compare models – that is, to compare modeling methods, species, sets of candidate predictors, and so forth – during the modeling process. In these comparisons it is very useful to have a measure that is independent of the prevalence of the event in the sample, especially when trying to draw conclusions based on models of different species or samples with varying frequencies.

AUC

A threshold-independent metric that is now frequently used to assess SDMs is the awkwardly-named "area under the curve" (AUC) of the receiver-operating characteristic (ROC) plot (Hanley & McNeil, 1982). This statistic is usually abbreviated AUC but is sometimes referred to as ROC. This metric was developed in the fields of medical diagnosis and signal processing. The ROC plot (Fig. 9.2) is a graph of the false-positive error rate on the x-axis (1 – Specificity; Table 9.3) versus the true positive rate on the y-axis (Sensitivity) based on each possible value of threshold probability – that is, every probability value predicted by the model for the data (for example, Table 9.5). The AUC is calculated by summing the area under the ROC curve. It can range from 0.5 to 1.0 where a value of 0.5 can be interpreted as random predictions and values above 0.5 indicate a performance better than random (Fig. 9.2). The value of the AUC indicates the proportion of the time a random selection from the positive group will score higher than a random selection from the negative group (Fielding & Bell, 1997). For binary data the AUC is identical to the two sample Mann–Whitney statistic (Hanley & McNeil, 1982). This statistic describes the overall ability of the model to discriminate between the two cases, i.e., species presence and absence. AUC values of 0.5–0.7 are considered low (poor model performance), 0.7–0.9 moderate, and >0.9 high (Swets, 1988; Manel et al., 2001).

Species frequency of occurrence in the training data (proportion of total observations that are presences) is called prevalence. This prevalence can represent the true species prevalence on the landscape (commonness) if the survey data include presences and absences from an unbiased, representative sample (Chapter 4). However, the proportion of presences and absences in the training sample can be manipulated by subsampling a presence/absence dataset. Further, these proportions are, by definition, artificial in the case of presence-only/background data where the size of the background sample is defined by the modeler, and the presence-only data do not represent the true prevalence of the species in a region. In order for a performance measure to be useful for comparing models for different species, it should be insensitive to factors such as species prevalence.

It has been stated that AUC is not affected by changes in species prevalence (Manel et al., 2001) and is therefore reliable for model comparisons (Cumming, 2000b). Some studies have shown that AUC does decrease with increasing species prevalence (Segurado & Araújo, 2004; Allouche et al., 2006), but due to the effect of species ecology on our ability to model their habitat requirements. It is more difficult to discriminate suitable from unsuitable habitat for habitat generalists, widespread species with high prevalence, than for habitat specialists (Brotons et al., 2004; Luoto et al., 2005; Elith et al., 2006; see Chapter 10). In contrast, a large comparison study manipulated the ratio of presence and absence observations in the training sample and concluded that the correlation between species frequency and model performance was an artifact of the effect of this ratio (sample prevalence) on model estimation. That study used logistic regression and related discriminative models (McPherson et al., 2004). In their study, more even samples of presence and absence taken from the same species data yielded higher AUC values. However, other studies have shown that manipulating the ratio of presences and absences in the training sample did not affect AUC (Franklin et al., 2009), that the effect of prevalence on AUC can be weak (Syphard & Franklin, in press-a), and that it is not an artifact of sample evenness but a result of species ecology (as discussed above).

The AUC, or some transformation of it (see Engler et al., 2004), is very widely used to evaluate SDM predictive performance, and it is very useful. However, when used as the only measure of SDM predictive performance, the AUC does not tell the whole story (Austin, 2007; Lobo et al., 2008; Peterson et al., 2008). Other aspects of model performance that might be important or informative include the magnitude of errors, model calibration, the spatial distribution of errors, or the differential cost of omission and commission errors. These aspects of model performance can be evaluated using additional threshold-independent methods described in the following subsections.

Correlation

The standard Pearson correlation coefficient has also been used as a threshold-independent measure of prediction accuracy in SDM, and is referred to as biserial correlation in the case where one of the variables is binary (Elith et al., 2006; Elith & Graham, 2009). The advantage of using this biserial correlation between observed presence/absence and the probability predicted from a model is that it takes into account the magnitude of the difference between the prediction and the observation, while AUC is only rank based. The Pearson correlation coefficient measures the degree to which predictions vary linearly with test data (observations), and therefore is likely to be sensitive to variations in sampling intensity in the training data (Phillips et al., 2009).

Rank correlation coefficients, such as the concordance index, have also been suggested as measures of SDM performance (Vaughan & Ormerod, 2005). The main difference between Pearson (applied to this problem as biserial) and rank correlation statistics is that the former is parametric and rank correlation or concordance measures are non-parametric. Rank correlation (Spearman's) has been used to evaluate SDMs (Phillips et al., 2009). A non-parametric correlation statistic would seem appropriate to this application to avoid the assumptions of normally distributed data and linear correlation.

Calibration

Pearce and Ferrier (2000) and Vaughan and Ormerod (2005) also discussed another aspect of model performance which they referred to as calibration. Calibration indicates whether predictions of 0.6 have a 60% chance of being occupied, and whether these are twice as likely to be occupied, or twice as suitable, as predictions of 0.3 (Vaughan & Ormerod 2005). A measure of model calibration is very useful if the SDM is used to designate critical habitat, or for other applications requiring quantitative interpretation of the likelihood of habitat occupancy. The average

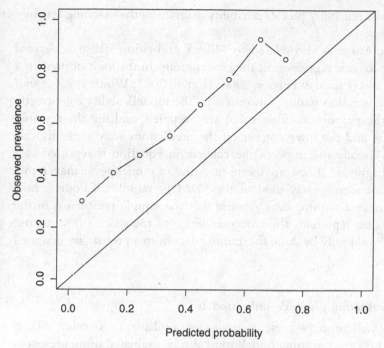

Fig. 9.3. A calibration plot (*sensu* Vaughan & Ormerod, 2005) comparing the average probability of species presence predicted from a model (for the data partially shown in Table 9.5) versus the average observed prevalence of the species among the observations predicted present at that threshold, TP/(TP + FP), that is, the proportion of positives correctly predicted. In this example, observations were averaged by decile. The Observed Prevalence is higher than the predicted probability (the 1:1 line is plotted for reference), suggesting a systematic bias in which the model underestimates species prevalence.

probability of species presence predicted from a model (predicted probability in Fig. 9.3) can be compared to the average observed prevalence of the species among the observations predicted present at that threshold, calculated as TP/(TP+FP) (see Table 9.2), that is, the proportion of positives correctly predicted. Averages can be calculated for regular probability intervals, or by defining equally sized groups of observations. When predicted probability is plotted against observed prevalence, as in Fig. 9.3, values that differ from the 1:1 line (45° angle) suggest an uncalibrated model. Values above the 1:1 line indicate that the model has underestimated species prevalence (observed prevalence is higher than estimated), while values below the line indicate overestimation (observed prevalence is lower than estimated). This type of calibration plot is also known as a reliability plot or reliability diagram in the machine learning literature.

This measure is also related to Miller's calibration statistics, derived from the logistic regression of the observations on the logit of predicted probabilities (Pearce & Ferrier, 2000; Harrell, 2001; Wintle *et al.*, 2005). Miller's calibration statistics also measure the model's ability to correctly predict the proportion of sites that are occupied, and the slope would equal one and the intercept zero if the predictions were perfectly calibrated. Usually the slope of the calibration equation is reported as a model diagnostic. These are useful measures of model performance that I have not seen widely used in the SDM literature. Of course, calibrating against training data assumes that the sample prevalence in the training data represents the true prevalence of the species in a region, and so should only be done for training data from a probability designed sample.

9.3.4 Evaluating presence-only models

In cases where only presence data are available to develop SDMs (Chapter 8), the resulting models must also be evaluated using presence-only data (Johnson *et al.*, 2006; Pearce & Boyce, 2006). This is an interesting challenge because the predictive performance measures described thus far evaluate the ability of a model to discriminate presence from absence. All of these measures can, however, be applied to presence and pseudo-absence or background data (describing available habitat), although they must be interpreted somewhat differently. For example, AUC calculated from presence and background data describes the probability that the model scores a random presence site higher than a random background site (Phillips *et al.*, 2009). However, some have interpreted species distribution models conditioned on presence-only data as models of potential distributions that should only be evaluated using occurrences and not pseudo-absences, and alternative formulations of the AUC have even been proposed (Peterson *et al.*, 2008).

Without absence data, threshold-dependent accuracy measures that require absences, e.g., specificity, cannot be calculated. One approach that has been used in this case to select a threshold of predicted probability, in order to delimit suitable habitat, is to choose a threshold that minimizes omission errors, but at the same time minimizes the area of the map predicted to be suitable (e.g., Loiselle *et al.*, 2008; Peterson *et al.*, 2008). Without absence observations a maximum threshold of predicted probability, 1, would minimize omission errors, but would also predict suitable habitat to be everywhere, and this is a trivial solution. Therefore, minimizing the area of predicted habitat can be used as a criterion for choosing a threshold probability to apply to an SDM when absence data are unavailable. In this approach, a threshold would be selected that encompasses a predefined proportion of observed species occurrences, say, 90% or 95%. This threshold criterion, which has been called minimum predicted area (MPA, Engler et al., 2004; Peterson et al., 2008), can also be used to evaluate and select among models when absence data are lacking. Models that yield a lower MPA would be considered more parsimonious and, therefore, superior. Some customized SDM software, e.g., Maxent (Phillips et al., 2006), has excellent tools for visualizing and selecting thresholds based on some balance between maximizing Sensitivity and minimizing predicted area.

Boyce et al. (2002) developed another SDM evaluation method based on the pattern of predicted habitat suitability values for a sample of presence-only test data. The frequency of presence locations within ranked (ordinal) classes or bins of predicted suitability is expected to increase with bin number (monotonically). That is to say, more of the presence-only test data points would be expected to coincide with high values, than with low values, of predicted habitat suitability if the model is valid. The frequency of occupied locations is adjusted for the area of the bin by dividing by the area of the landscape predicted to have the range of scores within that bin. The area-adjusted frequency of validation points is expected to have positive rank correlation with the bin rank. In other words, the observed frequency of evaluation points per bin is compared to expected (proportional to area). The variation in this measure is estimated by k-fold cross-validation. Boyce et al. proposed calculating Spearman's rank correlation coefficient for these frequency/bin rank data as a measure of model performance, and this measure has been referred to as the "Boyce index" (Hirzel et al., 2006).

Hirzel et al. (2006) carried out an extensive comparison of presence/absence and presence-only prediction performance measures. The Boyce index, the absolute validation index (AVI; proportion of presence evaluation points falling above some specified threshold of predicted suitability), and variations on these indices were compared to Kappa and AUC (calculated using background data). They concluded that the Boyce Index and a continuous (moving window) version of it were robust and informative measures of presence-only model performance if an adequate number of bins were defined. Ten bins seemed to work well. Too many bins increased the variance among the cross-validation partitions.

9.3.5 Spatial distribution of model uncertainty and error

All of the model evaluation metrics described thus far are global or non-spatial. Spatially explicit representations of model uncertainty or error are very informative in SDM. A simple way to examine the spatial pattern of model performance is to map the model predictions along with the actual observations of species presence and absence (Fig. 9.4). Clusters or specific locations of omission and commission errors may suggest missing explanatory variables, interactions between variables that have not been accounted for (Fig. 9.4), differences between models, as in Fig. 9.5, or simply observations that are unusual (see also, for example, Miller, 2005). By doing this, the distribution of model errors or residuals can be viewed in a discrete or qualitative way, by examining the pattern of true and false positive and negative prediction errors. Mapping the errors can also help to identify spatially-structured predictors that may not have been accounted for in the model (e.g., Rahbek & Graves, 2001; Diniz-Filho *et al.*, 2003).

One source of model uncertainty is in the estimation of parameters, for example in a logistic regression model (Chapter 6). It has been noted that maps of the upper and lower confidence intervals of predicted occurrence, as in Fig. 9.6, are useful spatially explicit indicators of the uncertainty associated with a GLM, and yet these are rarely published (Elith *et al.*, 2002). For a recent example of a map of model uncertainty shown for a Bayesian hierarchical model, see Ibáñez *et al.* (2009).

When evaluating predictions from different models applied to the same species data, typically to compare different modeling methods, it can be useful and informative to visually and quantitatively compare the amount of overlap between the maps. For example, two models could have very similar accuracy estimates, based on the AUC for test data, but they could predict suitable habitat in different places – the models could make their mistakes and their correct predictions in different places. There are a number of ways that agreement between habitat maps has been described. For example the percent of cells in the map that have the same label (presence, absence) could be calculated (e.g., Miller & Franklin, 2002; Tsoar *et al.*, 2007). Kappa can be used to describe the agreement between two maps, on a cell by cell basis. Kulczynski's coefficient (discussed in Hennig & Hausdorf, 2006) has also been used as a

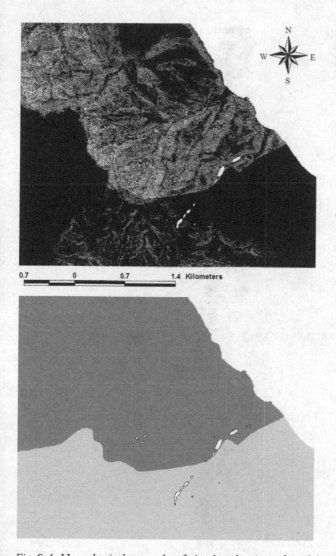

Fig. 9.4. Hypothetical example of visual evaluation of model predictions versus species distribution data for an endemic oak species, *Quercus tomentella*, shown for a small portion of Santa Catalina Island, California, USA. The predicted probability of presence based on a GLM is shown in the top panel where lighter tones indicate higher likelihood of presence. The white polygons are mapped stands of this rare species. For this model, a random sample of "pseudo-absence" locations were used, and so those points are not shown. In the top panel an area of omission error can be seen, where the model predicts low probability of occurrence but the species is present. The bottom panel shows that this corresponds to a boundary in the map of geologic substrate. The geology class in the lower part of the map is associated with the absence of this species in other parts of the island, and so these areas have lower predicted probability of occurrence based on the model, but in this part of the island it appears that the species does occur on that geologic substrate, resulting in an omission error.

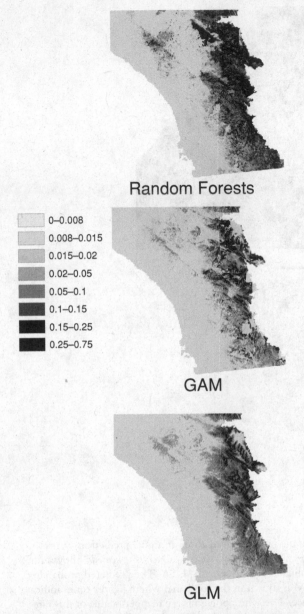

Random Forests

	0–0.008
	0.008–0.015
	0.015–0.02
	0.02–0.05
	0.05–0.1
	0.1–0.15
	0.15–0.25
	0.25–0.75

GAM

GLM

Fig. 9.5. Maps displaying predicted probability of presence from a generalized linear model (GLM), generalized additive model (GAM), and Random Forest model (RF) for a perennial herb, *Penstemon spectabilis* (sample prevalence 2%, AUC = 0.72–0.81) in southwestern California, USA (modified from Syphard and Franklin, submitted). Data are from the vegetation survey described in Fig. 4.1, and the study region and environmental predictors are shown in Fig. 5.1. Darker tones represent higher probability of occurrence in these maps. This species generally occurs in the mountainous, eastern portion of the study area, but differences can be seen among the predicted occurrence maps.

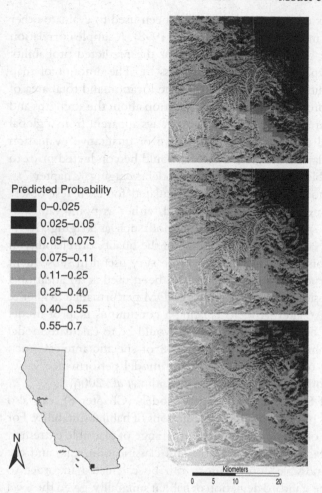

Fig. 9.6. Example of mapping the lower (top map) and upper (bottom map) confidence intervals of the predicted probability of species occurrence (middle map) based on a generalized linear model (GLM), shown for *Rhus integrifolia*, a woody shrub, in an area in southwestern California, USA. Data are from the vegetation survey described in Fig. 4.1, and the study region and environmental predictors are shown in Fig. 5.1. Lighter values indicate higher probability of occurrence.

measure of map agreement for habitat suitability maps (Ray & Burgman, 2006). Another measure of categorical map agreement, the so-called modified Lee-Sallee shape index (Lee & Sallee, 1970), the ratio of the intersection to the union of the predicted habitat area in two maps, could be used to compare predictive maps of species distributions. I have not seen this index used in SDM, but it has been used to evaluate other types of landscape models (Clarke & Gaydos, 1998). A simple correlation coefficient can be calculated, cell by cell, for the predicted probability of occurrence (Syphard & Franklin, in press-b). The amount of map overlap between different predictions, and the location and total area of predicted habitat provides important information about the strengths and limitations of different models that are not always apparent from a global measure of model performance. Quantitative or qualitative evaluation of the spatial overlap of alternative SDMs should be conducted prior to employing ensemble modeling methods (model averaging; Chapter 7).

Further, the model residuals can be calculated for each test obser-vation and these residuals could be mapped, either as point values or using an optimal spatial interpolation method such as kriging (Kyri-akidis & Dungan, 2001). This yields a map of the quality or reliability of the predictions from the SDM that might be very useful in subsequent applications. Interpolated residuals have also been used as an additional predictor in regression kriging, to improve SDM performance (Miller & Franklin, 2002). An alternative to creating a continuous map of model error in the form of interpolated residuals would be to calculate model error for geographical subareas that may be of specific interest (such as ecoregions), in order to determine if the model performance varies systematically among those regions (e.g., Randin et al., 2006).

Other types of SDMs, such as expert models (Chapter 8), can also have error bounds placed around the predictions of habitat suitability. For example, two sets of values, representing the range of plausible extremes, can be used for each predictor variable and classification rule, and the variability of all possible combinations can be calculated in order to display a map of the standard deviation of habitat suitability, given the level of subjective uncertainty in the modeling (described in Ray & Burgman, 2006). The same approach could potentially be used to map variability or uncertainty in predictions from different models, perhaps resulting from different modeling methods or candidate predictor variables, in order to identify geographical areas where confidence in predicted species occurrence is high, and where it is not.

Any method that describes and maps model uncertainty in its vari-ous forms is extremely useful to the subsequent users of SDM predic-tions, because that uncertainty will "propagate" through other models or procedures using predicted species distributions. Wilson et al. (2005) describe an interesting case study showing that the uncertainty in SDMs affected the subsequent design of a nature reserve network, and that those uncertainties can compound when predictions for several species are used. While users of prediction maps from SDM are always implicitly taking a risk of making a bad decision because of uncertainties that are inherent in any model projections, it is best if those risks are explicit and quantified, if possible, so that tradeoffs between risk and efficiency can be considered (Burgman, 2005).

9.4 Summary

Recently, a great deal of SDM literature has focused on models that predict a likelihood of species presence, e.g., the probability of a binary outcome (presence, absence). Therefore, evaluation of model accuracy has made extensive use of threshold-independent measures of prediction accuracy, in particular the AUC. This is an appropriate and useful way to summarize model performance especially when comparing different modeling methods, combinations of predictors, or species. However, this chapter has shown that prediction accuracy is just one measure of model validity. Calibration is another measure that should become more widely used in SDM. Threshold-dependent measures such as sensitivity, specificity and kappa are also very useful in those cases where a categorical prediction (suitable/unsuitable habitat) is required for subsequent applications of SDM products, and where omission and commission errors have different costs, for example in predicting the risk of species invasions versus assessing the protected status of rare species. Rule-based models with rules derived from expert assignment of factor levels to habitat suitability can also be evaluated for uncertainty by generating multiple plausible sets of rules. Even simple visual evaluation of the spatial pattern of model errors can help identify potential problems with an SDM such as missing environmental predictors, and suggest where more complex models may be warranted (e.g., autologistic regression, Chapter 6).

Further, it is important to keep in mind that the ability to evaluate the predictive performance of an SDM is limited by the data available for evaluation. Adequate data for model development can be the limiting component in SDM; independent and adequate data for evaluation can be very difficult to come by (Lobo, 2008). Some of the most important applications of SDM are very challenging to evaluate, for example, future projections of species distributions under scenarios of environmental change. These applications require extrapolation from the conditions used to estimate the model, and ways to evaluate models used for extrapolation are critically needed (Elith & Graham, 2009).

When evaluating an SDM for its intended purpose, the trade-off between bias and variance (Hastie *et al.*, 2001), that is, specificity (over-fitting) and generality (parsimony), is important. In general, SDMs that are intended for interpolation, that is, filling in the gaps in survey data and describing current species distributions, may benefit from greater complexity and specificity, can make fruitful use of local (spatial) modeling methods (Chapter 6; Osborne *et al.*, 2007), and it is probably safe to evaluate them using some kind of resampling of the training data, that is, *k*-fold cross validation or bootstrapping. On the other hand, SDMs that are required to extrapolate in space or time may, by necessity, be simpler, global models, in which case they are likely to make more general but robust predictions. It is very useful to validate them using data from novel places, for example by dividing a study area (e.g., Randin *et al.*, 2006), or times, e.g., using retrospective distribution data (Araújo *et al.*, 2005a; Vallecillo *et al.*, 2009).

Even when focusing on predictive performance as an evaluation criterion, this chapter and this book have heavily emphasized species occurrence data, because they are so widely used for SDM. There are excellent examples of calibration and evaluation of SDM using some continuous measure of species abundance or fitness (e.g., Iverson & Prasad, 1998; Leathwick, 2001; Moisen & Frescino, 2002; Potts & Elith, 2006; Leathwick *et al.*, 2006a; Bellis *et al.*, 2008); however, modeling quantitative measures of abundance or fitness is generally less well developed in SDM than modelling occurrence. In cases where some measure of species abundance is modeled, a wide variety of existing quantitative measures of observed versus predicted values is available for model evaluation.

Each step of the SDM process should be evaluated, including the data, modeling methods, and predictions, using criteria appropriate for the intended use of the resulting maps. Recommendations for an evaluation framework will be discussed in Chapter 10 (Fig. 10.1). Not only statistical criteria but also ecological realism criteria should be applied to evaluate correct selection of predictors, accurate description of response curves, fitting of ecologically rational relationships, and calibrated prediction of abundance if data are available to evaluate this (Austin *et al.*, 2006).

10 · *Implementation of species distribution models*

10.1 Introduction

Species distribution modeling involves the development of formal, usually quantitative rules linking species occurrence or abundance to environmental variables. This book has emphasized the use of species distribution models for spatial prediction of species occurrence – that is, applying the rules to maps of the environmental predictors in order to derive a map of the potential distribution (likelihood of occurrence) of, or habitat suitability for, a species. Although it is the models themselves, their parameters and validation, that are often emphasized in the literature, the predictive maps are the "data products" that get used, often in combination with other models or spatial data, for all of the purposes outlined in Chapter 1. These applications include conservation prioritization, planning and reserve design, environmental impact assessment, predicting the impacts of global change on ecosystems, predicting the risk of pathogens and exotic species invasion in new regions, and ecological restoration and species reintroductions.

A 2007 paper noted the profusion of recently published SDM studies – they had discovered, using a keyword search, 42 papers in the prior seven years (Peterson *et al.*, 2007). However, this has been eclipsed by a recent ISI Web of Science search (on 22 October 2008), using the fairly restrictive parameters Topic = ("species distribution model*" OR "ecological niche model*" OR "climat* niche model*"), that resulted in a list of 45 publications for 2007 alone, and 36 published papers recorded in the first nine months of 2008. An additional 44 papers on this topic were published in the first four months of 2009. These numbers, and the hundreds of studies discussed in this book, show that species distribution modeling activities are now reflected in a growing number of published studies in addition to an unknown number of unpublished applications by governmental and non-governmental organizations.

Many of the published studies have emphasized the statistical, machine learning, and other methods used for modeling (Chapters 6–8), as well as additional methodological issues including sampling (Chapter 4), data characteristics (Chapters 4, 5), and model evaluation (Chapter 9). It is important that these issues be rigorously evaluated in peer-reviewed studies, but the evaluation process frequently involves boiling down the entire complex SDM process, as outlined in Austin's (2002) framework (Chapter 1), to a single number – for example, the AUC, Kappa (Chapter 9), or some other measure of model performance. This is done so that comparisons can be made across species, modeling methods, sets of environmental predictors, species traits, sampling schemes, or combinations of these. Of course a model's predictive performance is an important measure of its usefulness, but as discussed in Chapter 9, and below in Section 10.6, there are a multitude of ways to evaluate a model.

My purpose in writing this book was to describe a general framework for spatial prediction of species distributions, building from the concepts of Austin (2002). The framework is intended to link ecological models of species distributions (theory) to the spatial data and statistical models used in empirical studies. Although I hope that this book has provided a useful summary of the state-of-the-art for experienced practitioners, I especially hope that it has provided practical guidelines to novice species distribution modelers and users of the resulting models, including students, researchers, natural resource managers and policy makers.

I proposed in Chapter 1 that the characteristics of the species (their life histories, biogeographic distributions, and other attributes), the scale of the analysis, and the data available for modeling and validation, together determine the "best" modeling framework to address in any particular question. I had the following specific expectations:

(1) The most suitable or effective modeling method will be determined, in large part, by the spatio-temporal characteristics of the species occurrence data available for modeling – the extent (relative to species range), relevant level of resolution, spatial pattern and density of the survey data, patchiness and detectability of the species, and species commonness, which may be related to habitat breadth and strength of species-environment correlations. Other species traits may also correlate with our ability to discriminate suitable from unsuitable habitat using this predictive framework.

(2) The availability of relevant mapped environmental predictors, or their surrogates, will affect the success of species distribution modeling more than the type of model used. In other words, the strength of the relationship between available predictors and species distributions will affect all types of models in the same way for a given species dataset, and will not determine which modeling method is best.

In the following sections I will summarize what is known, particularly from the growing number of comparative SDM studies, into a framework for implementation. Specifically, I will examine the impacts of species attributes, species occurrence data, environmental data, and modeling methods, on our ability to successfully model species distributions for a multiplicity of purposes. If there is a relationship that can be established between species traits, data traits, methods or other factors, and the performance of SDMs, then these models can be used more effectively, that is, with a better understanding of how reliable they are likely to be for a given application. Then I will present an overview of end-to-end steps for SDM (Fig. 10.1) that includes evaluation of the ecological, data, and statistical models (Chapter 1), not only the predictive performance of the SDM.

10.2 Species attributes

There have now been a number of comparative studies, some quite impressive in their scope, that have focused on the relationship between species traits and the accuracy of species distribution models. In particular, I would like to emphasize those studies that have focused on characteristics of the species niche (in environmental space) and species range (in geographical space). The majority of these studies have addressed the following questions: are the distributions of species with narrower environmental tolerances, or more geographically restricted ranges, easier to model than those of widespread generalists? Is this confounded by sampling issues – the inevitable problem of small numbers of observations for rarer species?

In order to frame this discussion, it is useful to start with a conceptual model of species commonness and rarity. Rabinowitz (1981) identified three factors that define species rarity: range size, habitat specificity and local density. The three factors have eight possible combinations (Table 10.1); six correspond to classes expressing degrees of species rarity

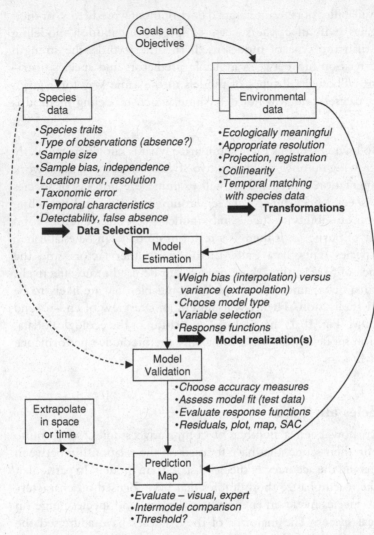

Fig. 10.1. Steps in species distribution modeling showing sources of uncertainty and decision steps in choosing data and methods to match modeling objectives. Modification of Fig. 1.2 (see also Beaumont, 2008; Elith & Leathwick, 2009). SAC = spatial autocorrelation.

or commonness, and two are considered to be unlikely or uncommon in nature. This framework addresses both species niche breadth in environmental space (habitat specificity), and species range in geographical space. It also incorporates local abundance as a factor to define rarity. Common species are those with large ranges, broad environmental

Table 10.1. *Classification of species commonness and rarity*

Geographical range		Wide		Narrow	
Habitat specificity		Broad	Restricted	Broad	Restricted
Local abundance	Somewhere Large	Common	Predictable, (widespread)	(Unlikely)	Endemic
	Everywhere Small	Widespread Sparse	Habitat Specialists	(Unlikely)	Rare on all counts

(From Rabinowitz, 1981.) See also Gurevitch *et al.* (2006).

tolerances, and high local abundance (at least somewhere within their range). Rare species have small ranges, high habitat specificity, and low local abundances everywhere they are found. Endemic species also have small ranges and high habitat specificity, but have high local abundances at least somewhere. (Note that this is somewhat different than the standard meaning of endemic, which is a species that is restricted to a particular geographic region.) Species with large ranges and broad ecological tolerances but low local abundance can be called widespread sparse, while those with large ranges and narrow ecological tolerances and any level of local abundance are referred to as predictable (or widespread habitat specialists). The combination of small range but low habitat specificity is considered to be unlikely in nature.

Table 10.2 summarizes studies that addressed the effect of range characteristics and breadth of ecological tolerance on SDM performance. This summary emphasizes those studies that examined more than just the effect of sample size or species sample frequency (prevalence) on various measures of model performance or accuracy (Chapter 9). Although these studies did not directly invoke the rarity classification scheme presented above, the observed trends can be summarized with respect to this continuum of species commonness to rarity based on environmental and geographical breadth or narrowness. In general, these studies have found that, all other things being equal, SDMs for rarer species, those that have smaller ranges, narrower ecological tolerances, or both, tend to have comparable or higher prediction accuracy than common, generalist species. Where the opposite pattern, less accurate models for rarer species, has been predicted or observed, a confounding factor may have been small sample size and low species frequency. By definition, SDMs for rare species are based on small numbers of species occurrence

Table 10.2. *Recent studies (in chronological order) that have examined the effect of species range size and habitat specificity on SDM performance*

Study No. species Taxa Source data	Environmental Space (Habitat Specificity)		Geographical Space (Range)			Findings
	Niche differentiation	Niche width	Range extent	Range size	Range overlap	
(Pearce & Ferrier, 2000), 25 species plants and animals, survey		Sample prevalence		(Sample prevalence)		More narrowly distributed species, more accurate models for plants only, opposite for animals.
(Karl et al., 2002), 7 bird species, survey		Sample prevalence		(Sample prevalence)		Controlling for sample size, models for rare species may be as accurate as those for common species.
(Hepinstall et al., 2002), 28 bird species, survey		Number of land cover types used				Narrower niche width, more accurate models.
(Stockwell & Peterson, 2002), 103 bird species, collections		Ecological breadth		Range size		Smaller range higher accuracy; Slight negative effect of niche width on models. (Species traits assigned by expert.)

Reference					
(Brotons et al., 2004), 30 bird species, atlas	Weighted ENFA Marginality (+) and Tolerance (−)	Weighted sample prevalence (+) and ENFA tolerance (+)			More marginal species higher accuracy; no significant effect of niche width.
(Segurado & Araújo, 2004), 44 species herpetofauna, atlas	ENFA marginality	ENFA tolerance	Extent of occurrence: distance between most distant cells	Area of occupancy = Prevalence (# grid cells)	Niche: specialists (high marginality, low tolerance) best models, generalists worst. Range: narrowly distributed (low occupancy, low extent) best models, widespread worst.
(Luoto et al., 2005), 98 butterfly species, atlas	Spatial autocovariate	Latitudinal range from N- to S-most record	Prevalence	(Latitudinal range)	Good predictions for species at margins of range where restricted to certain habitats, and specialists (clumped); worse predictions for widespread species.
(Elith et al., 2006), 6 major data sets, 226 species, survey and collections	Marginality: avg. distance presences to random points	"Range" overlap: mean overlap range of environmental variables	Maximum distance among pairs presence points	Area of convex polygon encompassing presences	Species that are rarer because they have environmentally or geographically restricted distributions are modeled with greater accuracy than common, generalist species.

(cont.)

Table 10.2. (cont.)

Study No. species Taxa Source data	Environmental Space (Habitat Specificity)		Geographical Space (Range)			Findings
	Niche differentiation	Niche width	Range extent	Range size	Range overlap	
(Zimmermann et al., 2007), 19 tree species, survey	Prevalence, abundance (core vs. satellite)	Prevalence, abundance (rural vs. urban)		Prevalence (urban, satellite vs. core, rural)		Core (widespread, common) highest accuracy; no difference in AUC among groups.
(McPherson & Jetz, 2007), 1300 bird species, atlas		Habitat tolerance: # habitats used		Extent (polygon area) in atlas data	Endemism: range within study area	Larger range size lower accuracy, perhaps due to local adaptation; higher Endemism higher accuracy because species ranges entirely within study area (covaries inversely with range size); broader habitat tolerance, slightly lower accuracy (covaries with range size).
(Tsoar et al., 2007), 42 species: land snails, birds and bats, collections	ENFA tolerance; sample prevalence			(Sample prevalence)		Tolerance strongly negatively affected accuracy (no effect of prevalence).

(Evangelista et al., 2008), 2 invasive plant species	A known specialist vs. generalist		Greater degree of agreement between specialist and modeled distribution than generalist.
(Loiselle et al., 2008), 76 plant species, collections		Two levels: Wide, Narrow	Wide-ranging species lower accuracy and greater commission error, especially with small samples.
(Franklin et al., 2009), 30 species herpetofauna, survey	Low (common, sparse) or high (habitat specialist, endemic)	Large (common, widespread) or small (endemic)	More accurate models for species with narrow habitat specificity and/or small range. (Species traits assigned by expert.)

For each study, number of species, taxonomic group, and source of species data are given. Range size refers to area while range extent is usually measured as distance. Range overlap is the degree of overlap between the entire species range and the study area. ENFA = Ecological niche factor analysis (Hirzel et al., 2002); collections = natural history collections; survey = biological field survey; atlas = species atlas data (Chapter 2).

records. Overall, it appears that the negative effect of small sample size on SDM performance for rare species may be offset, in many cases, by the well-defined range of environmental tolerances (narrow niche width, high niche differentiation) found within any given study area and corresponding empirical dataset (species occurrence data and environmental predictors). This often makes it possible to distinguish suitable from unsuitable habitat even with a small sample. Further, widespread species may encompass populations showing local adaptations.

These conclusions have usually been drawn based on the AUC which is a frequency-independent measure of SDM prediction accuracy (Chapter 9). Among these studies (Table 10.2), those that explicitly examined tradeoffs between omission (false negative predictions) and commission error (false positives) all observed an effect of sample size on the results. In particular, common species have higher prevalence in the sample, and therefore SDMs for these species have higher commission error (or lower specificity; Chapter 9), while models for rare species with low sample frequency have higher omission error or lower sensitivity (Manel *et al.*, 2001).

Both sampling and biological issues can make interpretation of some comparison studies challenging, for example, if the study has used sample frequency from the species survey to define range size or species commonness (confounding niche breadth and range size). Further, environmental tolerance tends to covary with range size – common species tend to have large ranges (as in McPherson & Jetz, 2007). Therefore, not all the boxes in Rabinowitz' classification scheme (Table 10.1) would be expected to contain equal numbers of species for any given community. In fact, ecological communities are often composed of a few common species and many rare ones, and the relationship between species dominance and community diversity has long been the subject of theoretical consideration (Preston, 1948; Whittaker, 1965; Hubbell, 2001). These sampling and biological issues have been addressed in creative and rigorous ways by developing independent measures of environmental tolerance, such as niche differentiation and niche width, and of biogeographical range, such as extent (largest distance between records) versus area (polygon encompassing records) (Table 10.2). The conclusions seem robust – given an adequate minimum number of observations for modeling, it is often possible to develop useful SDMs for rare species because they tend to have narrow and distinct environmental tolerances. SDMs for generalist species tend to have low specificity, presumably a true reflection of wide ecological tolerances and local adaptation. Generalist species may not occupy all potentially suitable habitat at any given time. Further, species tend to be restricted to particular habitats at the limits of their range and have broader habitat tolerances near their "optimal" niche conditions.

Other species attributes and aspects of species life history that have been considered include: taxonomic group, trophic level (Pearce & Ferrier, 2000; Huntley *et al.*, 2004; Elith *et al.*, 2006) migratory status, body size, affinity with fine scale habitat features (Stockwell & Peterson, 2002; McPherson & Jetz, 2007), and disturbance response (Syphard & Franklin, in review). Not surprisingly, it is more difficult to develop SDMs for species that are migratory or that are associated with fine-scale habitat features (those features that are not reflected in the set of environmental predictors used for modeling because of lack of suitable maps). Encouragingly, there does not appear to be a strong systematic difference in the performance of SDMs among different taxonomic groups of organisms or trophic levels – equally good and bad SDMs have been developed for plants and animals, invertebrates and microbial pathogens, birds and lizards, herbivores and top carnivores, and so forth, in a multitude of SDM studies.

The correlation of species traits with the performance of SDMs is an important area of continuing research both because it provides insights into basic questions about the factors determining species distribution and range, and because it helps guide the judicious use of SDMs in applied research. In general, comparative studies have shown that the performance of SDMs varies much more among species than along any other dimension of the SDM framework: model type, scale of investigation, or species or environmental data characteristics. Therefore, if general patterns can be found, they can guide the use of SDMs and the resulting spatial predictions in subsequent applications by suggesting which SDMs are likely to perform the best based on species traits. So, my first prediction was supported – species commonness or rarity is related to SDM performance. In fact, this view of species traits in relation to SDMs was probably too narrow and should be broadened to include other fundamental factors such as mobility or dispersal distance, longevity, size and so forth.

Species range traits and ecological traits are not only related to SDM performance (our ability to model their distributions), but also to their vulnerability to global environmental changes such as anthropogenic global warming. For example, species distribution on environmental gradients, niche breadth, and proximity of their distributions to migration barriers (such as oceans for terrestrial species) have been related to their sensitivity to climate change impacts (Broennimann *et al.*, 2006).

When using SDMs for extrapolation in time or space, e.g., to project the impacts of climate change on species distributions or predict the impact of an introduced species, probably the most important species trait predicting SDM performance is niche conservatism (Pearman *et al.*, 2008). Niche conservatism can be defined as the tendency of a species niche to remain unchanged over space or time (Chapter 3), and can be viewed as an untested assumption of SDM. Unfortunately, niche conservatism, as an attribute, is unknown for most species. In fact SDM is frequently used as one tool to examine the niche stability of species over short (species invasions, e.g., Broennimann *et al.*, 2007) or long (evolutionary) time periods (see Chapter 3).

10.3 Species data

Sample size, sample design, data resolution, and sample extent are important species data issues affecting the performance of SDMs (Chapter 4). Sample size encompasses both the total number of observations, and how many observations of species presence (prevalence) are available, in the case of presence/absence data. Sample design addresses the issue of whether the species data are an unbiased and representative set of observations, well distributed throughout the environmental gradients found within the species range or the study area. So many published SDM studies are conducted using presence/absence data, or presence-only observations, that relatively less attention has been paid to modeling other measures of species occurrence such as abundance or fitness.

These species data issues, sample size, sample design, resolution and sample extent, are interrelated. The geographical extent of SDMs can vary from local to global, but we can divide those large-scale studies that encompass entire species ranges from ecological- or landscape-scale studies whose extent may be smaller than the biogeographic range of the species being modeled. The first group includes studies based on large-extent species data often from natural history collections or atlas data (which themselves, in turn, are often based on collections records). The total number of observations tends to be large and the extent typically encompasses entire species ranges. However, often only species presence records are available for modeling, and even for very large datasets, the small number of observations for rare species can be a problem. Therefore, efforts have been directed at developing modeling methods for presence-only data (Chapter 8) that can effectively overcome the curse of dimensionality (Chapter 6), be used to mine large datasets, and describe non-linear patterns in multivariate data space. Additionally, point records of species occurrence have sometimes been aggregated to coarse-scale grid cells suitable for large-area investigations and allowing "absence" data to be generated via this aggregation (Section 4.6).

The second group of studies uses species location data from typical field-based survey methods including plant community sampling, forest inventory, breeding bird surveys, point counts, mist netting, trapping, trip cameras, radiotelemetry, and so forth (Chapter 4). Some of these are community-based (rather than single species) survey methods, yielding information on multiple species, and therefore also recording both presence and absence. While some surveys may cover only a limited area or number of species, some of these datasets are impressive in their size and extent. For example, they can include tens or even hundreds of thousands of forest inventory plots over hundreds of thousands of square kilometers (Leathwick, 1995; Iverson & Prasad, 1998), thousands of continent-wide multi-year species monitoring transects, e.g., the North American Breeding Bird Survey (Peterson *et al.*, 2007), or thousands of large-area survey locations systematically collected to monitor species and develop atlases within regions spanning tens of thousands of square kilometers (Segurado & Araújo, 2004; Brotons *et al.*, 2007).

The resolution of species data from ecological surveys is typically, but not always, fine-scaled relative to most mapped environmental predictors used in SDM. In other words, if species data are collected over 1-ha areas or smaller, interactions between broad-scale climate predictors and their local terrain modifiers are taken into account, and biotic interactions (e.g., competition between plants) can actually occur. On the other hand, if the resolution of the species data is coarse, because observations are made over a large area, or locational uncertainty is high, or aggregation to coarse grid cells has taken place, then the ability to model species–environment relations that occur at a finer spatial scale is lost. SDMs may then only be able to delimit general climatic range limits of species.

With regard to extent, an important issue to consider is how the species response functions might be expected to differ for those studies that encompass entire species ranges, versus smaller-extent studies that do not. It is possible and often useful to use SDM to interpolate species distribution information for a region within a portion of the species range if it is acknowledged that response functions may be linear instead of unimodal (the limits of the species range are not included) or otherwise non-linear, and that environmental correlates may be insignificant if the species distribution is not limited by any of the measured environmental predictors within the study region (Brotons et al., 2007). However, because predicting the impacts of climate change has become a prominent application of SDM, it is important to note that some authors have strongly recommended that SDMs used to project climate change impacts must be developed using species distribution data from throughout the entire species range (Thuiller et al., 2004b).

When species data were not all collected at the same time, for example, in the case of natural history collections, older data may have to be screened if they do not represent contemporary habitat conditions. Excluding historical collection locations from potential pseudo-absence data can also improve SDMs (Lutolf et al., 2006). Alternatively, historical and paleoecological species data can be matched with concurrent representations of environmental conditions (Stigall & Lieberman, 2006; Kremen et al., 2008; Peterson & Nyari, 2008). The dynamics of suitable habitat can also be modeled in cases where species monitoring has been conducted over time and predictor variables related to habitat dynamics are available (Alexander et al., 2006; Osborne & Suarez-Seoane, 2007).

In summary, with regard to the species data used in SDM, a number of features of the data need to be considered in order to: (a) evaluate and screen the species data for detectability, and taxonomic, locational and other errors; (b) select an appropriate study area extent; (c) potentially aggregate the data to a particular resolution; (d) select suitable predictors and spatially and temporally match them to the species data; and, (e) choose a modeling method that is robust to the biases of the data (Fig. 10.1). These decisions should be made in light of the objectives of the study, most importantly, and also with consideration of species traits and how they might affect the reliability of resulting models.

10.4 Environmental data and scale

While it is true that the choice of environmental predictors used in SDMs is often based on data availability (Dormann, 2007c), I argue that this is not because we lack an understanding of the proximal factors affecting species distributions, but because we often lack the data to describe them and need to rely on surrogates (Chapter 5).

As noted, when focusing on the use of SDMs to predict the impacts of climate change on species at broad scales (Pearson & Dawson, 2003), some insist that such projections should only be made using models based on climate variables as predictors. However, others maintain that additional factors that limit species distributions, and that may be relatively static at the time scale of climate change predictions, e.g., substrate, topography, should also be incorporated (Dormann, 2007c), as they have been in a number of climate change studies (e.g., Midgley *et al.*, 2006; Keith *et al.*, 2008).

Chapter 3 discussed a conceptual model and hierarchical framework of environmental factors affecting species distributions, and many SDM studies have included environmental predictors that, implicitly or explicitly, address this hierarchy of factors: climate, substrate, topography, gaps, and patches. Table 10.3 shows an example of a number of recent SDM studies that incorporate climate, substrate, topographical, patch (represented by vegetation or land cover data and landscape metrics) and locational (geographical coordinates) data. This is obviously an ad hoc list; for example, none of the studies listed used the distribution of other species as predictors, and yet Chapter 5 discussed several studies that have done so. With this limitation mind, Table 10.3, in combination with the results of the meta-analysis of animal SDMs by Meyer and Thuiller (2006), suggests that many SDM studies include predictors representing several levels in the spatial hierarchy of factors controlling the primary environmental regimes of light, heat, moisture, and mineral nutrients. These include climate, substrate, terrain, and land cover. In other words, even when the data model is not stated explicitly, SDM studies choose predictors based on an underlying general model of the factors affecting species distributions. Almost all of the examples shown in Table 10.3 used climate and terrain variables as predictors. Another generalization that is evident is that patch characteristics, described by vegetation or land cover categories and landscape metrics, are used much more frequently in SDM for animals than for plants. In contrast, substrate characteristics are used more frequently to model plants versus animals.

However, another pattern can be seen in Table 10.3. GIS-based SDM and remote-sensing-based SDM have developed along somewhat separate lines, in different literatures and different journals, rather than being tightly integrated, as it had been recognized that they should be since the early 1980s (Strahler, 1981; Hutchinson, 1982; Lees & Ritman, 1991; Stoms & Estes, 1993). Only two of the studies included in the table emphasized the remotely sensed predictors while the other relied primarily on digital maps of climate, geology, soils, vegetation and terrain. Again, this list is meant to illustrate a point, and Chapter 5 described many other studies that did use remotely sensed predictors in SDM. However,

Table 10.3. *A summary of the number and types of environmental predictors used in SDM shown for a range of studies in chronological order*

Citation	Location	Resolution (km²)	Sample size	Number species	Taxonomic group	Total number predictors	Climate	Elevation, terrain	Substrate, geology, soil	Location cover	Land cover	Remotely sensed
(Franklin, 1998)	Southern California, USA	0.0009	906	20	Shrubs	8	3	3		2		
(Iverson & Prasad, 1998)	Eastern USA	County	2500	80	Trees	33	7	3	18			
(Vayssiéres et al., 2000)	California, USA	NA	4010	3	Trees: oaks	25	16	1	8			
(Cumming, 2000b)	Southern Africa	648	4020	74	Ticks	49	36	1				12
(Araújo & Williams, 2000)	Europe	2500	4419	174	Trees	6	5	1				
(Leathwick & Austin, 2001)	New Zealand		14540	4	*Nothofagus* tree species	8	6	1	1			
(Pearce et al., 2001)	New South Wales, Australia	0.04	4572	5, 9, 64	Arboreal marsupials, reptiles, birds	24	3	3	3	1		14

Reference	Location				Taxon						
(Pearce et al., 2001)	New South Wales, Australia	0.04	2223	75	Overstory plants	10	3	3	3	1	1
(Elith & Burgman, 2002)	Victoria, Australia	0.0625	3522	8	Plants	35	23	7	2	2	1
(Detmers et al., 2002)	Southern Appalachia, USA	not given	215	6	Birds	26	13	13			13
(Stockwell & Peterson, 2002)	Mexico	25	1–100	103	Birds	8	2	1		3	2
(Miller & Franklin, 2002)	Mojave desert, California, USA	0.0009	3819	11	Plant assemblages	12	4	6	2		2
(Peterson et al., 2003)	Maine, USA	0.0081 (from 3.6)	8–541	30	Birds	24	21	3	2		
(Thuiller et al., 2003)	Catelonia, Spain	0.0324		4	Trees	19	16	2	1		
(Thuiller et al., 2003)	Portugal	100	933	4	Trees	9	2	7			
(Thuiller et al., 2003)	Europe	2500	4419	4	Trees	7	6	1			
(Brotons et al., 2004)	Catalan, Spain	1	1500	30	Birds	34	3	2	2	2	27
(Segurado & Araújo, 2004)	Portugal	100	993	44	Herptiles	16	5		2		9
(McPherson et al., 2004)	Southern Africa	648	4275	32	Birds	61	1	1			60

(cont.)

Table 10.3. (cont.)

Citation	Location	Resolution (km²)	Sample size	Number species	Taxonomic group	Total number predictors	Climate	Elevation, terrain	Substrate, geology, soil	Location	Land cover	Remotely sensed
(Huntley et al., 2004)	Europe	2500		306	Plants, insects, birds	3	3					
(Edwards et al., 2005)	Oregon, Washington, USA	0.0081	299	4	Lichens	9	2	3			4	
(Drake et al., 2006)	Switzerland	0.000625	550	106	Trees and shrubs	5	3	2				
(Elith et al., 2006)	Wet tropics, Australia	0.0064	9–74, 32–265	20, 20	Plants, birds	13	10	3				
(Elith et al., 2006)	Switzerland	0.01	36–5822	30	Trees	13	7	2	2		2	
(Elith et al., 2006)	New Zealand	0.01	18–211	52	Plants	13	8	3	2			
(Elith et al., 2006)	New South Wales, Australia	0.01	2–426	54	Plants, birds, reptiles, mammals	13	5	4	2		2	
(Elith et al., 2006)	Ontario, Canada	1	16–749	20	Birds	11	6	3		1	1	
(Prasad et al., 2006)	Eastern USA	400	9782	4	Trees	36	5	6	22		3	

The studies were selected to be illustrative, and it is not implied that they are a representative sample. Land cover refers to vegetation or land cover data.

better integration of novel remote sensing data into SDM remains an important research challenge, and a number of recent studies address this topic and suggest a way forward (see Chapter 5).

Table 10.3 also illustrates that the total number of predictors ranges from three or four to several dozen, and is typically around 8–40. This table only indicates the number of candidate predictors used in SDMs in these studies, and does not show which and how many predictors were included in the final models (which would vary among species). Clearly, those studies using many candidate predictors, especially if they are in the same category (e.g., climate), would have to be especially concerned about screening predictors for multicollinearity. A more thorough and systematic examination of the selection of environmental predictors for SDM would be very useful. Such an examination could more directly address questions such as: what environmental process models were used to derive the predictors? Which predictors represented direct, indirect and resource gradients? Were hierarchical interactions among predictors considered explicitly? Future research in this area could help guide the effective choice of predictors for SDM.

10.5 Modeling methods

Modeling methods commonly used in SDM (Chapters 6–8) differ in the types of data required, complexity and nature of the estimated functions, and predictive performance (Elith & Leathwick, 2009). However, among those methods that have been extensively tested and compared using presence–absence data, differences in performance among different types of models tend to be smaller than differences among species. Further, inter-model comparison studies have sometimes yielded conflicting results about relative performance of different methods, which is unsurprising as each study uses different data and often compares different groups of modeling methods. However, there are some generalizations emerging (Table 10.4), and it appears that learning to use innovative but complex methods is worth the investment of time.

Machine learning methods, especially those that incorporate model averaging such as bagged or boosted classification trees or random forests (Chapter 7), as well as non-parametric curve-fitting regression methods such as GAMs, tend to have better performance than single classification trees or parametric statistical methods such as GLMs (Segurado & Araújo, 2004; Lawler et al., 2006; Prasad et al., 2006; Elith & Graham, 2009). However, sometimes machine learning methods yielded lower

Table 10.4. *Summary of modeling methods used in SDM, and comments on their performance*

Method	Description	Response	Comments on SDM performance
Statistical			
Generalized linear models (GLM)	Flexible, modern multiple regression, copes with non-normal distributions of the response variable using link and variance functions	Continuous, ordinal, binary (P/A)	Effective global modeling method; performs well with adequate data, even when pseudo-absences are used. Global model may be prudent when projecting in time and space.
Generalized additive models (GAM)	Multiple regression but with curve fitting using splines or other methods	Continuous, ordinal, binary (P/A)	Performs well when not over-fit to data (controlled through d.f.); similar to and slightly better performance than GLM. Useful visualization of response curves.
Multivariate adaptive regression splines (MARS)	Generalization of stepwise linear regression, using piecewise linear splines	Continuous, ordinal, binary (P/A)	Performs similarly to and slightly better than GLM; good performance given adequate sample size. Computationally faster than GAMs for complex data. Fitting to categorical response requires special software.
Bayesian modeling	Estimates probability degree of belief in the likelihood of an event, incorporating prior and conditional probabilities	Continuous, ordinal, binary (P/A)	A number of SDM applications but not widely compared to other methods.
Spatial autoregressive models (SAR)	Autoregressive lag (response regressed on a spatial lag of itself) and error (spatially structured error) models	Continuous, ordinal, binary (P/A)	Where used, tends to perform better than non-spatial models, but limited by spatial density of data available for calibration. Autocovariate may mask importance of environmental predictors.

Decision trees (DT); Classification and regression trees (CRT)	Divisive, monothetic, supervised classifiers	Continuous, ordinal, categorical (P/A)	Single DTs perform poorly compared to other methods. Useful graphical visualization of classification rules.
Ensemble DTs: bagging, boosted regression trees (BRT) (stochastic gradient boosting, SGB), random forests	Estimate a large number of tree models based on subsets of the data and averaging the results	Continuous, ordinal, categorical (P/A)	Ensemble DTs tend to have good predictive performance given adequate sample size. Tools available to describe predictor importance and response functions.
Artificial neural networks (ANN)	Derive linear combinations of the predictors and predict using a non-linear function of features	Categorical (P/A)	Good performance when used by skilled practitioners. Performance sometimes worse than for statistical methods.
Genetic algorithms (GA), GARP (Genetic algorithm for rule-set production)	Population of rules generated using four methods; rules selected using genetic algorithm	GARP developed for P-only; generates pseudo-absence	Poor performance in comparison with other methods. Tends to overestimate current distribution (interpolation), underestimate distributions in novel conditions (extrapolation).
Maxent	Maximum entropy	Maxent software can generate pseudo-absence	Performs well in data-poor situations and when projecting in time and space.
Support vector machines (SVM)	Estimates the "support" of the statistical distribution of predictor variables where the species occurs	One- (P) or multiple (P/A) class	Only a few SDM applications to data (no extensive comparisons).

(cont)

Table 10.4. (cont.)

Method	Description	Response	Comments on SDM performance
Similarity and expert rules			
Envelope: BIOCLIM	Box (parallelepiped) classifier	P	P-only methods generally perform more poorly than statistical models. BIOCLIM underestimates distributions in novel conditions.
Similarity: Mahalonobis distance	Multivariate dissimilarity measure scaled by covariances of predictors	P	Performance sometimes comparable to statistical methods, better than BIOCLIM
Similarity: DOMAIN	Based on Gower multivariate distance	P	P-only methods underestimate distributions in novel conditions.
Environmental niche factor analysis (ENFA)	Multivariate difference between species mean and range of environmental conditions	P	Tends to overestimate current distributions compared to statistical methods.
Habitat suitability index (HSI)	Sum of mapping functions of predictors to habitat suitability, based on expert knowledge or data	None required	Tends to perform more poorly than statistical models, but sometimes only marginally so.

P = presence; A = absence; d.f. = degrees of freedom.

performance (Dormann *et al.*, 2008). Further, the incremental performance improvement of complex methods can be small, especially for those species for which reasonably high-performing models can be fit by most techniques. In particular, GLMs, GAMs and MARS often yield fairly similar predictions and have roughly similar accuracy (Leathwick *et al.*, 2006b; Elith & Leathwick, 2009). In those cases, this regression family of methods may be quite useful because they allow the modeler to examine and interpret the response curves estimated by the model. The gains in accuracy from using "black box" machine learning models may be cancelled by the loss of ability to interpret to model parameters for ecological meaning. Single classification trees have lower accuracy than other methods (but see Muñoz & Felicisimo, 2004), and in particular tend to underestimate species distributions – they are prone to omission error (Miller & Franklin, 2002; Lawler *et al.*, 2006). GARP tends to overestimate distributions, that is, it is prone to commission errors (Lawler *et al.*, 2006), especially when using small samples (Wisz *et al.*, 2008b).

Many model comparisons have been based almost exclusively on measures of predictive performance such as AUC or Kappa. A number of studies have shown that models with very similar performance metrics can yield very different spatial predictions, especially when used to extrapolate into new areas of environmental "space," e.g., novel combinations of environmental variables that did not occur in the training data. For example, one study found that classification trees, GLMs and GAMs gave similar spatial predictions, while environmental envelopes, distance-based methods and GARP (all developed for presence-only data) yielded distinct spatial predictions (Pearson *et al.*, 2006). Further, the presence-only methods that were tested tended to predict habitat losses when confronted with novel environmental conditions (e.g., climate change). Another recent study found that GAMs and GLMs gave similar spatial predictions, while random forests and single classification trees made distinct spatial predictions, and predicted distribution maps differed most when prediction accuracy was low for a given species (Syphard & Franklin, submitted). I recommend that greater emphasis should be given to examining the spatial pattern of predictions.

One of the greatest challenges in SDM is the need to develop models in data-poor situations, typically using biased, small samples of presence-only data. Methods relying on presence-only data generally perform more poorly than those using presence and (pseudo-) absence data (Elith *et al.*, 2006). Further, as noted above, presence–absence versus presence-only methods differ in the kinds of projections they make when used to extrapolate to future climates or novel environments, with a tendency for presence-only methods to predict greater losses of suitable habitat (Pearson *et al.*, 2006). However, among those SDM modeling methods applied to presence-only data, there are some clear trends. Multivariate distance and envelop-based methods (DOMAIN, BIOCLIM) tend to perform more poorly than other methods (Segurado & Araújo, 2004; Elith *et al.*, 2006). GARP and DOMAIN appear to underestimate distributions for novel environmental "spaces," e.g. non-analog climates (Hijmans & Graham, 2006; Pearson *et al.*, 2006). Maxent performs quite well (Elith *et al.*, 2006; Phillips *et al.*, 2006; Elith & Graham, 2009), especially in data-poor situations (e.g., Hernandez *et al.*, 2006; Pearson *et al.*, 2007; Wisz *et al.*, 2008b). It has been recommended that Maxent or one of the regression methods (Chapter 6) using pseudo-absence (background) data be used in those situations (Elith & Leathwick, 2009). However, it was noted in one study that, when Maxent was used to predict distributions for future and past climates, its predicted distributions were more variable than those from GAMs (Hijmans & Graham, 2006).

Several decisions must be made when choosing a modeling method and fitting the model. These include (a) weighing the trade-offs between bias (and precision) versus variance (and generality) based on whether the primary aim of modeling is ecological interpretation of factors controlling species distributions, interpolation to unsurveyed locations, or extrapolation into novel environments; (b) estimating the response functions; and, (c) addressing multicollinearity in the predictors (Fig. 10.1). Model-driven, parametric (low variance, high bias) approaches are more suitable when the aim is to estimate species response functions, interpret which factors control species distributions, or extrapolate into novel environments. Data-driven, non-parametric or machine learning approaches (high variance, low bias) may be more useful in situations where accurate interpolation to unsurveyed locations is the primary goal.

It is very useful to use more than one modeling method. This is likely to yield insights into the underlying species patterns as well as the model structure and assumptions. Different methods with varying data inputs yield predictions that can vary either subtly or wildly. Should results of difference models be averaged? Several ensemble methods described in Chapter 7 estimate many models as an automated step in the modeling process, and then their predictions are averaged in some way. An ensemble of predictions can also result from different realizations of the same type of model (varying inputs or parameters), or from different types of models (Section 7.8). A number of consensus methods exist for estimating the agreement among different model outputs, including simple averaging. Consensus predictions are not always an improvement over single models, but are useful in cases where there are not adequate data to evaluate competing models, as with species invading new areas, projecting future distributions under climate change, and in very data-poor settings (Section 7.8).

10.6 Model evaluation

Evaluation of any kind of ecological model takes many forms, from verifying the logic of the formulations, to assessing its fit to data or ability to forecast. As discussed in Chapter 9, there is a well-developed literature and widely implemented set of performance evaluation procedures applied to SDMs (Fielding & Bell, 1997). The AUC of the ROC plot has become a common currency for characterizing threshold-independent prediction accuracy applied to models of species presence/absence. The Kappa coefficient is another commonly used and important measure of the degree of agreement (above chance agreement) between predicted and observed values of a categorical variable. It is frequently used in map accuracy assessment, and also SDM evaluation. There are also standard and well-established measures of categorical omission and commission error, such as sensitivity and specificity, that have been widely applied in SDM. These are very useful when there are practical trade-offs between the costs of the two types of errors. All of these measures, and many variations on them, are easy to calculate given a set of reference data and predictions, and it is advisable to use multiple measures in order to characterize different aspects of prediction error or uncertainty. In particular, both a threshold-independent measure of model discrimination, such as AUC, and threshold-dependent measures of omission, commission and overall error should be examined. Given that many applications of SDMs require binary or ordinal classifications of habitat suitability, a suitable threshold to apply to probabilistic SDM predictions should be selected using the criteria and methods described in Chapter 9. Further, if only presence data are available, one of the presence-only evaluation methods described in Chapter 9 may be more appropriate. For example, the rank correlation coefficient of the frequency of presences versus ranked bins of predicted suitability, sometimes called the "Boyce index," can be used.

A current challenge in SDM is to address fully the spatial nature of the data, the models and their predictions. This is addressed to some degree in the SDM literature, but not universally. Mapping the residuals of model predictions, and calculating their spatial autocorrelation (Chapter 9), is an important step in model evaluation (Fig. 10.1), and can lead to additional model development, e.g. using methods that address spatial autocorrelation (Chapter 6). Developing different models (by varying the input data, model type, and so forth) is an important way to characterize scientific uncertainty in general, including in SDM. Some SDM studies do compare the spatial patterns of alternative predictions (habitat suitability, likelihood of species occurrence) using measures such as Kappa for categorical map agreement, or a correlation coefficient to compare continuous predictions. However, spatial agreement is emphasized less than the non-spatial performance measures in the literature. It is the spatial prediction maps of potential distribution or habitat suitability that will be used in risk assessment, planning and management, and so the spatial characteristics of the predictions deserve more attention.

In summary, when evaluating an SDM, a number of decisions must be made (Fig. 10.1). The data used for evaluation must be identified. Model evaluation measures should be chosen based on the objective of SDM and the nature of the evaluation data. Environmental response functions and the relative importance of different predictors should be evaluated for ecological realism. Further, spatial patterns of model residuals should be examined for spatial autocorrelation. Finally, following spatial prediction, the spatial configuration of the predicted distributions should be evaluated and compared among model realizations.

10.7 Summary – beyond species distribution modeling

Species distribution modeling, especially when used for spatial prediction, is rarely an end in itself, and almost always a means to an end (Chapter 1). SDM has been used to test ecological hypotheses about factors limiting species distributions and the nature of species responses to environmental gradients (Austin *et al.*, 1990), and to test phylogenetic hypotheses about niche stability versus divergence (e.g., Ackerly, 2004; Kozak & Wiens, 2006; Peterson & Nyari, 2008). It has been widely used to interpolate sparse biological survey in space in order to use predicted distributions for conservation planning, reserve design and impact and risk analysis. It is being used increasingly to extrapolate species distributions in time and space in order to predict the risk of species invasions, the spread of disease organisms, and the impacts of climate change on biodiversity. As such, the limitations of this static, empirical modeling approach are well documented and extensively discussed. Nonetheless, SDM has proven broadly useful for estimating first-order relationships between species distributions and environmental factors. The methods themselves are quite flexible and can accommodate biotic factors, spatial effects, and so forth. Of course, SDM is not process-based and cannot simulate the transient dynamics of environmental processes.

Further, to paraphrase Austin (2002), mechanistic models of ecological processes should be consistent with rigorous, quantitative empirical descriptions of the same system. A way forward is to use SDMs to interpolate plausible spatial realizations of species distributions to develop baseline species distributions for more mechanistic simulations (e.g., Mladenoff & He, 1999; Keith et al., 2008). A recent review paper by Thuiller et al. (2008), focusing on climate change impacts, reiterated a point often made with regard to models; there will always be trade-offs between using complex, mechanistic versus simple, empirical models to forecast environmental change. They suggested that adding mechanistic realism about species dispersal or migration to the empirical SDM might be an effective first step to improving forecasts about climate impacts on species distributions (Thuiller et al., 2008). A number of recent studies are doing just that – combining SDMs with dispersal kernels (Meentemeyer et al., 2008b; Williams et al., 2008), spatially explicit simulations (Iverson et al., 1999; Iverson et al., 2004b), and other approaches, to study species invasions and predict species range shifts in response to global environmental change. This type of specific recommendation for combining SDM with other data and approaches (e.g., Botkin et al., 2007) will help guide the judicious use of SDMs as well as future research to improve methods available to evaluate environmental change.

References

Abdulla, F. A. & Lettenmaier, D. P. (1997) Development of regional parameter estimation equations for a macroscale hydrologic model. *Journal of Hydrology*, **197**, 230–257.

Accad, A. & Neil, D. T. (2006) Modelling pre-clearing vegetation distribution using GIS-integrated statistical, ecological and data models: a case study from the wet tropics of Northeastern Australia. *Ecological Modelling*, **198**, 85–100.

Ackerly, D. D. (2003) Community assembly, niche conservatism, and adaptive evolution in changing environments. *International Journal of Plant Sciences*, **164**, S165–S184.

Ackerly, D. D. (2004) Adaptation, niche conservatism, and convergence: comparative studies of leaf evolution in the California chaparral. *American Naturalist*, **163**, 654–671.

Adjemian, J. C. Z., Girvetz, E. H., Beckett, L., & Foley, J. E. (2006) Analysis of Genetic Algorithm for Rule-Set Production (GARP) modeling approach for predicting distributions of fleas implicated as vectors of plague, *Yersinia pestis*, in California. *Journal of Medical Entomology*, **43**, 93–103.

Agresti, A. (1996) *An Introduction to Categorical Data Analysis*. New York: John Wiley and Sons.

Ahlqvist, O. (2005) Using uncertain conceptual spaces to translate between land cover categories. *International Journal of Geographical Information Science*, **19**, 831–857.

Akçakaya, H. R. (2001) Linking population-level risk assessment with landscape and habitat models. *Science of the Total Environment*, **274**, 283–291.

Akçakaya, H. R. (2000) Viability analyses with habitat-based metapopulation models. *Population Ecology*, **42**, 45–53.

Akçakaya, H. R. & Atwood, J. L. (1997) A habitat based metapopulation model of the California gnatcatcher. *Conservation Biology*, **11**, 422–434.

Akçakaya, H. R., Butchart, S. H. M., Mace, G. M., Stuart, S. N., & Hilton-Taylor, C. (2006) Use and misuse of the IUCN Red List Criteria in projecting climate change impacts on biodiversity. *Global Change Biology*, **12**, 2037–2043.

Akçakaya, H. R., Franklin, J., Syphard, A. D., & Stephenson, J. R. (2005) Viability of Bell's Sage Sparrow (*Amphispiza belli* ssp. *belli*) under altered fire regimes. *Ecological Applications*, **15**, 521–531.

Akçakaya, H. R., Radeloff, V. C., Mladenoff, D. J., & He, H. S. (2004) Integrating landscape and metapopulation modeling approaches: viability of the sharp-tailed grouse in a dynamic landscape. *Conservation Biology*, **18**, 526–537.

Albert, P. S. & McShane, L. M. (1995) A generalized estimating equations approach for spatially correlated binary data: applications to the analysis of neuroimaging data. *Biometrics*, **51**, 627–638.

Aldridge, C. L. & Boyce, M. S. (2007) Linking occurrence and fitness to persistence: habitat-based approach for endangered Greater Sage-Grouse. *Ecological Applications*, **17**, 508–526.

Alexander, S. M., Logan, T. B., & Paquet, P. C. (2006) Spatio-temporal co-occurrence of cougars (*Felis concolor*) and wolves (*Canis lupis*) and their prey during winter: a comparison of two analytical methods. *Journal of Biogeography*, **33**, 2001–2012.

Allouche, O., Tsoar, A., & Kadmon, R. (2006) Assessing the accuracy of species distribution models: prevalence, kappa and the true skill statistic (TSS). *Journal of Applied Ecology*, **43**, 1223–1232.

Andersen, M. C., Adams, H., Hope, B., & Powell, M. (2004) Risk analysis for invasive species: general framework and research needs. *Risk Analysis*, **24**, 893–900.

Anderson, D. R. & Burnham, K. P. (2002) Avoiding pitfalls when using information-theoretic methods. *Journal of Wildlife Management*, **66**, 912–918.

Anderson, R. P., Lew, D., & Peterson, A. T. (2003) Evaluating predictive models of species distributions: criteria for selecting optimal models. *Ecological Modelling*, **162**, 211–232.

Anselin, A., Meire, P. M., & Anselin, L. (1989) Multicriteria techniques in ecological evaluation: an example using the analytical hierarchy process. *Biological Conservation*, **49**, 215–229.

Anselin, L. (2002) Under the hood – Issues in the specification and interpretation of spatial regression models. *Agricultural Economics*, **27**, 247–267.

Anselin, L. & Rey, S. J. (1991) Properties of tests for spatial dependence in linear regression models. *Geographical Analysis*, **23**, 112–131.

Anselin, L., Florax, R. J. G. M., & Rey, S. J. (2004) *Advances in Spatial Econometrics: Methodology, Tools and Applications*. Berlin: Springer.

Araújo, M. B. & Guisan, A. (2006) Five (or so) challenges for species distribution modeling. *Journal of Biogeography*, **33**, 1677–1688.

Araújo, M. B. & Luoto, M. (2007) The importance of biotic interactions for modelling species distributions under climate change. *Global Ecology and Biogeography*, **16**, 743–753.

Araújo, M. B. & New, M. (2007) Ensemble forecasting of species distributions. *Trends in Ecology & Evolution*, **22**, 42–47.

Araújo, M. B. & Williams, P. H. (2000) Selecting areas for species persistence using occurrence data. *Biological Conservation*, **96**, 331–345.

Araújo, M. B., Pearson, R. G., Thuiller, W., & Erhard, M. (2005a) Validation of species-climate impact models under climate change. *Global Change Biology*, **11**, 1504–1513.

Araújo, M. B., Thuiller, W., & Pearson, R. G. (2006) Climate warming and the decline of amphibians and reptiles in Europe. *Journal of Biogeography*, **33**, 1712–1728.

Araújo, M. B., Thuiller, W., Williams, P. H., & Reginster, I. (2005b) Down-scaling European species atlas distributions to a finer resolution: implications for conservation planning. *Global Ecology and Biogeography*, **14**, 17–30.

Armstrong, J. S. & Collopy, F. (1992) Error measures for generalizing about forecasting methods: empirical comparisons. *International Journal of Forecasting*, **8**, 69–80.

Ashcroft, M. B. (2006) A method for improving landscape scale temperature predictions and the implications for vegetation modelling. *Ecological Modelling*, **197**, 394–404.

Asher, J., Warren, M., Fox, R., Harding, P., Jeffcoate, G., & Jeffcoat, S. (Eds.) (2001) *The Millennium Atlas of Butterflies in Britain and Ireland*. Oxford, UK: Oxford University Press.

Asner, G. P., Martin, R. E., Ford, A. J., Metcalfe, D. J., & Liddell, M. J. (2009) Leaf chemical and spectral diversity in Australian tropical forests. *Ecological Applications*, **19**, 236–253.

Aspinall, R. & Veitch, N. (1993) Habitat mapping from satellite imagery and wildlife survey data using a Bayesian modeling procedure in GIS. *Photogrammetric Engineering and Remote Sensing*, **59**, 537–543.

Aspinall, R. J. & Pearson, D. M. (1996) Data quality and spatial analysis: analytical use of GIS for ecological modeling. In Goodchild, M. F., Steyaert, L. T., Parks, B. O. et al. (Eds.) *GIS and Environmental Modeling: Progress and Research Issues*. Fort Collins, CO: GIS World Books, pp. 35–38.

Augustin, N., Mugglestone, M., & Buckland, S. (1996) An autologistic model for the spatial distribution of wildlife. *Journal of Applied Ecology*, **33**, 339–347.

Augustin, N., Mugglestone, M., & Buckland, S. (1998) The role of simulation in modelling spatially correlated data. *Environmetrics*, **9**, 175–196.

Austin, M. (1971) The role of regression analysis in plant ecology. *Proceedings of the Ecological Society of Australia*, **6**, 63–75.

Austin, M. P. (1980) Searching for a model for use in vegetation analysis. *Vegetatio*, **42**, 11–21.

Austin, M. P. (1985) Continuum concept, ordination methods, and niche theory. *Annual Review of Ecology and Systematics*, **16**, 39–61.

Austin, M. P. (1990) Community theory and competition in vegetation. In Grace, J. B. & Tilman, D. (Eds.) *Perspectives on Plant Competition*. San Diego, CA: Academic Press, pp. 215–238.

Austin, M. P. (1992) Modelling the environmental niche of plants: some implications for plant community response to elevation CO_2 levels. *Australian Journal of Botany*, **40**, 615–630.

Austin, M. P. (1998) An ecological perspective on biodiversity investigations: examples from Australian eucalypt forests. *Annals of the Missouri Botanical Garden*, **85**, 2–17.

Austin, M. (1999) A silent clash of paradigms: some inconsistencies in community ecology. *Oikos*, **86**, 170–178.

Austin, M. P. (2002) Spatial prediction of species distribution: an interface between ecological theory and statistical modelling. *Ecological Modelling*, **157**, 101–118.

Austin, M. P. (2007) Species distribution models and ecological theory: a critical assessment and some possible new approaches. *Ecological Modelling*, **200**, 1–19.

Austin, M. & Cunningham, R. (1981) Observational analysis of environmental gradients. *Proceedings of the Ecological Society of Australia*, **11**, 109–119.

Austin, M. P. & Heyligers, P. C. (1989) Vegetation survey design for conservation: gradsect sampling of forests in North-eastern New South Wales. *Biological Conservation*, **50**, 13–32.

Austin, M. P. & Heyligers, P. C. (1991) New approach to vegetation survey design: gradsect sampling. In Margules, C. R. & Austin, M. P. (Eds.) *Nature Conservation: Cost Effective Biological Surveys and Data Analysis*. East Melbourne, Australia: CSIRO, pp. 31–36.

Austin, M. P. & Meyers, J. (1996) Current approaches to modelling the environmental niche of eucalypts: implication for management of forest biodiversity. *Forest Ecology and Management*, **85**, 95–106.

Austin, M. P. & Smith, T. M. (1989) A new model for the continuum concept. *Vegetatio*, **83**, 35–47.

Austin, M. P., Belbin, L., Meyers, J. A., Doherty, M. D., & Luoto, M. (2006) Evaluation of statistical models used for predicting plant species distributions: Role of artificial data and theory. *Ecological Modelling*, **199**, 197–216.

Austin, M. P., Cunningham, R. B., & Fleming, P. M. (1984) New approaches to direct gradient analysis using environmental scalars and statistical curve-fitting procedures. *Vegetatio*, **55**, 11–27.

Austin, M. P., Nicholls, A. O., Doherty, M. D., & Meyers, J. A. (1994) Determining species response functions to an environmental gradient by means of a beta-function. *Journal of Vegetation Science*, **5**, 215–228.

Austin, M. P., Nicholls, A. O., & Margules, C. R. (1990) Determining the realised qualitative niche: environmental niches of five *Eucalyptus* species. *Ecological Monographs*, **60**, 161–177.

Austin, M. P., Pausas, J. G., & Nicholls, A. O. (1996) Patterns of tree species richness in relation to environment in southeastern New South Wales, Australia. *Australian Journal of Ecology*, **21**, 154–164.

Bach, M., Breuer, L., Frede, H. G., Huisman, J. A., Otte, A., & Waldhardt, R. (2006) Accuracy and congruency of three different digital land-use maps. *Landscape and Urban Planning*, **78**, 289–299.

Bailey, L. L., Simons, T. R., & Pollock, K. H. (2004) Estimating site occupancy and species detection probability parameters for terrestrial salamanders. *Ecological Applications*, **14**, 692–702.

Bailey, T. C. & Gatrell, A. C. (1995) *Interactive spatial data analysis*, Harlow, England, Longman.

Band, L. E. (1986) Topographic partitioning of watersheds with digital elevation models. *Water Resources Research*, **22**, 15–24.

Banks, W. E., D'errico, F., Peterson, A. T. *et al.* (2008) Human ecological niches and ranges during the LGM in Europe derived from an application of eco-cultural niche modeling. *Journal of Archaeological Science*, **35**, 481–491.

Barnes, T. K., Volety, A. K., Chartier, K., Mazzotti, F. J., & Pearlstine, L. (2007) A habitat suitability index model for the eastern oyster (*Crassostrea virginica*), a tool for restoration of the Caloosahatchee Estuary, Florida. *Journal of Shellfish Research*, **26**, 949–959.

Barnett, V. (2002) *Sample Survey: Principles and Methods*. London: Arnold.

Barrows, C. W. (1997) Habitat relationships of the Coachella Valley fringe-toed lizard (*Uma inornata*). *Southwestern Naturalist*, **42**, 218–223.

Barrows, C. W., Swartz, M. B., Hodges, W. L. *et al.* (2005) A framework for monitoring multiple-species conservation plants. *Journal of Wildlife Management*, **69**, 1333–1345.

Barry, S. & Elith, J. (2006) Error and uncertainty in habitat models. *Journal of Applied Ecology*, **43**, 413–423.

Barry, S. C. & Welsh, A. H. (2002) Generalized additive modelling and zero inflated count data. *Ecological Modelling*, **157**, 179–188.

Bazzaz, F. A. & Pickett, S. T. A. (1980) Physiological ecology of tropical succession: A comparative review. *Annual Review of Ecology and Systematics*, **11**, 287–310.

Beale, C. M., Lennon, J. J., Elston, D. A., Brewer, M. J., & Yearsley, J. M. (2007) Red herrings remain in geographical ecology: a reply to Hawkins *et al.* (2007). *Ecography*, **30**, 845–847.

Beaumont, L. J. (2008) Why is the choice of future climate scenarios for species distribution modelling important? *Ecology Letters*, **11**, 1135–1146.

Beck-Worner, C., Raso, G., Vounatsou, P. *et al.* (2007) Bayesian spatial risk prediction of *Schistosoma mansoni* infection in western Côte d'Ivoire using a remotely-sensed digital elevation model. *American Journal of Tropical Medicine and Hygiene*, **76**, 956–963.

Beerling, D., Huntley, B., & Bailey, J. (1995) Climate and the distribution of *Fallopia japonica*: use of an introduced species to test the predictive capacity of response surfaces. *Journal of Vegetation Science*, **6**, 269–282.

Beger, M. & Possingham, H. P. (2008) Environmental factors that influence the distribution of coral reef fishes: modeling occurrence data for broad-scale conservation and management. *Marine Ecology-Progress Series*, **361**, 1–13.

Beissinger, S. R. & McCullough, D. R. (Eds.) (2002) *Population Viability Analysis*. Chicago: University of Chicago Press.

Bell, J. F. (1996) Application of classification trees to the habitat preference of upland birds. *Journal of Applied Statistics*, **23**, 349–359.

Bellis, L. M., Pidgeon, A. M., Radeloff, V. C., St-Louis, V., Navarro, J. L., & Martella, M. B. (2008) Modeling habitat suitability for greater rheas based on satellite image texture. *Ecological Applications*, **18**, 1956–1966.

Benedict, M. Q., Levine, R. S., Hawley, W. A., & Lounibos, L. P. (2007) Spread of the tiger: global risk of invasion by the mosquito *Aedes albopictus*. *Vector-Borne and Zoonotic Diseases*, **7**, 76–85.

Benediktsson, J. A., Swain, P. H., & Ersoy, O. K. (1993) Conjugate-gradient neural networks in classification of multisource and very-high-dimensional remote sensing data. *International Journal of Remote Sensing*, **14**, 2883–2903.

Benito Garzón, M., Blazek, R., Neteler, M., Sánchez de Dios, R., Ollero, H. S., & Furlanello, C. (2006) Predicting habitat suitability with machine learning models: the potential area of *Pinus sylvestris* L. in the Iberian Peninsula. *Ecological Modelling*, **197**, 383–393.

Benito Garzón, M., Sanchez de Dios, R., & Sainz Ollero, H. (2008) Effects of climate change on the distribution of Iberian tree species. *Applied Vegetation Science*, **11**, 169–178.

Besag, J. (1972) Nearest-neighbour systems and the autologistic model for binary data. *Journal of the Royal Statistical Society B*, **34**, 75–83.

Besag, J. (1974) Spatial interaction and the statistical analysis of lattice systems. *Journal of the Royal Statistical Society, Series B*, **36**, 192–236.

Bickford, S. A. & Laffan, S. W. (2006) Multi-extent analysis of the relationship between pteridophyte species richness and climate. *Global Ecology and Biogeography*, **15**, 588–601.

Bio, A. M. F., Alkemade, R., & Barendregt, A. (1998) Determining alternative models for vegetation response analysis: a non-parametric approach. *Journal of Vegetation Science*, **9**, 5–16.

Borcard, D., Legendre, P., & Drapeau, P. (1992) Partialling out the spatial component of ecological variation. *Ecology*, **73**, 1045–1055.

Boitani, L., Sinibaldi, I., Corsi, F. *et al.* (2008) Distribution of medium- to large-sized African mammals based on habitat suitability models. *Biodiversity and Conservation*, **17**, 605–621.

Bojorquez-Tapia, L. A., Brower, L. P., Castilleja, G. *et al.* (2003) Mapping expert knowledge: redesigning the Monarch Butterfly Biosphere Reserve. *Conservation Biology*, **17**, 367–379.

Bolker, B. M., Brooks, M. E., Clark, C. J. *et al.* (2009) Generalized linear mixed models: a practical guide for ecology and evolution. *Trends in Ecology and Evolution*, **24**, 127–35.

Bolstad, P. V. & Lillesand, T. M. (1992) Improved classification of forest vegetation in northern Wisconsin through a rule-based combination of soils, terrain, and Landsat Thematic Mapper data. *Forest Science*, **38**, 5–20.

Bond, J. E. & Stockman, A. K. (2008) An integrative method for delimiting cohesion species: finding the population-species interface in a group of Californian trapdoor spiders with extreme genetic divergence in geographic structuring. *Systematic Biology*, **57**, 628–646.

Botkin, D. B., Saxe, H., Araújo, M. B. *et al.* (2007) Forecasting the effects of global warming on biodiversity. *Bioscience*, **57**, 227–236.

Bourgeron, P. S., Humphries, C. J., & Jensen, M. E. (2001) General data collection and sampling design considerations for integrated ecological assessment. In Bourgeron, P., Jensen, M. & Lessard, G. (Eds.) *An Integrated Ecological Assessment Protocols Guidebook*. New York: Springer-Verlag, pp. 92–107.

Box, E. (1981) *Macroclimate and Plant Forms: An Introduction to Predictive Modeling in Phytogeography*. The Hague: Dr. W. Junk.

Boyce, M. S. (2006) Scale for resource selection functions. *Diversity and Distributions*, **12**, 269–276.

Boyce, M. S., Vernier, P. R., Nielsen, S. E., & Schmiegelow, F. K. A. (2002) Evaluating resource selection functions. *Ecological Modelling*, **157**, 281–300.

Bradbury, R. B., Kyrkos, A., Morris, A. J., Clark, S. C., Perkins, A. J., & Wilson, J. D. (2000) Habitat associations and breeding success of yellowhammers on lowland farmland. *Journal of Applied Ecology*, **37**, 789–805.

Bradley, B. A. & Fleishman, E. (2008) Can remote sensing of land cover improve species distribution modelling? *Journal of Biogeography*, **35**, 1158–1159.

Braunisch, V., Bollmann, K., Graf, R. F., & Hirzel, A. H. (2008) Living on the edge – modelling habitat suitability for species at the edge of their fundamental niche. *Ecological Modelling*, **214**, 153–167.

Breiman, L. (1996) Bagging predictors. *Machine Learning*, **26**, 123–140.

Breiman, L. (2001a) Statistical modeling: the two cultures. *Statistical Science*, **16**, 199–215.

Breiman, L. (2001b) Random forests. *Machine Learning*, **45**, 15–32.

Breiman, L., Friedman, J., Olshen, R., & Stone, C. (1984) *Classification and Regression Trees*. Belmont, CA: Wadsworth.

Brigham, C. A., Thomson, D. M., Brigham, C. A., & Schwartz, M. W. (2003) Approaches to modeling population viability in plants: an overview. *Population viability in plants: Conservation, management and modeling of rare plants*, 145–171.

Brodley, C. & Utgoff, P. (1995) Multivariate decision trees. *Machine Learning*, **19**, 45–77.

Broennimann, O. & Guisan, A. (2008) Predicting current and future biological invasions: both native and invaded ranges matter. *Biology Letters*, **4**, 585–589.

Broennimann, O., Thuiller, W., Hughes, G., Midgley, G. F., Alkemade, J. M. R., & Guisan, A. (2006) Do geographic distribution, niche property and life form explain plants' vulnerability to global change? *Global Change Biology*, **12**, 1079–1093.

Broennimann, O., Treier, U. A., Muller-Scharer, H., Thuiller, W., Peterson, A. T., & Guisan, A. (2007) Evidence of climatic niche shift during biological invasion. *Ecology Letters*, **10**, 701–709.

Brotons, L., Herrando, S., & Pla, M. (2007) Updating bird species distribution at large spatial scales: applications of habitat modelling to data from long-term monitoring programs. *Diversity and Distributions*, **13**, 276–288.

Brotons, L., Thuiller, W., Araújo, M. B., & Hirzel, A. H. (2004) Presence-absence versus presence-only modelling methods for predicting bird habitat suitability. *Ecography*, **27**, 437–448.

Brown, D. (1994) Predicting vegetation types at treeline using topography and biophysical disturbance variables. *Journal of Vegetation Science*, **5**, 641–656.

Brown, D. B. & Bara, T. J. (1994) Recognition and reduction of systematic error in elevation and derivative surfaces from 7 1/2-minute DEMs. *Photogrammetric Engineering and Remote Sensing*, **60**, 189–194.

Browning, D. M., Beaupre, S. J., & Duncan, L. (2005) Using partitioned Mahalanobis D-2(K) to formulate a GIS-based model of timber rattlesnake hibernacula. *Journal of Wildlife Management*, **69**, 33–44.

Brzeziecki, B., Kienast, F., & Wildi, O. (1993) A simulated map of the potential natural forest vegetation of Switzerland. *Journal of Vegetation Science*, **4**, 499–508.

Buckley, L. B. & Roughgarden, J. (2004) Biodiversity conservation: effects of changes in climate and land use. *Nature*, **430**, 2 pp. following 33; discussion following 33.

Buermann, W., Saatchi, S., Smith, T. B. *et al.* (2008) Predicting species distributions across the Amazonian and Andean regions using remote sensing data. *Journal of Biogeography*, **35**, 1160–1176.

Burgman, M. A. (2005) *Risks and Decisions for Conservation and Environmental Management.* Cambridge, UK: Cambridge University Press.

Burgman, M. A. & Fox, J. C. (2003) Bias in species range estimates from minimum convex polygons: implications for conservation and options for improved planning. *Animal Conservation,* **6**, 19–28.

Burgman, M. A., Lindenmayer, D. B., & Elith, J. (2005) Managing landscapes for conservation under uncertainty. *Ecology,* **86**, 2007–2017.

Burnham, K. P. & Anderson, D. R. (1998) *Model Selection and Multimodel Inference: A Practical Information – Theoretic Approach.* New York, USA: Springer-Verlag.

Burrough, P. & McDonnell, R. (1998) *Principles of Geographical Information Systems,* Oxford: Oxford University Press.

Busby, J. R. (1986) A biogeoclimatic analysis of *Nothofagus cunninghamii* (Hook.) Oerst. in southeastern Australia. *Australian Journal of Ecology,* **11**, 1–7.

Busby, J. R. (1991) BIOCLIM – a bioclimatic analysis and prediction system. In Margules, C. R. & Austin, M. P. (Eds.) *Nature Conservation: Cost Effective Biological Surveys and Data Analysis.* East Melbourne, Australia: CSIRO, pp. 64–68.

Callaway, R. M. & Pennings, S. C. (2000) Facilitation may buffer competitive effects: Indirect and diffuse interactions among salt marsh plants. *American Naturalist,* **156**, 416–424.

Cantor, S. B., Sun, C. C., Tortolero-Luna, G., Richards-Kortum, R., & Follen, M. (1999) A comparison of C/B ratios from studies using receiver operating characteristic curve analysis. *Journal of Clinical Epidemiology,* **52**, 885–892.

Card, D. H. (1982) Using known map category marginal frequencies to improve estimates of thematic map accuracy. *Photogrammetric Engineering and Remote Sensing,* **48**, 431–439.

Carlson, K. M., Asner, G. P., Hughes, R. F., Ostertag, R., & Martin, R. E. (2007) Hyperspectral remote sensing of canopy biodiversity in Hawaiian lowland rainforests. *Ecosystems,* **10**, 536–549.

Carmel, Y. & Flather, C. H. (2006) Constrained range expansion and climate change assessments. *Frontiers in Ecology and the Environment,* **4**, 178–179.

Carpenter, G., Gillison, A. N., & Winter, J. (1993) DOMAIN: a flexible modeling procedure for mapping potential distributions of plants and animals. *Biodiversity and Conservation,* **2**, 667–680.

Carpenter, G. A., Gopal, S., Macomber, S., Martens, S., Woodcock, C. E., & Franklin, J. (1999) A neural network method for efficient vegetation mapping. *Remote Sensing of Environment,* **70**, 326–338.

Carroll, C. & Johnson, D. S. (2008) The importance of being spatial (and reserved): Assessing Northern Spotted Owl habitat relationships with hierarchical Bayesian Models. *Conservation Biology,* **22**, 1026–1036.

Carroll, C., Phillips, M. K., Schumaker, N. H., & Smith, D. W. (2003) Impacts of landscape change on wolf restoration success: Planning a reintroduction program based on static and dynamic spatial models. *Conservation Biology,* **17**, 536–548.

Carroll, C., Zielinski, W. J., & Noss, R. F. (1999) Using presence-absence data to build and test spatial habitat models for the fisher in the Klamath Region, U.S.A. *Conservation Biology,* **13**, 1344–1359.

Carstens, B. C. & Richards, C. L. (2007) Integrating coalescent and ecological niche modeling in comparative phylogeography. *Evolution,* **61**, 1439–1454.

Carter, G. M., Stolen, E. D., & Breininger, D. R. (2006) A rapid approach to modeling species–habitat relationships. *Biological Conservation*, **127**, 237–244.

Cawsey, E. M., Austin, M. P., & Baker, B. L. (2002) Regional vegetation mapping in Australia: a case study in the practical use of statistical modelling. *Biodiversity and Conservation*, **11**, 2239–2274.

Chambers, J. M. & Hastie, T. J. (1992) *Statistical Models in S.* Pacific Grove, CA: Wadsworth.

Chase, J. M. & Leibold, M. A. (2003) *Ecological Niches: Linking Classical and Contemporary Approaches.* Chicago: Chicago University Press.

Chatfield, C. (1995) Model uncertainty, data mining and statistical inference. *Journal of the Royal Statistical Society*, **158**, 419–466.

Chefaoui, R. M. & Lobo, J. M. (2008) Assessing the effects of pseudo-absences on predictive distribution model performance. *Ecological Modelling*, **210**, 478–486.

Chen, D. M. & Stow, D. (2003) Strategies for integrating information from multiple spatial resolutions into land-use/land-cover classification routines. *Photogrammetric Engineering and Remote Sensing*, **69**, 1279–1287.

Ciarniello, L. M., Boyce, M. S., Seip, D. R., & Heard, D. C. (2007) Grizzly bear habitat selection is scale dependent. *Ecological Applications*, **17**, 1424–1440.

Civco, D. L. (1993) Artificial neural networks for land-cover classification and mapping. *International Journal of Geographic Information Systems*, **7**, 173–186.

Clark, J. D., Dunn, J. E., & Smith, K. G. (1993) A multivariate model of female black bear habitat use for a geographic information system. *Journal of Wildlife Management*, **57**, 519–526.

Clark, L. A. & Pregibon, D. (1992) Tree-based models. In Chambers, J. & Hastie, T. J. (Eds.) *Statistical Models in S.* Pacific Grove, CA: Wadsworth and Brooks/Cole Advanced Books and Software, pp. 377–419.

Clark, M. L., Roberts, D. A., & Clark, D. B. (2005) Hyperspectral discrimination of tropical rain forest tree species at leaf to crown scales. *Remote Sensing of Environment*, **96**, 375–398.

Clarke, K. C. (1997) *Getting Started with Geographic Information Systems.* Upper Saddle River, NJ: Prentice Hall.

Clarke, K. C. & Gaydos, L. J. (1998) Loose-coupling a cellular automaton model and GIS: long-term urban growth prediction for San Francisco and Washington/Baltimore. *International Journal of Geographic Information Science*, **12**, 699–714.

Clevenger, A. P., Wierzchowski, J., Chruszcz, B., & Gunson, K. (2002) GIS-generated, expert-based models for identifying wildlife habitat linkages and planning mitigation passages. *Conservation Biology*, **16**, 503–514.

Cochran, W. G. (1977) *Sampling Techniques.* New York: Wiley.

Collins, S. L., Glenn, S. M., & Roberts, D. W. (1993) The hierarchical continuum concept. *Journal of Vegetation Science*, **4**, 149–156.

Congalton, R. G. (1991) A review of assessing the accuracy of classifications of remotely sensed data. *Remote Sensing of Environment*, **37**, 35–46.

Cohen, W. B. & Goward, S. N. (2004) Landsat's role in ecological applications of remote sensing. *Bioscience*, **54**, 535–545.

Cooperrider, A. Y., Boyd, R. J., & Stuart, H. R. (Eds.) (1986) *Inventory and Monitoring of Wildlife Habitat.* Denver, CO: US Department of the Interior, Bureau of Land Management, Service Center.

Cord, P. L. (2008) Using species-environmental amplitudes to predict pH values from vegetation. *Journal of Vegetation Science*, **19**, 437–444.

Cordellier, M. & Pfenninger, M. (2009) Inferring the past to predict the future: climate modelling predictions and phylogeography for the freshwater gastropod *Radix balthica* (Pulmonata, Basommatophora). *Molecular Ecology*, **18**, 534–544.

Corsi, F., Dupre, E., & Biotani, L. (1999) A large-scale model of wolf distribution in Italy for conservation planning. *Conservation Biology*, **13**, 150–159.

Costa, J., Peterson, A. T., & Beard, C. B. (2002) Ecologic niche modeling and differentiation of populations of *Triatoma brasiliensis neiva*, 1911, the most important Chagas' disease vector in northeastern Brazil (Hemiptera, Reduviidae, Triatominae). *American Journal of Tropical Medicine and Hygiene*, **67**, 516–520.

Coudun, C. & Gégout, J.-C. (2007) Quantitative prediction of the distribution and abundance of *Vaccinium myrtillus* with climatic and edaphic factors. *Journal of Vegetation Science*, **18**, 517–524.

Coudun, C., Gégout, J.-C., Piedallu, C., & Rameau, J.-C. (2006) Soil nutritional factors improve models of plant species distribution: an illustration with *Acer campestre* (L.) in France. *Journal of Biogeography*, **33**, 1750–1763.

Cramer, J. S. (2003) *Logit Models from Economics and Other Fields*. Cambridge, UK: Cambridge University Press.

Cramer, W., Bondeau, A., Woodward, F. I. *et al.* (2001) Global response of terrestrial ecosystem structure and function to CO_2 and climate change: results from six dynamic global vegetation models. *Global Change Biology*, **7**, 357–373.

Crawley, M. J. (2005) *Statistics: An Introduction Using R*. Chichester, UK: John Wiley & Sons, Ltd.

Cressie, N. (1991) *Statistics for Spatial Data*. New York: John Wiley and Sons.

Crossman, N. D. & Bass, D. A. (2008) Application of common predictive habitat techniques for post-border weed risk management. *Diversity and Distributions*, **14**, 213–224.

Culvenor, D. S. (2002) TIDA: an algorithm for the delineation of tree crowns in high spatial resolution remotely sensed imagery. *Computers & Geosciences*, **28**, 33–44.

Cumming, G. S. (2000a) Using between-model comparisons to fine-tune linear models of species ranges. *Journal of Biogeography*, **27**, 441–455.

Cumming, G. S. (2000b) Using habitat models to map diversity: pan-African species richness of ticks (Acari: Ixodida). *Journal of Biogeography*, **27**, 425–440.

Cumming, G. S. (2002) Comparing climate and vegetation as limiting factors for species ranges of African ticks. *Ecology*, **83**, 255–268.

Cunningham, H. R., Rissler, L. J., & Apodaca, J. J. (2009) Competition at the range boundary in the slimy salamander: using reciprocal transplants for studies on the role of biotic interactions in spatial distributions. *Journal of Animal Ecology*, **78**, 52–62.

Curran, P. (1980) Multispectral remote sensing of vegetation amount. *Progress in Physical Geography*, **4**, 315–341.

Currie, D. J. (1991) Energy and large-scale patterns of animal-species and plant-species richness. *American Naturalist*, **137**, 27–49.

Cutler, D. R., Edwards Jr., T. C., Beard, K. H. *et al.* (2007) Random forests for classification in ecology. *Ecology*, **88**, 2783–2792.

Daly, C. (2006) Guidelines for assessing the suitability of spatial climate data sets. *International Journal of Climatology*, **26**, 707–721.

Daly, C. & Taylor, G. H. (1996) Development of a new Oregon precipitation map using the PRISM model. In Goodchild, M. F., Steyaert, L. T., Park, B. O. *et al.* (Eds.) *GIS and Environmental Modeling: Progress and Research issues.* Fort Collins, CO: GIS World Books, pp. 91–92.

Daly, C., Halbleib, M., Smith, J. I. *et al.* (2008) Physiographically sensitive mapping of climatological temperature and precipitation across the conterminous United States. *International Journal of Climatology*, **28**, 2031–2064.

Daly, C., Neilson, R. P., & Phillips, D. L. (1994) A statistical-topographic model for mapping climatological precipitation over mountainous terrain. *Applied Meteorology*, **33**, 140–158.

Das, A., Lele, S. R., Glass, G. E., Shields, T., & Patz, J. (2002) Modeling as a discrete spatial response using generalized linear mixed models: application to Lyme disease. *International Journal of Geographic Information Science*, **16**, 152–166.

Davis, F. W. & Goetz, S. (1990) Modeling vegetation pattern using digital terrain data. *Landscape Ecology*, **4**, 69–80.

Davis, F. W., Seo, C., & Zielinski, W. J. (2007) Regional variation in home-range-scale habitat models for fisher (*Martes pennanti*) in California. *Ecological Applications*, **17**, 2195–2213.

Davis, F. W., Stine, P. A., Stoms, D. M., Borchert, M. I., & Hollander, A. (1995) Gap analysis of the actual vegetation of California: 1. The southwestern region. *Madroño*, **42**, 40–78.

De'ath, G. (2002) Multivariate regression trees: a new technique for modeling species–environment relationships. *Ecology*, **83**, 1105–1117.

De'ath, G. (2007) Boosted trees for ecological modeling and prediction. *Ecology*, **88**, 243–251.

De'ath, G. & Fabricius, K. E. (2000) Classification and regression trees: a powerful yet simple technique for ecological data analysis. *Ecology*, **81**, 3178–3192.

Dechka, J. A., Franklin, S. E., Watmough, M. D., Bennett, R. P., & Ingstrup, D. W. (2002) Classification of wetland habitat and vegetation communities using multi-temporal Ikonos imagery in southern Saskatchewan. *Canadian Journal of Remote Sensing*, **28**, 679–685.

Dedecker, A. P., Goethals, P. L. M., Gabriels, W., & De Pauw, N. (2004) Optimization of Artificial Neural Network (ANN) model design for prediction of macroinvertebrates in the Zwalm river basin (Flanders, Belgium). *Ecological Modelling*, **174**, 161–173.

DeFries, R. S. & Townshend, J. R. G. (1999) Global land cover characterization from satellite data: from research to operational implementation? *Global Ecology and Biogeography*, **8**, 367–379.

DeMatteo, K. E. & Loiselle, B. A. (2008) New data on the status and distribution of the bush dog (*Speothos venaticus*): evaluating its quality of protection and directing research efforts. *Biological Conservation*, **141**, 2494–2505.

DeMers, M. (1991) Classification and purpose in automated vegetation maps. *Geographical Review*, **81**, 267–280.

DeMers, M. (2009) *Fundamentals of Geographic Information Systems*, Hoboken, NJ: Wiley.

Deng, Y., Wilson, J. P., & Bauer, B. O. (2007) DEM resolution dependencies of terrain attributes across a landscape. *International Journal of Geographical Information Science*, **21**, 187–213.

Dettmers, R., Buehler, D. A. & Bartlett, J. B. (2002) A test and comparison of wildlife-habitat modeling techniques for predicting bird occurrence at a regional scale. In Scott, J. M., Heglund, P. J., Morrison, M., Raphael, M., Haufler, J., & Wall, B. (Eds.) *Predicting Species Occurrences: Issues of Accuracy and Scale*. Covelo, CA: Island Press, pp. 607–615.

Diggle, P. (1983) *Statistical Analysis of Spatial Point Patterns*. London: Academic Press.

Di Luzio, M., Johnson, G. L., Daly, C., Eischeid, J. K., & Arnold, J. G. (2008) Constructing retrospective gridded daily precipitation and temperature datasets for the conterminous United States. *Journal of Applied Meteorology and Climatology*, **47**, 475–497.

Diniz-Filho, J. A. F. & Bini, L. M., (2005) Modelling geographical patterns in species richness using eigenvector-based spatial filters. *Global Ecology and Biogeography*, **14**, 177–185.

Diniz-Filho, J. A. F., Bini, L. M., & Hawkins, B. A. (2003) Spatial autocorrelation and red herrings in geographical ecology. *Global Ecology & Biogeography*, **12**, 53–64.

Diniz-Filho, J. A. F., Hawkins, B. A., Bini, L. M., De Marco, P., & Blackburn, T. M. (2007a) Are spatial regression methods a panacea or a Pandora's box? A reply to Beale *et al.* (2007). *Ecography*, **30**, 848–851.

Diniz-Filho, J. A. F., Rangel, T. F. L. V. B., Bini, L. M., & Hawkins, B. A. (2007b) Macroevolutionary dynamics in environmental space and the latitudinal diversity gradient in New World birds. *Proceedings of the Royal Society Biological Sciences Series B*, **274**, 43–52.

Dirnböck, T., Dullinger, S., Gottfried, M., Ginzler, C., & Grabherr, G. (2003) Mapping alpine vegetation based on image analysis, topographic variables and Canonical Correspondence Analysis. *Applied Vegetation Science*, **6**, 85–96.

Donald, P. F. & Fuller, R. J. (1998) Ornithological atlas data: a review of uses and limitations. *Bird Study*, **45**, 129–145.

Dormann, C. F. (2007a) Assessing the validity of autologistic regression. *Ecological Modelling*, **207**, 234–242.

Dormann, C. F. (2007b) Effects of incorporating spatial autocorrelation into the analysis of species distribution data. *Global Ecology and Biogeography*, **16**, 129–138.

Dormann, C. F. (2007c) Promising the future? Global change projections of species distributions. *Basic and Applied Ecology*, **8**, 387–397.

Dormann, C. F., McPherson, J. M., Araújo, M. B. *et al.* (2007) Methods to account for spatial autocorrelation in the analysis of species distributional data: a review. *Ecography*, **30**, 609–628.

Dormann, C. F., Purschke, O., García-Marquez, J., Lautenbach, S., & Schröder, B. (2008) Components of uncertainty in species distribution analysis: a case study of the Great Grey Shrike. *Ecology* **89**, 3371–3386.

Doswald, N., Zimmermann, F., & Breitenmoser, U. (2007) Testing expert groups for a habitat suitability model for the lynx *Lynx lynx* in the Swiss Alps. *Wildlife Biology*, **13**, 430–446.

Dozier, J. (1980) A clear-sky spectral solar radiation model for snow-covered mountainous terrain. *Water Resources Research*, **16**, 709–718.

Dozier, J. & Frew, J. (1990) Rapid calculation of terrain parameters for radiation modeling from digital elevation data. *IEEE Transactions on Geoscience and Remote Sensing*, **28**, 963–969.

Drake, J. M., Randin, C., & Guisan, A. (2006) Modelling ecological niches with support vector machines. *Journal of Applied Ecology*, **43**, 424–432.

Dray, S., Legendre, P., & Peres-Neto, P. R. (2006) Spatial modelling: a comprehensive framework for principal coordinate analysis of neighbour matrices (PCNM). *Ecological Modelling*, **196**, 483–493.

Dreisbach, T. A., Smith, J. A., & Molina, R. (2002) Challenges of modeling fungal habitats: when and where do you find chanterelles? In Scott, J. M., Heglund, P. J., Morrison, M. L. *et al.* (Eds.) *Predicting Species Occurrences: Issues of Accuracy and Scale*. Covelo, CA: Island Press, pp. 475–481.

Dubayah, R. & Rich, P. M. (1995) Topographic solar radiation for GIS. *International Journal of Geographic Information Systems*, **9**, 405–419.

Dubayah, R. & Rich, P. M. (1996) GIS-based solar radiation modeling. In Goodchild, M. F., Steyaert, L. T., Parks, B. O. *et al.* (Eds.) *GIS and Environmental Modeling: Progress and Research Issues*. Fort Collins, CO: GIS World Books, pp. 129–134.

Dudík, M., Phillips, S. J., & Schapire, R. E. (2007) Maximum entropy density estimation with generalized regularization and an application to species distribution modeling. *Journal of Machine Learning Research*, **8**, 1217–1260.

Duncan, B. W., Breininger, D. R., Schmalzer, P. A., & Larson, V. L. (1995) Validating a Florida scrub jay habitat suitability model using demographic data on Kennedy Space Center. *Photogrammetric Engineering & Remote Sensing*, **61**, 1361–1370.

Dunk, J. R., Zielinski, W. J., & Preisler, H. K. (2004) Predicting the occurrence of rare mollusks in northern California forests. *Ecological Applications*, **14**, 713–729.

Dutilleul, P. (1993) Modifying the *t*-Test for assessing the correlation between two spatial processes. *Biometrics*, **49**, 305–314.

Early, R., Anderson, B., & Thomas, C. D. (2008) Using habitat distribution models to evaluate large-scale landscape priorities for spatially dynamic species. *Journal of Applied Ecology*, **45**, 228–238.

Echarri, F., Tambussi, C., & Hospitaleche, C. A. (2009) Predicting the distribution of the crested tinamous, *Eudromia* spp. (Aves, Tinamiformes). *Journal of Ornithology*, **150**, doi:10.1007/s10336–008–0319–5.

Edwards, T. C., Jr., Cutler, D. R., Zimmermann, N. E., Geiser, L., & Alegria, J. (2005) Model-based stratifications for enhancing the detection of rare ecological events. *Ecology*, **86**, 1081–1090.

Edwards, T. C., Jr., Cutler, D. R., Zimmermann, N. E., Geiser, L., & Moisen, G. G. (2006) Effects of sample survey design on the accuracy of classification tree models in species distribution models. *Ecological Modelling*, **199**, 132–141.

Eisen, R. J., Reynolds, P. J., Ettestad, P. *et al.* (2007) Residence-linked human plague in New Mexico: a habitat-suitability model. *American Journal of Tropical Medicine and Hygiene*, **77**, 121–125.

Elder, J. F. (2003) The generalization paradox of ensembles. *Journal of Computational and Graphical Statistics*, **12**, 853–864.

Elith, J. & Burgman, M. A. (2002) Predictions and their validation: rare plants in the Central Highlands, Victoria, Australia. In Scott, J. M., Heglund, P. J., Morrison, M. L. *et al.* (Eds.) *Predicting Species Occurrences: Issues of Accuracy and Scale.* Covelo, CA: Island Press, pp. 303–313.

Elith, J. & Graham, C. (2009) Do they? How do they? WHY do they differ? On finding reasons for differing performances of species distribution models. *Ecography,* **32,** 1–12.

Elith, J., Graham, C. H., Anderson, R. P. *et al.* (2006) Novel methods improve prediction of species distributions from occurrence data. *Ecography,* **29,** 129–151.

Elith, J. & Leathwick, J. (2007) Predicting species distributions from museum and herbarium records using multiresponse models fitted with multivariate adaptive regression splines. *Diversity and Distributions,* **13,** 265–275.

Elith, J., Burgman, M. A., & Regan, H. M. (2002) Mapping epistemic uncertainties and vague concepts in predictions of species distribution. *Ecological Modelling,* **157,** 313–329.

Elith, J. & Leathwick, J. (2009) Conservation prioritisation using species distribution modelling. In Moilanen, A., Wilson, K. A., & Possingham, H. (Eds.) *Spatial Conservation Prioritization: Quantitative Methods and Computational Tools.* Oxford, UK: Oxford University Press, pp. 70–93.

Elith, J., Leathwick, J. R., & Hastie, T. (2008) A working guide to boosted regression trees. *Journal of Animal Ecology,* **77,** 802–813.

Ellenberg, H. (1988) *Vegetation Ecology of Central Europe.* Cambridge, UK: Cambridge University Press.

Ellenberg, H., Weber, H. E., Duell, R., Wirth, V., Werner, W., & Paulissen, D. (1991) *Indicator Values of Plants in Central Europe.* Goettingen (Germany, F.R.): Goltze.

Ellison, A. M. (2004) Bayesian inference in ecology. *Ecology Letters,* **7,** 509–520.

Elton, C. (1927) *Animal Ecology.* Chicago: University of Chicago Press.

Engler, R., Guisan, A., & Rechsteiner, L. (2004) An improved approach for predicting the distribution of rare and endangered species from occurrence and pseudo-absence data. *Journal of Applied Ecology,* **41,** 263–274.

Epstein, H. E., Lauenroth, W. K., Burke, I. C., & Coffin, D. P. (1997) Productivity patterns of C-3 and C-4 functional types in the US Great Plains. *Ecology,* **78,** 722–731.

Erasmus, B. F. N., van Jaarsveld, A. S., Chown, S. L., Kshatriya, M., & Wessels, K. J. (2002) Vulnerability of South African animal taxa to climate change. *Global Change Biology,* **8,** 679–693.

Ertsen, A. C. D., Alkemade, J. R. M., & Wassen, M. J. (1998) Calibrating Ellenberg indicator values for moisture, acidity, nutrient availability and salinity in the Netherlands. *Plant Ecology,* **135,** 113–124.

Estes, J. E. & Mooneyhan, D. W. (1994) Of maps and myths. *Photogrammetric Engineering and Remote Sensing,* **60,** 517–524.

Estes, L. D., Okin, G. S., Mwangi, A. G., & Shugart, H. H. (2008) Habitat selection by a rare forest antelope: a multi-scale approach combining field data and imagery from three sensors. *Remote Sensing of Environment,* **112,** 2033–2050.

Evangelista, P. H., Kumar, S., Stohlgren, T. J. *et al.* (2008) Modelling invasion for a habitat generalist and a specialist plant species. *Diversity and Distributions*, **14**, 808–817.

Fairbanks, D. H. K. & McGwire, K. C. (2004) Patterns of floristic richness in vegetation communities of California: regional scale analysis with multi-temporal NDVI. *Global Ecology and Biogeography*, **13**, 221–235.

Farber, O. & Kadmon, R. (2003) Assessment of alternative approaches for bioclimatic modeling with special emphasis on the Mahalanobis distance. *Ecological Modelling*, **160**, 115–130.

Fels, J. E. (1994) Modeling and mapping potential vegetation using digital terrain data. Unpublished Ph.D. Thesis, College of Forest Resource, North Carolina State University.

Feranec, J., Hazeu, G., Christensen, S., & Jaffrain, G., (2007) CORINE land cover change detection in Europe (case studies of the Netherlands and Slovakia). *Land Use Policy*, **24**, 234–247.

Fernandez, N., Delibes, M., & Palomares, F. (2006) Landscape evaluation in conservation: Molecular sampling and habitat modeling for the Iberian lynx. *Ecological Applications*, **16**, 1037–1049.

Ferrer-Castan, D. & Vetaas, O. R. (2005) Pteridophyte richness, climate and topography in the Iberian Peninsula: comparing spatial and nonspatial models of richness patterns. *Global Ecology and Biogeography*, **14**, 155–165.

Ferrier, S. (2002) Mapping spatial pattern in biodiversity for regional conservation planning: where to from here? *Systematic Biology*, **51**, 331–363.

Ferrier, S. & Guisan, A. (2006) Spatial modelling of biodiversity at the community level. *Journal of Applied Ecology*, **43**, 393–404.

Ferrier, S., Drielsma, M., Manion, G., & Watson, G. (2002a) Extended statistical approaches to modelling spatial pattern in biodiversity in northeast New South Wales. II. Community-level modeling. *Biodiversity and Conservation*, **11**, 2309–2338.

Ferrier, S., Manion, G., Elith, J., & Richardson, K. (2007) Using generalized dissimilarity modelling to analyse and predict patterns of beta diversity in regional biodiversity assessment. *Diversity and Distributions*, **13**, 252–264.

Ferrier, S., Watson, G., Pearce, J., & Drielsma, M. (2002b) Extended statistical approaches to modelling spatial pattern in biodiversity in northeast New South Wales. I. Species-level modelling. *Biodiversity and Conservation*, **11**, 2275–2307.

Fertig, W. & Reiners, W. A. (2002) Predicting Presence/Absence of plant species for range mapping: a case study from Wyoming. In Scott, J. M., Heglund, P. J., Morrison, M. L. *et al.* (Eds.) *Predicting Species Occurrences: Issues of Accuracy and Scale.* Covelo, CA: Island Press, pp. 483–489.

Ficetola, G. F., Thuiller, W., & Miaud, C. (2007) Prediction and validation of the potential global distribution of a problematic alien invasive species – the American bullfrog. *Diversity and Distributions*, **13**, 476–485.

Ficetola, G. F., Thuiller, W., & Padoa-Schioppa, E. (2009) From introduction to the establishment of alien species: bioclimatic differences between presence and reproduction localities in the slider turtle. *Diversity and Distributions*, **15**, 108–116.

Fielding, A. H. & Bell, J. F. (1997) A review of methods for the assessment of prediction errors in conservation presence/absence models. *Environmental Conservation*, **24**, 38–49.

Fielding, A. H. (Ed.) (1999) *Machine learning methods for ecological applications*, Boston, Kluwer.

Fielding, A. H. & Haworth, P. F. (1995) Testing the generality of bird-habitat models. *Conservation Biology*, **9**, 1466–1481.

Fischer, H. (1990) Simulating the distribution of plant communities in an alpine landscape. *Coenoses*, **5**, 37–43.

Fisher, R. N., Suarez, A. V., & Case, T. J. (2002) Spatial patterns in the abundance of the coastal horned lizard. *Conservation Biology*, **16**, 205–215.

Fitzgerald, R. W. & Lees, B. G. (1992) The application of neural networks to floristic classification of remote sensing and GIS data in complex terrain. *Proceedings of the XVII Congress ISPRS*. Bethesda, MD, American Society of Photogrammetry and Remote Sensing.

Fitzpatrick, M. C., Gove, A. D., Sanders, N. J., & Dunn, R. R. (2008) Climate change, plant migration, and range collapse in a global biodiversity hotspot: the *Banksia* (Proteaceae) of Western Australia. *Global Change Biology*, **14**, 1337–1352.

Fleishman, E., Murphy, D. D., & Sjögren-Gulve, P. (2002) Modeling species richness and habitat suitability for taxa of conservation interest. In Scott, J. M., Heglund, P. J., Morrison, M. L. *et al.* (Eds.) *Predicting Species Occurrences: Issues of Accuracy and Scale*. Covelo, CA: Island Press, pp. 507–517.

Fleming, G., Van Der Merwe, M., & McFerren, G. (2007) Fuzzy expert systems and GIS for cholera health risk prediction in southern Africa. *Environmental Modelling & Software*, **22**, 442–448.

Flesch, A. D. & Hahn, L. A. (2005) Distribution of birds and plants at the western and southern edges of the Madrean Sky Islands in Sonora, Mexico. *U S Forest Service Rocky Mountain Research Station Proceedings RMRS-P*, **36**, 80–87.

Florinsky, I. V. (1998) Combined analysis of digital terrain models and remotely sensed data in landscape investigation. *Progress in Physical Geography*, **22**, 33–60.

Foley, J. A., DeFries, R., Asner, G. P. *et al.* (2005) Global consequences of land use. *Science*, **309**, 570–574.

Fonseca, R. L., Guimaraes, P. R., Morbiolo, S. R., Scachetti-Pereira, R., & Peterson, A. T. (2006) Predicting invasive potential of smooth crotalaria (*Crotalaria pallida*) in Brazilian national parks based on African records. *Weed Science*, **54**, 458–463.

Foody, G. M. (2004) Spatial nonstationarity and scale-dependency in the relationship between species richness and environmental determinants for the sub-Saharan endemic avifauna. *Global Ecology and Biogeography*, **13**, 315–320.

Foody, G. M. & Cutler, M. E. J. (2003) Tree biodiversity in protected and logged Bornean tropical rain forests and its measurement by satellite remote sensing. *Journal of Biogeography*, **30**, 1053–1066.

Fornes, G. L. (2004) Habitat use by loggerhead shrikes (*Lanius ludovicianus*) at Midewin National Tallgrass Prairie, Illinois: an application of Brooks and Temple's habitat suitability index. *American Midland Naturalist*, **151**, 338–345.

Fortin, M.-J. & Dale, M. R. T. (2005) *Spatial Analysis: A Guide for Ecologists*. Cambridge, UK: Cambridge University Press.

Fortin, M. J., Keitt, T. H., Maurer, B. A., Taper, M. L., Kaufman, D. M., & Blackburn, T. M. (2005) Species geographic ranges and distributional limits: pattern analysis and statistical issues. *Oikos*, **108**, 7–17.

Fotheringham, A. S., Brunsdon, C., & Charlton, M. (2002) *Geographically Weighted Regression: The Analysis of Spatially Varying Relationships*. Chichester, UK: Wiley & Sons.

Franklin, J. (1995) Predictive vegetation mapping: geographic modeling of biospatial patterns in relation to environmental gradients. *Progress in Physical Geography*, **19**, 474–499.

Franklin, J. (1998) Predicting the distribution of shrub species in southern California from climate and terrain-derived variables. *Journal of Vegetation Science*, **9**, 733–748.

Franklin, J. (2001) Geographic information science and ecological assessment. In Bourgeron, P., Jensen, M. & Lessard, G. (Eds.) *An Integrated Ecological Assessment Protocols Guidebook*. New York: Springer-Verlag, pp. 151–161.

Franklin, J. (2002) Enhancing a regional vegetation map with predictive models of dominant plant species in chaparral. *Applied Vegetation Science*, **5**, 135–146.

Franklin, J. (in press) Spatial point pattern analysis of plants. In Rey, S. J. & Anselin, L. (Eds.) *Perspectives on Spatial Data Analysis*. New York: Springer.

Franklin, J. & Woodcock, C. E. (1997) Multiscale vegetation data for the mountains of Southern California: spatial and categorical resolution. In Quattrochi, D. A. & Goodchild, M. F. (Eds.) *Scale in Remote Sensing and GIS*. Boca Raton, FL, CRC/Lewis Publishers Inc., pp. 141–168.

Franklin, J., Keeler-Wolf, T., Thomas, K. *et al.* (2001) Stratified sampling for field survey of environmental gradients in the Mojave Desert Ecoregion. In Millington, A., Walsh, S., & Osborne, P. (Eds.) *GIS and Remote Sensing Applications in Biogeography and Ecology*. Netherlands: Kluwer Academic Publishers, pp. 229–253.

Franklin, J., Logan, T., Woodcock, C. E., & Strahler, A. H. (1986) Coniferous forest classification and inventory using Landsat and digital terrain data. *IEEE Transactions on Geoscience and Remote Sensing*, **GE-24**, 139–149.

Franklin, J., McCullough, P., & Gray, C. (2000) Terrain variables used for predictive mapping of vegetation communities in Southern California. In Wilson, J. & Gallant, J. (Eds.) *Terrain Analysis: Principles and Applications*. New York: Wiley & Sons, pp. 331–353.

Franklin, J., Phinn, S. R., Woodcock, C. E., & Rogan, J. (2003) Rationale and conceptual framework for classification approaches to assess forest resources and properties. In Wulder, M. A. & Franklin, S. E. (Eds.) *Remote Sensing of Forest Environments: Concepts and Case Studies*. Boston, MA: Kluwer Academic Publishers, pp. 279–300.

Franklin, J., Syphard, A. D., He, H. S., & Mladenoff, D. J. (2005) The effects of altered fire regimes on patterns of plant succession in the foothills and mountains of southern California. *Ecosystems*, **8**, 885–898.

Franklin, J., Wejnert, K. E., Hathaway, S. A., Rochester, C. J., & Fisher, R. N. (2009) Effect of species rarity on the accuracy of species distribution models for reptiles and amphibians in southern California. *Diversity and Distributions* **15**, 167–177.

Franklin, S. E. & Wulder, M. A. (2002) Remote sensing methods in medium spatial resolution satellite data land cover classification of large areas. *Progress in Physical Geography*, **26**, 173–205.

Freeman, E. A. & Moisen, G. G. (2008a) A comparison of the performance of threshold criteria for binary classification in terms of predicted prevalence. *Ecological Modelling*, **217**, 48–58.

Freeman, E. A. & Moisen, G. (2008b) PresenceAbsence: An R package for presence absence analysis. *Journal of Statistical Software*, **23**.

Frescino, T. S., Edwards, T. C., Jr., & Moisen, G. G. (2001) Modeling spatially explicit forest structural attributes using Generalized Additive Models. *Journal of Vegetation Science*, **12**, 15–26.

Freund, Y. & Schapire, R. (1996) Experiments with a new boosting algorithm. *Machine Learning: Proceedings of the 13th International Conference*. San Francisco: Morgan Kauffman, pp. 148–156.

Friedman, J. H. (1991) Multivariate adaptive regression splines. *Annals of Statistics*, **19**, 1–141.

Friedman, J. H. (2001) Greedy function approximation: a gradient boosting machine. *Annals of Statistics*, **29**, 1189–1232.

Friedman, J. H. (2002) Stochastic gradient boosting. *Computational Statistics and Data Analysis*, **38**, 367–378.

Friedman, J. H. & Roosen, C. B. (1995) An introduction to multivariate adaptive regression splines. *Statistical Methods in Medical Research*, **4**, 197–217.

Gahegan, M. (2003) Is inductive machine learning just another wild goose (or might it lay the golden egg)? *International Journal of Geographical Information Science*, **17**, 69–92.

Gallagher, R. V., Beaumont, L. J., Downey, P. O., Hughes, L., & Leishman, M. R. (2008) Projecting the impact of climate change on bitou bush and boneseed distributions in Australia. *Plant Protection Quarterly*, **23**, 37.

Gamon, J. A. & Qiu, H.-L. (1999) Ecological applications of remote sensing at multiple scales. In Pugnaire, F. I. & Valladares, F. (Eds.) *Handbook of Functional Plant Ecology*. New York: Marcel Dekker, Inc., pp. 805–846.

Gasc, J. P., Cabela, A., Crnobrnja-Isailovic, J. *et al.* (1997) Atlas of amphibians and reptiles in Europe. *Atlas of amphibians and reptiles in Europe*, 1–496.

Gaston, K. J. (2003) *The Structure and Dynamics of Geographic Ranges*. Oxford, UK: Oxford University Press.

Gauch, H. G. & Whittaker, R. H. (1972) Coencline simulation. *Ecology*, **53**.

Gause, G. F. (1934) *The Struggle for Existence*. Baltimore, MD: Williams and Wilkins.

Gégout, J.-C., Coudun, C., Bailly, G., & Jabiol, B. (2005) EcoPlant: a forest site database to link floristic data with soil resources and climate conditions. *Journal of Vegetation Science*, **16**, 257–260.

German, G. & Gahegan, M. (1996) Neural network architectures for the classification of temporal image sequences. *Computers & Geosciences*, **22**, 969–979.

Gerrard, R., Stine, P., Church, R., & Gilpin, M., (2001) Habitat evaluation using GIS: a case study applied to the San Joaquin Kit Fox. *Landscape and Urban Planning*, **52**, 239–255.

Gibson, L., Barrett, B., & Burbidge, A. (2007) Dealing with uncertain absences in habitat modelling: a case study of a rare ground-dwelling parrot. *Diversity and Distributions*, **13**, 704–713.

Gibson, L. A., Wilson, B. A., Cahill, D. M., & Hill, J. (2004) Modelling habitat suitability of the swamp antechinus (*Antechinus minimus maritimus*) in the coastal heathlands of southern Victoria, Australia. *Biological Conservation*, **117**, 143–150.

Gillespie, T. W. (2005) Predicting woody-plant species richness in tropical dry forests: A case study from south Florida, USA. *Ecological Applications*, **15**, 27–37.

Gillespie, T. W., Foody, G. M., Rocchini, D., Giorgi, A. P., & Saatchi, S. (2008) Measuring and modelling biodiversity from space. *Progress in Physical Geography*, **32**, 203–221.

Gillison, A. N. & Brewer, K. R. W. (1985) The use of gradient directed transects or gradsect in natural resources survey. *Journal of Environmental Management*, **20**, 103–127.

Gleason, H. A. (1926) The individualistic concept of the plant association. *Bulletin of the Torrey Botanical Club*, **53**, 7–26.

Godinho, R., Teixeira, J., Rebelo, R. *et al.* (1999) Atlas of the continental Portuguese herpetofauna: an assemblage of published and new data. *Revista Espanola de Herpetologia*, **13**, 61–82.

Goetz, S., Steinberg, D., Dubayah, R., & Blair, B. (2007) Laser remote sensing of canopy habitat heterogeneity as a predictor of bird species richness in an eastern temperate forest, USA. *Remote Sensing of Environment*, **108**, 254–263.

Goodchild, M. F. (1988) Stepping over the line: technological constraints and the new cartography. *American Cartographer*, **15**, 311–319.

Goodchild, M. F. (1992) Geographical data modeling. *Computers & Geosciences*, **18**, 401–408.

Goodchild, M. F. (1994) Integrating GIS and remote sensing for vegetation analysis and modeling: methodological issues. *Journal of Vegetation Science*, **5**, 615–626.

Goodchild, M. F. (1996) The spatial data infrastructure of environmental modeling. In Goodchild, M. F., Steyaert, L. T., Parks, B. O. *et al.* (Eds.) *GIS and Environmental Modeling: Progress and Research Issues*. Fort Collins, CO: GIS World Books, pp. 11–15.

Goodchild, M. F. & Gopal, S. (1989) *The accuracy of spatial databases*, London, Taylor & Francis.

Goodchild, M. F. & Tate, N. J. (1992) Description of terrain as a fractal surface, and application to digital elevation model quality assessment. *Photogrammetric Engineering and Remote Sensing*, **58**, 1568–1570.

Goodchild, M. F., Parks, B. O., & Steyaert, L. T. (1993) *Environmental Modeling with GIS*: New York: Oxford University Press.

Goodchild, M. F., Steyaert, L. T., Parks, B. O. *et al.* (Eds.) (1996) *GIS and Environmental Modeling: Progress and Research Issues*. Fort Collins, CO: GIS World Books.

Gottschalk, T. K., Huettmann, F., & Ehlers, M. (2005) Thirty years of analysing and modelling avian habitat relationships using satellite imagery data: a review. *International Journal of Remote Sensing*, **26**, 2631–2656.

Gotway, C. & Stroup, W. (1997) A generalized linear model approach to spatial data analysis and prediction. *Journal of Agricultural, Biological, and Environmental Statistics*, **2**, 152–178.

Gould, W. (2000) Remote sensing of vegetation, plant species richness, and regional biodiversity hotspots. *Ecological Applications*, **10**, 1861–1870.

Graham, C. H. & Blake, J. G. (2001) Influence of patch- and landscape-level factors on bird assemblages in a fragmented tropical landscape. *Ecological Applications*, **11**, 1709–1721.

Graham, C. H. & Fine, P. V. A. (in press) Phylogenetic beta diversity: linking ecological and evolutionary processes across space in time. *Trends in Ecology & Evolution*.

Graham, C. H. & Hijmans, R. J. (2006) A comparison of methods for mapping species ranges and species richness. *Global Ecology and Biogeography*, **15**, 578–587.

Graham, C. H., Elith, J., Hijmans, R. J. *et al.* (2008) The influence of spatial errors in species occurrence data used in distribution models. *Journal of Applied Ecology*, **45**, 239–247.

Graham, C. H., Ferrier, S., Huettman, F., Moritz, C., & Peterson, A. T. (2004) New developments in museum-based informatics and applications in biodiversity analysis. *Trends in Ecology & Evolution*, **19**, 497–503.

Granadeiro, J. P., Andrade, J., & Palmeirim, J. M. (2004) Modelling the distribution of shorebirds in estuarine areas using generalised additive models. *Journal of Sea Research*, **52**, 227–240.

Greenberg, J. A., Dobrowski, S. Z., Ramirez, C. M., Tuil, J. L., & Ustin, S. L. (2006) A bottom-up approach to vegetation mapping of the Lake Tahoe Basin using hyperspatial image analysis. *Photogrammetric Engineering and Remote Sensing*, **72**, 581–589.

Greig-Smith, P. (1983) *Quantitative Plant Ecology*. Oxford, UK: Blackwell Scientific Publications.

Gret-Regamey, A. & Straub, D. (2006) Spatially explicit avalanche risk assessment linking Bayesian networks to a GIS. *Natural Hazards and Earth System Sciences*, **6**, 911–926.

Griffith, D. A. (2006) Assessing spatial dependence in count data: Winsorized and spatial filter specification alternatives to the auto-Poisson model. *Geographical Analysis*, **38**, 160–179.

Griffith, D. A. & Peres-Neto, P. R. (2006) Spatial modeling in ecology: The flexibility of eigenfunction spatial analyses. *Ecology*, **87**, 2603–2613.

Grinnell, J. (1917a) The niche-relationships of the California thrasher. *Auk*, **34**, 427–433.

Grinnell, J. (1917b) Field tests and theories concerning distributional control. *The American Naturalist*, **51**, 115–128.

Grossman, D. H., Faber-Langendoen, D., Weakley, A. S. *et al.* (1998) *International Classification of Ecological Communities: Terrestrial Vegetation of the United States Volume 1. The National Vegetation Classification System: Development, Status and Applications*. http://consci.tnc.org/library/pubs/class/index.html. Washington, DC: The Nature Conservancy.

Gu, W. D. & Swihart, R. K. (2004) Absent or undetected? Effects of non-detection of species occurrence on wildlife-habitat models. *Biological Conservation*, **116**, 195–203.

Guisan, A. & Harrell, F. E. (2000) Ordinal response regression models in ecology. *Journal of Vegetation Science*, **11**, 617–626.

Guisan, A. & Thuiller, W. (2005) Predicting species distributions: offering more than simple habitat models. *Ecology Letters*, **8**, 993–1009.

Guisan, A. & Zimmermann, N. E. (2000) Predictive habitat distribution models in ecology. *Ecological Modelling*, **135**, 147–186.

Guisan, A., Edwards, T. C., Jr., & Hastie, T. (2002) Generalized linear and generalized additive models in studies of species distributions: setting the scene. *Ecological Modelling*, **157**, 89–100.

Guisan, A., Lehmann, A., Ferrier, S. *et al.* (2006) Making better biogeographical predictions of species distributions. *Journal of Applied Ecology*, **43**, 386–392.

Guisan, A., Theurillat, J.-P., & Kienast, F. (1998) Predicting the potential distribution of plant species in an alpine environment. *Journal of Vegetation Science*, **9**, 65–74.

Guisan, A., Weiss, S., & Weiss, A. (1999) GLM versus CCA spatial modeling of plant species distributions. *Plant Ecology*, **143**, 107–122.

Guisan, A., Zimmermann, N. E., Elith, J., Graham, C. H., Phillips, S., & Peterson, A. T. (2007) What matters for predicting the occurrences of trees: Techniques, data, or species characteristics? *Ecological Monographs*, **77**, 615–630.

Gumpertz, M., Graham, J., & Ristaino, J. (1997) Autologistic model of spatial pattern of Phytophthora epidemic in bell pepper: Effects of soil variation on disease presence. *Journal of Agricultural, Biological and Environmental Statistics*, **2**, 131–156.

Gumpertz, M., Wu, C., & Pye, H. (2000) Logistic regression for southern pine beetle outbreaks with spatial and temporal autocorrelation. *Forest Science*, **46**, 95–107.

Guo, Q. H., Kelly, M., Gong, P., & Liu, D. S. (2007) An object-based classification approach in mapping tree mortality using high spatial resolution imagery. *GIScience & Remote Sensing*, **44**, 24–47.

Guo, Q. H., Kelly, M., & Graham, C. H. (2005) Support vector machines for predicting distribution of sudden oak death in California. *Ecological Modelling*, **182**, 75–90.

Gurevitch, J., Scheiner, S. M., & Fox, G. A. (2006) *The Ecology of Plants*, Sunderland, MA, Sinauer Associates.

Gustafson, E. J. (1998) Quantifying landscape spatial pattern: What is the state of the art? *Ecosystems*, **1**, 143–156.

Hagemeijer, W. J. M. & Blair, M. J. (1997) *The EBCC Atlas of European Breeding Birds. Their Distribution and Abundance*. London: Poyser.

Haining, R. P. (1990) *Spatial Data Analysis in the Social and Environmental Sciences*. Cambridge, UK: Cambridge University Press.

Hanley, J. A. & McNeil, B. J. (1982) The meaning and use of the area under a receiver operating characteristics curve. *Radiology*, **143**, 29–36.

Hannah, L. & Phillips, B. (2004) Extinction-risk coverage is worth inaccuracies. *Nature*, **430**, 141.

Hanski, I. (1999) *Metapopulation Ecology*. Oxford, UK: Oxford University Press.

Harrell, F. E. (2001) *Regression Modelling Strategies: with Applications to Linear Models, Logistic Regression and Survival Analysis*. New York: Springer.

Hastie, T. J. & Pregibon, D. (1992) Generalized linear models. In Chambers, J. & Hastie, T. (Eds.) In *Statistical Models in S*. Pacific Grove, CA: Wadsworth, pp. 195–247.

Hastie, T., Tibshirani, R. J., & Buja, A. (1994) Flexible discriminant analysis by optimal scoring. *Journal of the American Statistical Association*, **89**, 1255–1270.

Hastie, T., Tibshirani, R., & Friedman, J. (2009) *The Elements of Statistical Learning: Data Mining, Inference and Prediction*. 2nd edn., New York: Springer-Verlag.

Hastie, T., Tibshirani, R., & Friedman, J. (2001) *The Elements of Statistical Learning: Data Mining, Inference and Prediction*. New York: Springer-Verlag.

Hastie, T. J. & Tibshirani, R. J. (1990) *Generalized Additive Models*. London: Chapman and Hall.

Hathaway, S. A. (2000) An exploratory analysis of the biogeographic distribution of herpetofauna (reptiles and amphibians) and environmental variation in San Diego County using museum records and survey data. Unpublished Masters Thesis, Geography Department, San Diego State University.

Hawkins, B. A., Diniz-Filho, J. A. F., Bini, L. M., De Marco, P., & Blackburn, T. M. (2007) Red herrings revisited: spatial autocorrelation and parameter estimation in geographical ecology. *Ecography*, **30**, 375–384.

Hazin, H. & Erzini, K. (2008) Assessing swordfish distribution in the South Atlantic from spatial predictions. *Fisheries Research*, **90**, 45–55.

He, H. S. & Mladenoff, D. J. (1999) Spatially explicit and stochastic simulation of forest-landscape fire disturbance and succession. *Ecology*, **80**, 81–90.

Heath, L. S., Birdsey, R. A., & Williams, D. W. (2002) Methodology for estimating soil carbon for the forest carbon budget model of the United States, 2001. *Environmental Pollution*, **116**, 373–380.

Heikkinen, R. K., Luoto, M., Araújo, M. B., Virkkala, R., Thuiller, W., & Sykes, M. T. (2006) Methods and uncertainties in bioclimatic envelope modelling under climate change. *Progress in Physical Geography*, **30**, 751–777.

Heikkinen, R. K., Luoto, M., Virkkala, R., Pearson, R. G., & Korber, J. H. (2007) Biotic interactions improve prediction of boreal bird distributions at macro-scales. *Global Ecology and Biogeography*, **16**, 754–763.

Heikkinen, R. K., Luoto, M., Virkkala, R., & Rainio, K. (2004) Effects of habitat cover, landscape structure and spatial variables on the abundance of birds in an agricultural-forest mosaic. *Journal of Applied Ecology*, **41**, 824–835.

Hellgren, E. C., Bales, S. L., Gregory, M. S., Leslie, D. M., & Clark, J. D. (2007) Testing a Mahalanobis distance model of black bear habitat use in the Ouachita Mountains of Oklahoma. *Journal of Wildlife Management*, **71**, 924–928.

Henle, K., Lindenmayer, D. B., Margules, C. R., Saunders, D. A., & Wissel, C. (2004) Species survival in fragmented landscapes: where are we now? *Biodiversity and Conservation*, **13**, 1–8.

Hennig, C. & Hausdorf, B. (2006) Design of dissimilarity measures: a new dissimilarity between species distribution areas. In Batagelj, V., Bock, H.-H., Ferligoj, A., & Žiberna, A. (Eds.) *Data Science and Classification*. Berlin, Heidelberg: Springer, pp. 29–37.

Hepinstall, J. A., Krohn, W. B., & Sader, S. A. (2002) Effects of niche width on the performance and agreement of avian habitat models. In Scott, J. M., Heglund, P. J., Morrison, M. L. et al. (Eds.) Predicting Species Occurrences: Issues of Accuracy and Scale. Covelo, CA: Island Press, pp. 593–606.

Herborg, L. M., Jerde, C. L., Lodge, D. M., Ruiz, G. M., & Macisaac, H. J. (2007) Predicting invasion risk using measures of introduction effort and environmental niche models. Ecological Applications, 17, 663–674.

Hernandez, M., Miller, S. N., Goodrich, D. C. et al. (2000) Modeling runoff response to land cover and rainfall spatial variability in semi-arid watersheds. Environmental Monitoring and Assessment, 64, 285–298.

Hernandez, P. A., Franke, I., Herzog, S. K. et al. (2008) Predicting species distributions in poorly-studied landscapes. Biodiversity and Conservation, 17, 1353–1366.

Hernandez, P. A., Graham, C. H., Master, L. L., & Albert, D. L. (2006) The effect of sample size and species characteristics on performance of different species distribution modeling methods. Ecography, 29, 773–785.

Hershey, R. R. (2000) Modeling the spatial distribution of ten tree species in Pennsylvania. In Mowrer, H. T. & Congalton, R. G. (Eds.) Quantifying Spatial Uncertainty in Natural Resources: Theory and Applications for GIS and Remote Sensing. Chelsea, MI: Ann Arbor Press, pp. 119–136.

Hestir, E. L., Khanna, S., Andrew, M. E. et al. (2008) Identification of invasive vegetation using hyperspectral remote sensing in the California Delta ecosystem. Remote Sensing of Environment, 112, 4034–4047.

Hijmans, R. J. & Graham, C. H. (2006) The ability of climate envelope models to predict the effect of climate change on species distributions. Global Change Biology, 12, 2272–2281.

Hijmans, R. J., Cameron, S. E., Parra, J. L., Jones, P. G., & Jarvis, A. (2005) Very high resolution interpolated climate surfaces for global land areas. International Journal of Climatology, 25, 1965–1978.

Hijmans, R. J., Guarino, L., Cruz, M., & Rojas, E. (2001) Computer tools for spatial analysis of plant genetic resources data: 1. DIVA-GIS. Plant Genetic Resources Newsletter, 127, 15–19.

Hilbert, D. W. & Ostendorf, B. (2001) The utility of artificial neural networks for modelling the distribution of vegetation in past, present and future climates. Ecological Modelling, 146, 311–327.

Hilbert, D. W. & van den Muyzenberg, J. (1999) Using an artificial neural network to characterise the relative suitability of environments for forest types in a complex tropical vegetation mosaic. Diversity and Distributions, 5, 263–274.

Hill, M. O., Roy, D. B., Mountford, J. O., & Bunce, R. G. H. (2000) Extending Ellenberg's indicator values to a new area: an algorithmic approach. Journal of Applied Ecology, 37, 3–15.

Hirzel, A. & Guisan, A. (2002) Which is the optimal sampling strategy for habitat suitability modelling? Ecological Modelling, 157, 331–341.

Hirzel, A. H. & Arlettaz, R. (2003) Modeling habitat suitability for complex species distributions by environmental-distance geometric mean. Environmental Management, 32, 614–623.

Hirzel, A. H. & Le Lay, G. (2008) Habitat suitability modelling and niche theory. Journal of Applied Ecology, 45, 1372–1381.

Hirzel, A. H., Hausser, J., Chessel, D., & Perrin, N. (2002) Ecological-niche factor analysis: how to compute habitat-suitability maps without absence data? *Ecology*, **83**, 2027–2036.

Hirzel, A. H., Helfer, V., & Métral, F. (2001) Assessing habitat-suitability models with a virtual species. *Ecological Modelling*, **145**.

Hirzel, A. H., Le Lay, G., Helfer, V., Randin, C., & Guisan, A. (2006) Evaluating the ability of habitat suitability models to predict species presences. *Ecological Modelling*, **199**, 142–152.

Hirzel, A. H., Posse, B., Oggier, P. A., Crettenand, Y., Glenz, C., & Arlettaz, R. (2004) Ecological requirements of reintroduced species and the implications for release policy: the case of the bearded vulture. *Journal of Applied Ecology*, **41**, 1103–1116.

Hobbs, N. T. (2003) Challenges and opportunities in integrating ecological knowledge across scales. *Forest Ecology and Management*, **181**, 223–238.

Hoeting, J. A., Davis, R. A., Merton, A. A., & Thompson, S. E. (2006) Model selection for geostatistical models. *Ecological Applications*, **16**, 87–98.

Hoeting, J. A., Madigan, D., Raftery, A. E., & Volinsky, C. T. (2000) Bayesian model averaging: a tutorial. *Statistical Science*, **14**, 382–417.

Hoffer, R. M. E. A. (1975) Natural resource mapping in mountainous terrain by computer analysis of ERTS-1 satellite data, Purdue University, W. Lafayette, In *Agricultural Experimental Station Research Bulletin* 919, and LARS Contract Report 061575.

Holdridge, L. (1947) Determination of world plant formations from simple climatic data. *Science*, **105**, 367–368.

Holling, C. S. (1992) Cross-scale morphology, geometry, and dynamics of ecosystems. *Ecological Monographs*, **62**, 447–502.

Homann, P. S., Sollins, P., Fiorella, M., Thorson, T., & Kern, J. S. (1998) Regional soil organic carbon storage estimates for western Oregon by multiple approaches. *Soil Science Society of America Journal*, **62**, 789–796.

Hooten, M. B., Larsen, D. R., & Wikle, C. K. (2003) Predicting the spatial distribution of ground flora on large domains using a hierarchical Bayesian model. *Landscape Ecology*, **18**, 487–502.

Hosmer, D. & Lemeshow, S. (2000) *Applied Logistic Regression*, New York, Wiley.

Howell, J. E., Peterson, J. T., & Conroy, M. J. (2008) Building hierarchical models of avian distributions for the state of Georgia. *Journal of Wildlife Management*, **72**, 168–178.

Huang, C. Y. & Geiger, E. L. (2008) Climate anomalies provide opportunities for large-scale mapping of non-native plant abundance in desert grasslands. *Diversity and Distributions*, **14**, 875–884.

Hubbell, S. P. (2001) *The Unified Neutral Theory of Biodiversity and Biogeography*. Princeton: Princeton University Press.

Huberty, C. J. (1994) *Applied Discriminant Analysis*. New York, USA: Wiley Interscience.

Hugall, A., Moritz, C., Moussalli, A., & Stanisic, J. (2002) Reconciling paleodistribution models and comparative phylogeography in the Wet Tropics rainforest land snail *Gnarosophia bellendenkerensis* (Brazier 1875). *Proceedings of the National Academy of Sciences of the United States of America*, **99**, 6112–6117.

Hunsaker, C. T., Nisbet, R. A., Lam, D. C. L. *et al.* (1993) Spatial models of ecological systems and processes: the role of GIS. In Goodchild, M. F., Parks, B. O., & Steyaert, L. T. (Eds.) *Environmental Modeling with GIS*. New York, NY: Oxford University Press, pp. 248–264.

Hunter, G. J. & Goodchild, M. F. (1997) Modeling the uncertainty of slope and aspect estimates derived from spatial databases. *Geographical Analysis*, **29**, 35–49.

Huntley, B. & Webb, T., III (1988) *Vegetation History*. Dordrecht: Kluwer.

Huntley, B., Bartlein, P. J. & Prentice, I. C. (1989) Climate control of the distribution and abundance of beech (*Fagus* L.) in Europe and North America. *Journal of Biogeography*, **16**, 551–560.

Huntley, B., Green, R. E., Collingham, Y. C. *et al.* (2004) The performance of models relating species geographical distributions to climate is independent of trophic level. *Ecology Letters*, **7**, 417–426.

Hurlbert, A. H. & Haskell, J. P. (2003) The effect of energy and seasonality on avian species richness and community composition. *American Naturalist*, **161**, 83–97.

Hurtt, G. C., Rosentrater, L., Frolking, S., & Moore, B., III, (2001) Linking remote-sensing estimates of land cover and census statistics on land use to produce maps of land use of the conterminous United States. *Global Biogeochemical Cycles*, **15**, 673–685.

Huston, M. A. (2002) Introductory essay: critical issues for improving predictions. In Scott, J. M., Heglund, P. J., Morrison, M. L. *et al.* (Eds.) *Predicting Species Occurrences: Issues of Accuracy and Scale*. Covelo, CA: Island Press, pp. 7–21.

Hutchinson, C. F. (1982) Techniques for combining Landsat and ancillary data for digital classification improvement. *Photogrammetric Engineering and Remote Sensing*, **48**, 123–130.

Hutchinson, M. F. (1987) Methods for generation of weather sequences. In Bunting, A. H. (Ed.) *Agricultural environments: characterisation, classification and mapping*. Wallingford: CAB, International, 149–157.

Hutchinson, M. F. (1996) Thin plate spline interpolation of mean rainfall. In Goodchild, M. F., Steyaert, L. T. & Park, B. O. (Eds.) *GIS and Environmental Modeling: Progress and Research Issues*. Fort Collins, CO: GIS World Books, pp. 85–90.

Hutchinson, M. F. & Gallant, J. (2000) Digital elevation models and representation of terrain shape. In Wilson, J. & Gallant, J. (Eds.) *Terrain Analysis: Principles and Applications*. New York: John Wiley & Sons, pp. 29–50.

Hutchinson, M. F. & Gessler, P. E. (1994) Splines – more than just a smooth interpolator. *Geoderma*, **62**, 45–67.

Ibáñez, I., Silander, J. A., Jr., Wilson, A. M., Lafleur, N., Tanaka, N., & Tsuyama, I. (2009) Multivariate forecasts of potential distributions of invasive plant species. *Ecological Applications*, **19**, 359–375.

Inglis, G. J., Huren, H., Oldman, J., & Haskew, R. (2006) Using habitat suitability index and particle dispersion models for early detection of marine invaders. *Ecological Applications*, **16**, 1377–1390.

Iverson, L. R. & Prasad, A. M. (1998) Predicting abundance of 80 tree species following climate change in the eastern United States. *Ecological Monographs*, **68**, 465–485.

Iverson, L. R., Prasad, A. M., Matthews, S. N., & Peters, M. (2008) Estimating potential habitat for 134 eastern US tree species under six climate scenarios. *Forest Ecology and Management*, **254**, 390–406.

Iverson, L. R., Prasad, A., & Schwartz, M. W. (1999) Modeling potential future individual tree-species distributions in the eastern United States under a climate change scenario: a case study with *Pinus virginiana*. *Ecological Modelling*, **115**, 77–93.

Iverson, L. R., Schwartz, M. W., & Prasad, A. M. (2004a) How fast and far might tree species migrate in the eastern United States due to climate change? *Global Ecology and Biogeography*, **13**, 209–219.

Iverson, L. R., Schwartz, M. W., & Prasad, A. M. (2004b) Potential colonization of newly available tree-species habitat under climate change: an analysis for five eastern US species. *Landscape Ecology*, **19**, 787–799.

Jaberg, C. & Guisan, A. (2001) Modeling the influence of landscape structure on bat species distribution and community composition in the Swiss Jura Mountains. *Journal of Applied Ecology*, **38**, 1169–1181.

Jackson, S. T. & Overpeck, J. T. (2000) Responses of plant populations and communities to environmental changes of the late Quaternary. *Paleobiology*, **26**.

Jacobs, B. F., Romme, W. H., & Allen, C. D. (2008) Mapping "old" vs. "young" piñon-juniper stands with a predictive topo-climatic model. *Ecological Applications*, **18**, 1627–1641.

Jeffers, J. N. R. (1999) Genetic algorithms I: Ecological applications of the Beagle and Gafer genetic algorithms. In Fielding, A. H. (Ed.) *Machine Learning Methods for Ecological Applications*. Dordrecht: Kluwer Academic Publishing, pp. 107–121.

Jenny, H. (1941) *Factors of Soil Formation*. New York: McGraw Hill.

Jenson, S. K. & Domingue, J. O. (1988) Extracting topographic structure from digital elevation data for geographic information system analysis. *Photogrammetric Engineering & Remote Sensing*, **54**, 1593–1600.

Jetz, W. & Rahbek, C. (2002) Geographic range size and determinants of avian species richness. *Science*, **297**, 1548–1551.

Jetz, W., Rahbek, C. & Lichstein, J. W. (2005) Local and global approaches to spatial data analysis in ecology. *Global Ecology and Biogeography*, **14**, 97–98.

Jiménez-Valverde, A., Lobo, J. M., & Hortal, J. (2008) Not as good as they seem: the importance of concepts in species distribution modeling. *Diversity and Distributions*, **14**, 885–890.

Jiménez-Valverde, A., Ortuno, V. M., & Lobo, J. M. (2007) Exploring the distribution of *Sterocorax Ortuno*, 1990 (Coleoptera, Carabidae) species in the Iberian peninsula. *Journal of Biogeography*, **34**, 1426–1438.

Johnson, C. J. & Gillingham, M. P. (2005) An evaluation of mapped species distribution models used for conservation planning. *Environmental Conservation*, **32**, 117–128.

Johnson, C. M. & Krohn, W. B. (2002) Dynamics patterns of associations between environmental factors and island use by breeding seabirds. In Scott, J. M., Heglund, P. J., Morrison, M. L. *et al*. (Eds.) *Predicting Species Occurrences: Issues of Scale and Accuracy*. Covelo, CA: Island Press, pp. 171–182.

Johnson, C. J., Nielsen, S. E., Merrill, E. H., McDonald, T. L., & Boyce, M. S. (2006) Resource selection functions based on use-availability data: Theoretical motivation and evaluation methods. *Journal of Wildlife Management*, **70**, 347–357.

Johnson, C. M., Johnson, L. B., Richards, C., & Beasley, V. (2002) Predicting the occurrence of amphibians: an assessment of multiple-scale models. In Scott, J. M., Heglund, P. J., Morrison, M. L. *et al.* (Eds.) *Predicting Species Occurrences: Issues of Accuracy and Scale*. Covelo, CA: Island Press, pp. 157–170.

Johnson, D. H. (1980) The comparison of usage and availability measurements for evaluating resource preference. *Ecology*, **61**, 65–71.

Jones, J. P. & Cassetti, E. (1992) *Applications of the Expansion Method*. London: Routledge.

Jones, M. M., Olivas Rojas, P., Tuomisto, H., & Clark, D. B. (2007) Environmental and neighborhood effects on tree fern distributions in a neotropical lowland rain forest. *Journal of Vegetation Science*, **18**, 13–24.

Jones, M. T., Niemi, G. J., Hanowski, J. M., & Regal, R. R. (2002) Poisson regression: a better approach to modeling abundance data? In Scott, J. M., Heglund, P. J., Morrison, M. L. *et al.* (Eds.) *Predicting Species Occurrences: Issues of Accuracy and Scale*. Covelo, CA: Island Press, pp. 411–418.

Justice, C. O., Vermote, E., Townshend, J. R. G. *et al.* (1998) The Moderate Resolution Imaging Spectroradiometer (MODIS): land remote sensing for global change research. *IEEE Transactions on Geoscience and Remote Sensing*, **36**, 1228–1249.

Kadmon, R., Farber, O., & Danin, A. (2003) A systematic analysis of factors affecting the performance of climatic envelope models. *Ecological Applications*, **13**, 853–867.

Kadmon, R., Farber, O., & Danin, A. (2004) Effect of roadside bias on the accuracy of predictive maps produced by bioclimatic models. *Ecological Applications*, **14**, 401–413.

Kambhampati, S. & Peterson, A. T. (2007) Ecological niche conservation and differentiation in the wood-feeding cockroaches, *Cryptocercus*, in the United States. *Biological Journal of the Linnean Society*, **90**, 457–466.

Karl, J. W., Svancara, L. K., Heglund, P. J., Wright, N. M., & Scott, J. M. (2002) Species commonness and the accuracy of habitat-relationship models. In Scott, J. M., Heglund, P. J., Morrison, M. L. *et al.* (Eds.) *Predicting Species Occurrences: Issues of Accuracy and Scale*. Covelo, CA: Island Press, pp. 573–580.

Kearney, M. (2006) Habitat, environment and niche: what are we modelling? *Oikos*, **115**, 186–191.

Keating, K. A. & Cherry, S. (2004) Use and interpretation of logistic regression in habitat selection studies. *Journal of Wildlife Management*, **68**, 774–789.

Keith, D. A., Akçakaya, H. R., Thuiller, W. *et al.* (2008) Predicting extinction risks under climate change: coupling stochastic population models with dynamic bioclimatic habitat models. *Biology Letters*, **4**, 560–563.

Keitt, T. H., Bjørnstad, O. N., Dixon, P. M., & Citron-Pousty, S. (2002) Accounting for spatial pattern when modeling organism–environment interactions. *Ecography*, **25**, 616–625.

Kelly, M. & Meentemeyer, R. K. (2002) Landscape dynamics of the spread of sudden oak death. *Photogrammetric Engineering and Remote Sensing*, **68**, 1001–1009.

Kelly, M., Allen-Diaz, B., & Kobzina, N. (2005) Digitization of a historic dataset: the Wieslander California vegetation type mapping project. *Madroño*, **52**, 191–201.

Kelly, M., Guo, Q., Liu, D., & Shaari, D. (2007) Modeling the risk for a new invasive forest disease in the United States: An evaluation of five environmental niche models. *Computers Environment and Urban Systems*, **31**, 689–710.

Kerr, J. T. & Ostrovsky, M. (2003) From space to species: ecological applications for remote sensing. *Trends in Ecology & Evolution*, **16**, 299–305.

Kerr, J. T., Southwood, T. R. E., & Cihlar, J. (2001) Remotely sensed habitat diversity predicts butterfly species richness and community similarity. *Proceedings of the National Academy of Science, USA*, **98**, 11365–11370.

Kessell, S. R. (1976) Gradient modeling: a new approach to fire modeling and wilderness resource management. *Environmental Management*, **1**, 39–48.

Kessell, S. R. (1978) Forum: perspectives in fire research. *Environmental Management*, **2**, 291–312.

Kessell, S. R. (1979) *Gradient Modeling: Resource and Fire Management*. New York: Springer-Verlag.

Kindvall, O., Bergman, K.-O., Akçakaya, H. R. *et al.* (2004) Woodland brown butterfly (*Lopinga achine*) in Sweden: viability in a dynamic landscape maintained by grazing. *Species Conservation and Management: Case Studies*, 171–178.

Kissling, W. D. & Carl, G. (2008) Spatial autocorrelation and the selection of simultaneous autoregressive models. *Global Ecology and Biogeography*, **17**, 59–71.

Knick, S. T. & Dyer, D. L. (1997) Spatial distribution of black-tailed jackrabbit habitat determined by GIS in southwestern Idaho. *Journal of Wildlife Management*, **61**, 75–85.

Knick, S. T. & Rotenberry, J. T. (1998) Limitations to mapping habitat areas in changing landscapes using the Mahalanobis distance statistic. *Journal of Agricultural Biological and Environmental Statistics*, **3**, 311–322.

Knowles, L. L., Carstens, B. C., & Keat, M. L. (2007) Coupling genetic and ecological-niche models to examine how past population distributions contribute to divergence. *Current Biology*, **17**, 940–946.

Köppen, W. P. (1923) *Die Klimate der Erde : Grundriss der Klimakunde*. Berlin; Leipzig: Walter de Gruyter & So.

Kozak, K. H. & Wiens, J. J. (2006) Does niche conservatism promote speciation? A case study in North American salamanders. *Evolution*, **60**, 2604–2621.

Kozak, K. H., Graham, C. H., & Wiens, J. J. (2008) Integrating GIS-based environmental data into evolutionary biology. *Trends in Ecology & Evolution*, **23**, 141–148.

Kremen, C., Cameron, A., Moilanen, A. *et al.* (2008) Aligning conservation priorities across taxa in Madagascar with high-resolution planning tools. *Science*, **320**, 222–226.

Köchler, A. W. (1973) Problems in classifying and mapping vegetation for ecological regionalization. *Ecology*, **54**, 512–523.

Köchler, A. W. & Zonneveld, I. S. (Eds.) (1998) *Vegetation Mapping*. Boston: Kluwer Academic Publishers.

Kueppers, L. M., Snyder, M. A., Sloan, L. C., Zavaleta, E. S., & Fulfrost, B. (2005) Modeled regional climate change and California endemic oak ranges. *Proceedings of the National Academy of Sciences of the United States of America*, **102**, 16281–16286.

Kueppers, L. M., Zavaleta, E., Fulfrost, B., Snyder, M. A., & Sloan, L. C. (2006) Constrained range expansion and climate change assessments – The authors reply. *Frontiers in Ecology and the Environment*, **4**, 179–179.

Kühn, I. (2007) Incorporating spatial autocorrelation may invert observed patterns. *Diversity and Distributions*, **13**, 66–69.

Kuhnert, P. M., Do, K.-A., & McClure, R. (2000) Combining non-parametric models with logistic regression: an application to motor vehicle injury data. *Computational Statistics and Data Analysis*, **34**, 371–386.

Kupfer, J. A. & Farris, C. A. (2007) Incorporating spatial non-stationarity of regression coefficients into predictive vegetation models. *Landscape Ecology*, **22**, 837–852.

Kyriakidis, P. C. & Dungan, J. L. (2001) A geostatistical approach for mapping thematic classification accuracy and evaluating the impact of inaccurate spatial data on ecological model predictions. *Environmental and Ecological Statistics*, **8**, 311–330.

Ladle, R. J., Jepson, P., Araújo, M. B., & Whittaker, T. J. (2004) Dangers of crying wolf over risk of extinctions. *Nature*, **428**, 799–799.

La Morgia, V., Bona, F., & Badino, G. (2008) Bayesian modelling procedures for the evaluation of changes in wildlife habitat suitability: a case study of roe deer in the Italian Alps. *Journal of Applied Ecology*, **45**, 863–872.

Lane, S. N., Richards, K. S., & Chandler, J. H. (1998) *Landform Monitoring, Modelling and Analysis*. Chichester, UK: John Wiley and Sons.

Larson, M. A., Thompson, F. R., Millspaugh, J. J., Dijak, W. D., & Shifley, S. R. (2004) Linking population viability, habitat suitability, and landscape simulation models for conservation planning. *Ecological Modelling*, **180**, 103–118.

Lassalle, G., Beguer, M., Beaulaton, L., & Rochard, E. (2008) Diadromous fish conservation plans need to consider global warming issues: an approach using biogeographical models. *Biological Conservation*, **141**, 1105–1118.

Lassau, S. A. & Hochuli, D. F. (2007) Associations between wasp communities and forest structure: do strong local patterns hold across landscapes? *Austral Ecology*, **32**, 656–662.

Lassueur, T., Joost, S., & Randin, C. F. (2006) Very high resolution digital elevation models: do they improve models of plant species distribution? *Ecological Modelling*, **198**, 139–153.

Lathrop, R. G., Aber, J. D., & Bognar, J. A. (1995) Spatial variability of digital soil maps and its impact on regional ecosystem modeling. *Ecological Modelling*, **82**, 1–10.

Latifovic, R., Zhu, Z.-L., Cihlar, J., & Giri, C. (2002) Land cover of North America 2000. Natural Resources Canada, Canada Center for Remote Sensing. US Geological Survey, EROS Data Center.

Latimer, A. M., Wu, S. S., Gelfand, A. E., & Silander, J. A. (2006) Building statistical models to analyze species distributions. *Ecological Applications*, **16**, 33–50.

Laurent, E. J., Shi, H. J., Gatziolis, D., Lebouton, J. P., Walters, M. B., & Liu, J. G. (2005) Using the spatial and spectral precision of satellite imagery to predict wildlife occurrence patterns. *Remote Sensing of Environment*, **97**, 249–262.

Lawler, J. J., White, D., Neilson, R. P., & Blaustein, A. R. (2006) Predicting climate-induced range shifts: model differences and model reliability. *Global Change Biology*, **12**, 1568–1584.

Lawler, J. J., Shafer, S. L., White, D. (2009) Projected climate-induced faunal changes in the Western Hemisphere. *Ecology*, **90**, 588–597.

Leathwick, J. R. (1995) Climatic relationships of some New Zealand forest tree species. *Journal of Vegetation Science*, **6**, 237–248.

Leathwick, J. R. (1998) Are New Zealand's *Nothofagus* species in equilibrium with their environment? *Journal of Vegetation Science*, **9**, 719–732.

Leathwick, J. R. (2001) New Zealand's potential forest pattern as predicted from current species-environment relationships. *New Zealand Journal of Botany*, **39**, 447–464.

Leathwick, J. R. & Austin, M. P. (2001) Competitive interactions between tree species in New Zealand's old-growth indigenous forests. *Ecology*, **82**, 2560–2573.

Leathwick, J. R., Elith, J., Francis, M. P., Hastie, T., & Taylor, P. (2006a) Variation in demersal fish species richness in the oceans surrounding New Zealand: an analysis using boosted regression trees. *Marine Ecology-Progress Series*, **321**, 267–281.

Leathwick, J. R., Elith, J., & Hastie, T. (2006b) Comparative performance of generalized additive models and multivariate adaptive regression splines for statistical modelling of species distributions. *Ecological Modelling*, **199**, 188–196.

Leathwick, J. R., Rowe, D., Richardson, J., Elith, J., & Hastie, T. (2005) Using multivariate adaptive regression splines to predict the distributions of New Zealand's freshwater diadromous fish. *Freshwater Biology*, **50**, 2034–2052.

Lee, D. R. & Sallee, G. T. (1970) A method of measuring shape. *Geographical Review*, **60**, 555–563.

Lee, M., Fahrig, L., Freemark, K., & Currie, D. J. (2002) Importance of patch scale vs landscape scale on selected forest birds. *Oikos*, **96**, 110–118.

Lee, S. & Choi, J. (2004) Landslide susceptibility mapping using GIS and the weight-of-evidence model. *International Journal of Geographical Information Science*, **18**, 789–814.

Lees, B. G. & Ritman, K. (1991) Decision-tree and rule-induction approach to integration of remotely sensed and GIS data in mapping vegetation in disturbed or hilly environments. *Environmental Management*, **15**, 823–831.

Lefsky, M. A., Cohen, W. B., & Spies, T. A. (2001) An evaluation of alternate remote sensing products for forest inventory, monitoring, and mapping of Douglas-fir forests in western Oregon. *Canadian Journal of Forest Research*, **31**, 78–87.

Legates, D. R. & McCabe, G. J., Jr. (1998) Evaluating the use of "goodness-of-fit" measures in hydrologic and hydroclimatic model validation. *Water Resources Research*, **35**, 233.

Legendre, P. (1993) Spatial autocorrelation: problem or new paradigm. *Ecology*, **74**, 1659–1673.

Legendre, P. & Fortin, M. J. (1989) Spatial pattern and ecological analysis. *Vegetatio*, **80**, 107–138.

Legendre, P. & Legendre, L. (1998) *Numerical Ecology*. Amsterdam: Elsevier.

Legendre, P., Borcard, D., & Peres-Neto, P. R. (2005) Analyzing beta diversity: partitioning the spatial variation of community composition data. *Ecological Monographs*, **75**, 435–450.

Lehmann, A. (1998) GIS modeling of submerged macrophyte distribution using Generalized Additive Models. *Plant Ecology*, **139**, 113–124.

Lehmann, A., Overton, J. M., & Leathwick, J. R. (2002) GRASP: generalized regression analysis and spatial prediction. *Ecological Modelling*, **157**, 189–207.

Lek, S. & Guégan, J.-F. (2000) *Artificial Neural Networks: Application to Ecology and Evolution.* Berlin: Springer-Verlag.

Le Maitre, D. C., Thuiller, W., & Schonegevel, L. (2008) Developing an approach to defining the potential distributions of invasive plant species: a case study of *Hakea* species in South Africa. *Global Ecology and Biogeography*, **17**, 569–584.

Lenihan, J. M. (1993) Ecological response surfaces for North American boreal tree species and their use in forest classification. *Journal of Vegetation Science*, **4**, 667–680.

Lenihan, J. M. & Neilson, R. (1993) A rule-based vegetation formation model for Canada. *Journal of Biogeography*, **20**, 615–628.

Lennon, J. J. (2002) Red shifts and red herrings in geographical ecology. *Ecography*, **23**, 101–113.

Levin, N., Shmida, A., Levanoni, O., Tamari, H., & Kark, S. (2007) Predicting mountain plant richness and rarity from space using satellite-derived vegetation indices. *Diversity and Distributions*, **13**, 692–703.

Levin, S. A. (1992) The problem of pattern and scale in ecology. *Ecology*, **73**, 1943–1967.

Levine, R. S., Peterson, A. T., & Benedict, M. Q. (2004) Geographic and eco-logic distributions of the *Anopheles gambiae* complex predicted using a genetic algorithm. *American Journal of Tropical Medicine and Hygiene*, **70**, 105–109.

Leyequien, E., Verrelst, J., Slot, M., Schaepman-Strub, G., Heitkonig, I. M. A., & Skidmore, A. (2007) Capturing the fugitive: applying remote sensing to terrestrial animal distribution and diversity (vol 9, pg 1, 2007). *International Journal of Applied Earth Observation and Geoinformation*, **9**, 224–224.

Li, J., Hilbert, D. W., Parker, T., & Williams, S. (2009) How do species respond to climate change along an elevation gradient? A case study of the grey-headed robin (*Heteromyias albispecularis*). *Global Change Biology*, **15**, 255–267.

Li, W. J., Wang, Z. J., Ma, Z. J., & Tang, H. X. (1999) Designing the core zone in a biosphere reserve based on suitable habitats: Yancheng Biosphere Reserve and the red crowned crane (*Grus japonensis*). *Biological Conservation*, **90**, 167–173.

Lichstein, J. W., Simons, T. R., Shriner, S. A., & Franzberg, K. E. (2002) Spatial autocorrelation and autoregressive models in ecology. *Ecological Monographs*, **72**, 445–463.

Lindenmayer, D. B. & Possingham, H. P. (1996) Modelling the inter-relationships between habitat patchiness, dispersal capability and metapopulation persistence of the endangered species, Leadbeater's possum, in south-eastern Australia. *Landscape Ecology*, **11**, 79–105.

Lindenmayer, D. B., McCarthy, M. A., Parris, K. M., & Pope, M. L. (2000) Habitat fragmentation, landscape context, and mammalian assemblages in southeastern Australia. *Journal of Mammalogy*, **81**, 787–797.

Linderman, M. A., Liu, J., Qi, J. *et al.* (2004) Using artificial neural networks to map the spatial distribution of understory bamboo from remotely sensed data. *International Journal of Remote Sensing*, **25**, 1685–1700.

Lippitt, C. D., Rogan, J., Toledano, J. et al. (2008) Incorporating anthropogenic variables into a species distribution model to map gypsy moth risk. Ecological Modelling, 210, 339–350.

Liu, C. R., Berry, P. M., Dawson, T. P., & Pearson, R. G. (2005) Selecting thresholds of occurrence in the prediction of species distributions. Ecography, 28, 385–393.

Lobo, J. M. (2008) More complex distribution models or more representative data? Biodiversity Informatics, 5, 14–19.

Lobo, J. M., Jay-Robert, P., & Lumaret, J.-P. (2004) Modelling the species richness distribution for French Aphodiidae (Coleoptera, Scarabaeoidea). Ecography, 27, 145–156.

Lobo, J. M., Jimenez-Valverde, A., & Real, R. (2008) AUC: a misleading measure of the performance of predictive distribution models. Global Ecology and Biogeography, 17, 145–151.

Lobo, J. M., Lumaret, J.-P., & Jay-Robert, P. (2002) Modelling the species richness of French dung beetles (Coleoptera, Scarabaeidae) and delimiting the predictive capacity of different groups of explanatory variables. Global Ecology & Biogeography, 11, 265–277.

Lohse, K. A., Newburn, D. A., Opperman, J. J., & Merenlender, A. M. (2008) Forecasting relative impacts of land use on anadromous fish habitat to guide conservation planning. Ecological Applications, 18, 467–482.

Loiselle, B. A., Jorgensen, P. M., Consiglio, T. et al. (2008) Predicting species distributions from herbarium collections: does climate bias in collection sampling influence model outcomes? Journal of Biogeography, 35, 105–116.

Long, P. R., Zefania, S., Ffrench-Constant, R. H., & Szekely, T. (2008) Estimating the population size of an endangered shorebird, the Madagascar plover, using a habitat suitability model. Animal Conservation, 11, 118–127.

Loo, S. E., Mac Nally, R., & Lake, P. S. (2007) Forecasting New Zealand mudsnail invasion range: model comparisons using native and invaded ranges. Ecological Applications, 17, 181–189.

Lopez-Lopez, P., Garcia-Ripolles, C., Soutullo, A., Cadahia, L., & Urios, V. (2007) Identifying potentially suitable nesting habitat for golden eagles applied to 'important bird areas' design. Animal Conservation, 10, 208–218.

Louzao, M., Hyrenbach, K. D., Arcos, J. M., Abello, P., De Sola, L. G., & Oro, D. (2006) Oceanographic habitat of an endangered Mediterranean procellariiform: implications for marine protected areas. Ecological Applications, 16, 1683–1695.

Loveland, T. R. & Ohlen, D. O. (1993) Experimental AVHRR land data sets for environmental monitoring and modeling. In Goodchild, M. F., Parks, B. O., & Steyaert, L. T. (Eds.) Environmental Modeling with GIS. New York: Oxford University Press.

Ludeke, A. K., Maggio, R. C., & Reid, L. M. (1990) An analysis of anthropogenic deforestation using logistic regression and GIS. Journal of Environmental Management, 31, 247–259.

Luoto, M., Poyry, J., Heikkinen, R. K., & Saarinen, K. (2005) Uncertainty of bioclimate envelope models based on the geographical distribution of species. Global Ecology and Biogeography, 14, 575–584.

Lutolf, M., Kienast, F., & Guisan, A. (2006) The ghost of past species occurrence: improving species distribution models for presence-only data. *Journal of Applied Ecology*, **43**, 802–815.

Lynn, H., Mohler, C. L., DeGloria, S. D., & McCulloch, C. E. (1995) Error assessment in decision-tree models applied to vegetation analysis. *Landscape Ecology*, **10**, 323–335.

Lytle, D. J. (1993) Digital soils databases for the United States. In Goodchild, M. F., Parks, B. O., & Steyaert, L. T. (Eds.) *Environmental Modeling with GIS*. New York: Oxford University Press, pp. 386–391.

Lytle, D. J., Bliss, N. B. & Waltman, S. W. (1996) Interpreting the State Soil Geographic Database (STATSGO). In Goodchild, M. F., Steyaert, L. T., Parks, B. O. *et al.* (Eds.) *GIS and Environmental Modeling: Progress and Research Issues*. Fort Collins, CO: GIS World Books, pp. 49–52.

MacKenzie, D. I., Nichols, J. D., Lachman, G. B., Droege, S., Royle, A., & Langtimm, C. A. (2002) Estimating site occupancy rates when detection probabilities are less than one. *Ecology*, **83**, 2248–2255.

MacKenzie, D. I., Nichols, J. D., Sutton, N., Kawanishi, K., & Bailey, L. L. (2005) Improving inferences in population studies of rare species that are detected imperfectly. *Ecology*, **86**, 1101–1113.

Mackey, B. G. (1993a) Predicting the potential distribution of rain-forest structural characteristics. *Journal of Vegetation Science*, **4**, 43–54.

Mackey, B. G. (1993b) A spatial analysis of the environmental relations of rainforest structural types. *Journal of Biogeography*, **20**, 303–336.

Mackey, B. G. & Lindenmayer, D. B. (2001) Towards a hierarchical framework for modeling the spatial distribution of animals. *Journal of Biogeography*, **28**, 1147–1166.

Mackey, B. G. & Sims, R. A. (1993) A climatic analysis of selected boreal tree species, and potential responses to global climate change. *World Resource Review*, **5**, 469–487.

Maindonald, J. & Braun, J. (2003) *Data Analysis and Graphics Using R – An Example-based Approach*. Cambridge, UK: Cambridge University Press.

Manel, S., Dias, J. M., Buckton, S. T., & Ormerod, S. J. (1999a) Alternative methods for predicting species distribution: an illustration with Himalayan river birds. *Journal of Applied Ecology*, **36**, 734–747.

Manel, S., Dias, J. M., & Ormerod, S. J. (1999b) Comparing discriminant analysis, neural networks and logistic regression for predicting species distributions: a case study with a Himalayan river bird. *Ecological Modelling*, **120**, 337–347.

Manel, S., Williams, H. C., & Ormerod, S. J. (2001) Evaluating presence–absence models in ecology: the need to account for prevalence. *Journal of Applied Ecology*, **38**, 921–931.

Manly, B. F. J., McDonald, L. L., Thomas, D. L., McDonald, T. L., & Erickson, W. (2002) *Resource Selection by Animals: Statistical Design and Analysis for Field Studies*. New York, NY, USA: Kluwer Press.

Mann, L. K., King, A. W., Dale, V. H. *et al.* (1999) The role of soil classification in geographic information system modeling of habitat pattern: threatened calcareous ecosystems. *Ecosystems*, **2**, 524–538.

Margules, C. R. & Austin, M. P. (1994) Biological models for monitoring species decline: the construction and use of databases. *Philosophical Transactions of the Royal Society of London B*, **344**, 69–75.

Margules, C. R., Nicholls, A. O., & Austin, M. P. (1987) Diversity of *Eucalyptus* species predicted by a multivariable environmental gradient. *Oecologia*, **71**, 229–232.

Mark, D. M. (1979) Phenomenon-based data-structuring and digital terrain modeling. *Geo-Processing*, **1**, 27–36.

Marks, D., Dozier, J. & Frew, J. (1984) Automated basin delineation from digital terrain data. *Geo Processing*, **2**, 299–311.

Marmion, M., Parviainen, M., Luoto, M., Heikkinen, R. K., & Thuiller, W. (2009) Evaluation of consensus methods in predictive species distribution modelling. *Diversity and Distributions*, **15**, 59–69.

Martínez-Meyer, E. & Peterson, A. T., (2006) Conservatism of ecological niche characteristics in North American plant species over the Pleistocene-to-Recent transition. *Journal of Biogeography*, **33**, 1779–1789.

Martínez-Meyer, E., Peterson, A. T., & Hargrove, W. W. (2004) Ecological niches as stable distributional constraints on mammal species, with implications for Pleistocene extinctions and climate change projections for biodiversity. *Global Ecology and Biogeography*, **13**, 305–314.

Martínez-Meyer, E., Peterson, A. T., Servin, J. I., & Kiff, L. F. (2006) Ecological niche modelling and prioritizing areas for species reintroductions. *Oryx*, **40**, 411–418.

Massolo, A., Della Stella, R. M., & Meriggi, A. (2007) Zoning and wild boar management: a multi-criteria approach to planning. *Hystrix*, **18**, 57–68.

Mastrorillo, S., Lek, S., Dauba, F., & Belaud, A. (1997) The use of artificial neural networks to predict the presence of small-bodied fish in a river. *Freshwater Biology*, **38**, 237–246.

Mathew, J., Jha, V. K., & Rawat, G. S. (2007) Weights of evidence modelling for landslide hazard zonation mapping in part of Bhagirathi Valley, Uttarakhand. *Current Science*, **92**, 628–638.

Mau-Crimmins, T. M., Schussman, H. R., & Geiger, E. L. (2006) Can the invaded range of a species be predicted sufficiently using only native-range data? Lehmann lovegrass (*Eragrostis lehmanniana*) in the southwestern United States. *Ecological Modelling*, **193**, 736–746.

Mayer, K. E. & Laudenslayer, W. F., Jr. (1988) *A Guide to Wildlife Habitats in California*. Sacramento: The Resource Agency, State of California.

McAlpine, C. A., Rhodes, J. R., Bowen, M. E. *et al.* (2008) Can multiscale models of species distribution be generalized from region to region? A case study of the koala. *Journal of Applied Ecology*, **45**, 558–567.

McAlpine, C. A., Scarth, P., Phinn, S. P., & Eyre, T. J. (2002) Mapping glider habitat in dry eucalypt forests for Montreal Process indication 1.1e: fragmentation of forest types. *Australian Forestry*, **64**, 232–241.

McCullagh, P. & Nelder, J. A. (1989) *Generalized Linear Models*. London: Chapman and Hall.

McCune, B. (2006) Non-parametric habitat models with automatic interactions. *Journal of Vegetation Science*, **17**, 819–830.

McCune, B. (2007) Improved estimates of incident radiation and heat load using non-parametric regression against topographic variables. *Journal of Vegetation Science*, **18**, 751–754.

McGarigal, K. (2002) Landscape pattern metrics. In el-Shaarawi, A. H. & Piergorsch, W. W. (Eds.) *Encyclopedia of Environmetrics*. Chichester, Sussex, UK: John Wiley & Sons, pp. 1135–1142.

McGarigal, K. & Marks, B. J. (1995) FRAGSTATS: spatial pattern analysis program for quantifying landscape structure. http://www.umass.edu/landeco/research/fragstats/documents/Metrics/Metrics%20TOC.htm.

McHarg, I. L. (1969) *Design with Nature*. New York: Wiley.

McNab, W. H. (1989) Terrain shape index: quantifying effect of minor landforms on tree height. *Forest Science*, **35**, 91–104.

McNab, W. H. (1993) A topographic index to quantify the effect of mesoscale landform on site productivity. *Canadian Journal of Forest Research*, **23**, 1100–1107.

McPherson, J. M. & Jetz, W., (2007) Effects of species ecology on the accuracy of distribution models. *Ecography*, **30**, 135–151.

McPherson, J. M., Jetz, W., & Rogers, D. J. (2004) The effects of species range sizes on the accuracy of distribution models: ecological phenomenon or statistical artefact? *Journal of Applied Ecology*, **41**, 811–823.

McPherson, J. M., Jetz, W., & Rogers, D. J. (2006) Using coarse-grained occurrence data to predict species distributions at finer spatial resolution – possibilities and limitations. *Ecological Modelling*, **192**, 499–522.

McRea Jr., J. E., Greene, H. G., O'Connell, V. M., & Wakefield, W. W. (1999) Mapping marine habitats with high resolution sidescan sonar. *Oceanologica Acta*, **22**, 679–686.

Meentemeyer, R. K., Anacker, B. L., Mark, W., & Rizzo, D. M. (2008b) Early detection of emerging forest disease using dispersal estimation and ecological niche modeling. *Ecological Applications*, **18**, 377–390.

Meentemeyer, R. K., Moody, A., & Franklin, J. (2001) Landscape-scale patterns of shrub-species abundance in California chaparral: the role of topographically mediated resource gradients. *Plant Ecology*, **156**, 19–41.

Meentemeyer, R. K., Rank, N. E., Anacker, B. L., Rizzo, D. M., & Cushman, J. H. (2008a) Influence of land-cover change on the spread of an invasive forest pathogen. *Ecological Applications*, **18**, 159–171.

Melbourne, B. A., Davies, K. F., Margules, C. R. *et al.* (2004) Species survival in fragmented landscapes: where to from here? *Biodiversity and Conservation*, **13**, 275–284.

Menke, S. B., Fisher, R. N., Jetz, W., & Holway, D. A. (2007) Biotic and abiotic controls of argentine ant invasion success at local and landscape scales. *Ecology*, **88**, 3164–3173.

Meyer, C. B. & Thuiller, W. (2006) Accuracy of resource selection functions across spatial scales. *Diversity and Distributions*, **12**, 288–297.

Meynard, C. N. & Quinn, J. F. (2007) Predicting species distributions: a critical comparison of the most common statistical models using artificial species. *Journal of Biogeography*, **34**, 1455–1469.

Michaelsen, J., Davis, F. W., & Borchert, M. (1987) A non-parametric method for analyzing hierarchical relationships in ecological data. *Coenoses*, **2**, 39–48.

Michaelsen, J., Schimel, D. S., Friedl, M. A., Davis, F. W., & Dubayah, R. C. (1994) Regression tree analysis of satellite and terrain data to guide vegetation sampling and surveys. *Journal of Vegetation Science*, **5**, 673–686.

Midgley, G. F., Hannah, L., Millar, D., Rutherford, M. C., & Powrie, L. W. (2002) Assessing the vulnerability of species richness to anthropogenic climate change in a biodiversity hotspot. *Global Ecology and Biogeography*, **11**, 445–451.

Midgley, G. F., Hughes, G. O., Thuiller, W., & Rebelo, A. G. (2006) Migration rate limitations on climate change-induced range shifts in Cape Proteaceae. *Diversity and Distributions*, **12**, 555–562.

Miller, J. (2005) Incorporating spatial dependence in predictive vegetation models: residual interpolation methods. *The Professional Geographer*, **57**, 169–184.

Miller, J. & Franklin, J. (2002) Predictive vegetation modeling with spatial dependence – vegetation alliances in the Mojave Desert. *Ecological Modelling*, **57**, 227–247.

Miller, J., & Franklin, J. (2006) Explicitly incorporating spatial dependence in predictive vegetation models in the form of explanatory variables: a Mojave Desert case study. *Journal of Geographical Systems*, **8**, 411–435.

Miller, J., Franklin, J., & Aspinall, R. (2007) Incorporating spatial dependence in predictive vegetation models. *Ecological Modelling*, **202**, 225–242.

Miller, J. R., Turner, M. G., Smithwick, E. A. H., Dent, C. L., & Stanley, E. H. (2004) Spatial extrapolation: The science of predicting ecological patterns and processes. *Bioscience*, **54**, 310–320.

Mitchell-Jones, A. J., Amori, G., Bogdanowicz, W. *et al.* (1999) *Atlas of European Mammals*, i–xi, 1–484.

Mladenoff, D. J. & Baker, W. L. (1999) Development of forest and landscape modeling approaches. In Mladenoff, D. J. & Baker, W. L. (Eds.) *Spatial Modeling of Forest Landscape Change: Approaches and Applications*. Cambridge, UK: Cambridge University Press, pp. 1–13.

Mladenoff, D. J. & He, H. S. (1999) Design, behavior and application of LANDIS, as object-oriented model of forest landscape disturbance and succession. In Mladenoff, D. J. & Baker, W. L. (Eds.) *Spatial Modeling of Forest Landscape Change: Approaches and Applications*. Cambridge, UK: Cambridge University Press, pp. 125–162.

Mohamed, K. I., Papes, M., Williams, R., Benz, B. W., & Peterson, A. T. (2006) Global invasion potential of 10 parasitic witchweeds and related Orobanchaceae. *Ambio*, **35**, 281–288.

Moisen, G. G. & Frescino, T. S. (2002) Comparing five modelling techniques for predicting forest characteristics. *Ecological Modelling*, **157**, 209–225.

Moisen, G. G., Freeman, E. A., Blackard, J. A., Frescino, T. S., Zimmermann, N. E., & Edwards Jr., T. C. (2006) Predicting tree species presence and basal area in Utah: a comparison of stochastic gradient boosting, generalized additive models and tree-based methods. *Ecological Modelling*, **199**, 176–187.

Monestiez, P., Dubroca, L., Bonnin, E., Durbec, J.-P., & Guinet, C. (2006) Geostatistical modeling of spatial distribution of *Blaaenoterpa physalus* in the Northwestern

Mediterranean Sea from sparse count data and heterogeneous observation efforts. *Ecological Modelling*, **193**, 615–628.

Monteiro De Barros, F. S., De Aguiar, D. B., Rosa-Freitas, M. G. *et al.* (2007) Distribution summaries of malaria vectors in the northern Brazilian Amazon. *Journal of Vector Ecology*, **32**, 161–167.

Moore, D. M., Lees, B. G., & Davey, S. M. (1991a) A new method for predicting vegetation distributions using decision tree analysis in a geographic information system. *Environmental Management*, **15**, 59–71.

Moore, I. D., Grayson, R., & Ladson, A. (1991b) Digital terrain modelling: a review of hydrological, geomorphological, and biological applications. *Hydrological Processes*, **5**, 3–30.

Moore, I. D., Burch, G. J., & Mackenzie, D. H. (1988) Topographic effects on the distribution of surface soil water and the location of ephemeral gullies. *Transactions of the American Society of Agricultural Engineers*, **31**, 1098–1107.

Morrison, M. L., Marcot, B. G., & Mannan, R. W. (1992) *Wildlife–habitat Relationships*. Madison, WI: The University of Wisconsin Press.

Morrison, M. L., Marcot, B. G., & Mannan, R. W. (1998) *Wildlife-Habitat Relationships: Concepts and Applications*. 2nd edn. Madison, WI: The University of Wisconsin Press.

Mueller, T., Olson, K. A., Fuller, T. K., Schaller, G. B., Murray, M. G., & Leimgruber, P. (2008) In search of forage: predicting dynamic habitats of Mongolian gazelles using satellite-based estimates of vegetation productivity. *Journal of Applied Ecology*, **45**, 649–658.

Mueller-Dombois, D. & Ellenberg, H. (1974) *Aims and Methods of Vegetation Ecology*. New York: Wiley.

Mugglestone, M. A., Kenward, M. G., & Clark, S. J. (2002) Generalized estimating equations for spatially referenced binary data. In Gregori, D., Carmeci, G., Friedl, H., Ferligoj, A., & Wedlin, A. (Eds.) *Correlated Data Modeling*. Trieste: FrancoAngeli S.R.L.

Muñoz, J. & Felicisimo, A. M. (2004) Comparison of statistical methods commonly used in predictive modelling. *Journal of Vegetation Science*, **15**, 285–292.

Murphy, H. T. & Lovett-Doust, J. (2004) Context and connectivity in plant metapopulations and landscape mosaics: does the matrix matter? *Oikos*, **105**, 3–14.

Murray-Smith, C., Brummitt, N. A., Oliveira-Filho, A. T. *et al.* (2009) Plant diversity hotspots in the Atlantic coastal forests of Brazil. *Conservation Biology*, **23**, 151–163.

Nagendra, H. (2001) Using remote sensing to assess biodiversity. *International Journal of Remote Sensing*, **22**, 2377–2400.

Nams, V. O., Mowat, G., & Panian, M. A. (2006) Determining the spatial scale for conservation purposes – an example with grizzly bears. *Biological Conservation*, **128**, 109–119.

Neilson, R. & Running, S. (1996) Global dynamic vegetation modelling: coupling biogeochemistry and biogeography models. In Walker, B. & Steffen, W. (Eds.) *Global Change and Terrestrial Ecosystems*. Cambridge, UK: Cambridge University Press, pp. 451–465.

Neilson, R. P., King, G. A., & Koerper, G. (1992) Toward a rule-based biome model. *Landscape Ecology*, **7**, 27–43.

Nelder, V. J., Crosslet, D. C., & Cofinas, M. (1995) Using geographic information systems (GIS) to determine the adequacy of sampling in vegetation surveys. *Biological Conservation*, **73**.

Nelson, T. A. & Boots, B. (2008) Detecting spatial hot spots in landscape ecology. *Ecography*, **31**, 556–566.

Newton-Cross, G., White, P. C. L., & Harris, S. (2007) Modelling the distribution of badgers *Meles meles*: comparing predictions from field-based and remotely derived habitat data. *Mammal Review*, **37**, 54–70.

Nicholls, A. (1989) How to make biological surveys go further with generalised linear models. *Biological Conservation*, **50**, 51–75.

Niu, X. Z. & Duiker, S. W. (2006) Carbon sequestration potential by afforestation of marginal agricultural land in the Midwestern US. *Forest Ecology and Management*, **223**, 415–427.

Nogues-Bravo, D., Araújo, M. B., Romdal, T., & Rahbek, C. (2008) Scale effects and human impact on the elevational species richness gradients. *Nature*, **453**, 216–218.

O'Connor, R. J., Jones, M. T., White, D. *et al.* (1996) Spatial partitioning of environmental correlates of avian biodiversity in the conterminous United States. *Biodiversity Letters*, **3**, 97–110.

Ohmann, J. L. & Gregory, M. J. (2002) Predictive mapping of forest composition and structure with direct gradient analysis and nearest neighbor imputation in coastal Oregon, U.S.A. *Canadian Journal of Forest Research*, **32**, 725–741.

Ohmann, J. L. & Spies, T. A. (1998) Regional gradient analysis and spatial pattern of woody plant communities of Oregon forests. *Ecological Monographs*, **68**, 151–182.

Oindo, B. O. & Skidmore, A. K. (2002) Interannual variability of NDVI and species richness in Kenya. *International Journal of Remote Sensing*, **23**, 285–298.

Oksanen, J. & Minchin, P. R. (2002) Continuum theory revisited: what shape are species responses along ecological gradients? *Ecological Modelling*, **157**, 119–129.

Olden, J. D., Joy, M. K., & Death, R. G. (2006) Rediscovering the species in community-wide predictive modeling. *Ecological Applications*, **16**, 1449–1460.

Olden, J. D., Lawler, J. J., & Poff, N. L. (2008) Machine learning methods without tears: A primer for ecologists. *Quarterly Review of Biology*, **83**, 171–193.

Olivier, F. & Wotherspoon, S. J. (2006) Modelling habitat selection using presence-only data: Case study of a colonial hollow nesting bird, the snow petrel. *Ecological Modelling*, **195**, 187–204.

Olivier, F. & Wotherspoon, S. J. (2008) Nest selection by snow petrels *Pagodroma nivea* in East Antarctica – Validating predictive habitat selection models at the continental scale. *Ecological Modelling*, **210**, 414–430.

Orme, C. D. L., Davies, R. G., Burgess, M. *et al.* (2005) Global hotspots of species richness are not congruent with endemism or threat. *Nature*, **436**, 1016–1019.

Ortega-Huerta, M. A. & Peterson, A. T. (2004) Modelling spatial patterns of biodiversity for conservation prioritization in North-eastern Mexico. *Diversity and Distributions*, **10**, 39–54.

Osborne, P. E. & Suarez-Seoane, S. (2002) Should data be partitioned spatially before building large-scale distribution models? *Ecological Modelling*, **157**, 249–259.

Osborne, P. E. & Suarez-Seoane, S. (2007) Identifying core areas in a species range using temporal suitability analysis: an example using little bustards *Tetrax tetrax* L. in Spain. *Biodiversity and Conservation*, **16**, 3505–3518.

Osborne, P. E. & Tigar, B. J. (1992) Interpreting bird atlas data using logistic models: an example from Lesotho, southern Africa. *Journal of Applied Ecology*, **29**, 55–62.

Osborne, P. E., Alonso, J. C., & Bryant, R. G. (2001) Modelling landscape-scale habitat use using GIS and remote sensing: a case study with great bustards. *Journal of Applied Ecology*, **38**, 458–471.

Osborne, P. E., Foody, G. M., & Suarez-Seoane, S. (2007) Non-stationarity and local approaches to modelling the distribution of wildlife. *Diversity and Distributions*, **13**, 313–323.

Ozesmi, S. L. & Ozesmi, U. (1999) An artificial neural network approach to spatial habitat modelling with interspecific interaction. *Ecological Modelling*, **116**, 15–31.

Palmer, M. W., Earls, P. G., Hoagland, B. W., White, P. S., & Wohlgemuth, T. (2002) Quantitative tools for perfecting species lists. *Environmetrics*, **13**, 121–137.

Parisien, M.-A. & Moritz, M. A. (2009) Environmental controls on the distribution of wildfire at multiple spatial scales. *Ecological Monographs*, **79**, 127–154.

Parmesan, C. (2006) Ecological and evolutionary responses to recent climate change. *Annual Review of Ecology Evolution and Systematics*, **37**, 637–669.

Pausas, J. G. & Austin, M. P. (2001) Patterns of plant species richness in relation to different environments: An appraisal. *Journal of Vegetation Science*, **12**, 153–166.

Pausas, J. G. & Carreras, J. (1995) The effect of bedrock type, temperature and moisture on species richness of Pyrenean Scots Pine (*Pinus sylvestris* L.) Forests. *Vegetatio*, **116**, 85–92.

Pausas, J. G., Braithwaite, L. W., & Austin, M. P. (1995) Modeling habitat quality for arboreal marsupials in the south coastal forests of New-South-Wales, Australia. *Forest Ecology and Management*, **78**, 39–49.

Pausas, J. G., Carreras, J., Ferre, A., & Font, X. (2003) Coarse-scale plant species richness in relation to environmental heterogeneity. *Journal of Vegetation Science*, **14**, 661–668.

Pearce, J. & Ferrier, S. (2000) An evaluation of alternative algorithms for fitting species distribution models using logistic regression. *Ecological Modelling*, **128**, 127–147.

Pearce, J. & Lindenmayer, D. (1998) Bioclimatic analysis to enhance reintroduction biology of the endangered helmeted honeyeater (*Lichenostomus melanops cassidix*) in southeastern Australia. *Restoration Ecology*, **6**, 238–243.

Pearce, J., Ferrier, S., & Scotts, D. (2001) An evaluation of the predictive performance of distributional models for flora and fauna in north-east New South Wales. *Journal of Environmental Management*, **62**, 171–184.

Pearce, J. L. & Boyce, M. S. (2006) Modelling distribution and abundance with presence-only data. *Journal of Applied Ecology*, **43**, 405–412.

Pearman, P. B., Guisan, A., Broennimann, O., & Randin, C. F. (2008) Niche dynamics in space and time. *Trends in Ecology & Evolution*, **23**, 149–158.

Pearson, R. G. & Dawson, T. P. (2003) Predicting the impacts of climate change on the distribution of species: are bioclimatic envelope models useful? *Global Ecology & Biogeography*, **12**, 361–371.

Pearson, R. G., Dawson, T. P., Berry, P. M., & Harrison, P. A., (2002) SPECIES: a spatial evaluation of climate impact on the envelope of species. *Ecological Modelling*, **154**, 289–300.

Pearson, R. G., Dawson, T. P., & Liu, C. (2004) Modelling species distribution in Britain: a hierarchical integration of climate and land cover. *Ecography*, **27**, 285–298.

Pearson, R. G., Raxworthy, C. J., Nakamura, M., & Peterson, A. T. (2007) Predicting species distributions from small numbers of occurrence records: a test case using cryptic geckos in Madagascar. *Journal of Biogeography*, **34**, 102–117.

Pearson, R. G., Thuiller, W., Araújo, M. B. *et al.* (2006) Model-based uncertainty in species range prediction. *Journal of Biogeography*, **33**, 1704–1711.

Pearson, S. M. (1993) The spatial extent and relative influence of landscape-level factors on wintering bird populations. *Landscape Ecology*, **8**, 3–18.

Pereira, J. M. & Itami, R. M. (1991) GIS-based habitat modeling using logistic multiple regression: a study of the Mt. Graham Red Squirrel. *Photogrammetric Engineering and Remote Sensing*, **57**, 1475–1486.

Pereira, J. M. C. & Duckstein, L. (1993) A multiple criteria decision-making approach to GIS-based land suitability evaluation. *International Journal of Geographical Information Systems*, **7**, 407–424.

Perring, F. (1958) A theoretical approach to a study of chalk grassland. *Journal of Ecology*, **46**, 665–679.

Perring, F. (1959) Topographical gradients of chalk grassland. *Journal of Ecology*, **47**, 447–481.

Perring, F. (1960) Climatic gradients of chalk grassland. *Journal of Ecology*, **48**, 415–442.

Peters, D. P. C., Herrick, J. E., Urban, D. L., Gardner, R. H., & Breshears, D. D. (2004) Strategies for ecological extrapolation. *Oikos*, **106**, 627–636.

Peterson, A. T. (2003) Predicting the geography of species invasions via ecological niche modeling. *Quarterly Review of Biology*, **78**, 419–433.

Peterson, A. T. (2006) Ecologic niche modeling and spatial patterns of disease transmission. *Emerging Infectious Diseases*, **12**, 1822–1826.

Peterson, A. T. & Holt, R. D. (2003) Niche differentiation in Mexican birds: using point occurrences to detect ecological innovation. *Ecology Letters*, **6**, 774–782.

Peterson, A. T. & Kluza, D. A. (2003) New distributional modelling approaches for gap analysis. *Animal Conservation*, **6**, 47–54.

Peterson, A. T. & Nyari, A. S. (2008) Ecological niche conservatism and Pleistocene refugia in the thrush-like mourner, *Schiffornis* sp., in the neotropics. *Evolution*, **62**, 173–183.

Peterson, A. T. & Robins, C. R. (2003) Using ecological-niche modeling to predict Barred Owl invasions with implications for Spotted Owl conservation. *Conservation Biology*, **17**, 1161–1165.

Peterson, A. T. & Vieglais, D. A. (2001) Predicting species invasions using ecological niche modeling: New approaches from bioinformatics attack a pressing problem. *Bioscience*, **51**, 363–371.

Peterson, A. T., Ortega-Huerta, M. A., Bartley, J. *et al.* (2002) Future projections for Mexican faunas under global climate change scenarios. *Nature*, **416**, 626–629.

Peterson, A. T., Papes, M., & Eaton, M. (2007) Transferability and model evaluation in ecological niche modeling: a comparison of GARP and Maxent. *Ecography*, **30**, 550–560.

Peterson, A. T., Papes, M., & Kluza, D. A. (2003) Predicting the potential invasive distributions of four alien plant species in North America. *Weed Science*, **51**, 863–868.

Peterson, A. T., Papes, M., & Soberón, J., (2008) Rethinking receiver operating characteristic analysis applications in ecological niche modeling. *Ecological Modelling*, **213**, 63–72.

Peterson, A. T., Sánchez-Cordero, V., Martínez-Meyer, E., & Navarro-Sigüenza, A. G. (2006) Tracking population extirpations via melding ecological niche modeling with land cover information. *Ecological Modelling*, **195**, 229–236.

Peterson, A. T., Soberón, J., & Sanchez-Cordero, V. (1999) Conservatism of ecological niches in evolutionary time. *Science*, **285**, 1265–1267.

Pettorelli, N., Vik, J. O., Mysterud, A., Gaillard, J. M., Tucker, C. J., & Stenseth, N. C. (2005) Using the satellite-derived NDVI to assess ecological responses to environmental change. *Trends in Ecology & Evolution*, **20**, 503–510.

Phillips, S. J. (2008) Transferability, sample selection bias and background data in presence-only modelling: a response to Peterson *et al.* (2007). *Ecography*, **31**, 272–278.

Phillips, S. J., Anderson, R. P., & Schapire, R. E. (2006) Maximum entropy modeling of species geographic distributions. *Ecological Modelling*, **190**, 231–259.

Phillips, S. J. & Dudík, M. (2008) Modeling of species distributions with Maxent: new extensions and a comprehensive evaluation. *Ecography*, **31**, 161–175.

Phillips, S. J., Dudík, M., Elith, J. *et al.* (2009) Sample selection bias and presence-only distribution models: implications for background and pseudo-absence data. *Ecological Applications*, **19**, 181–197.

Phinn, S. R. (1998) A framework for selecting appropriate remotely sensed data dimensions for environmental monitoring and management. *International Journal of Remote Sensing*, **19**, 3457–3463.

Phinn, S. R., Stow, D. A., & Zedler, J. B. (1996) Monitoring wetland habitat restoration in southern California using airborne multispectral video data. *Restoration Ecology*, **4**, 412–422.

Phinn, S. R., Stow, D. S., Franklin, J., Mertes, L. A. K., & Michaelsen, J. (2003) Optimizing remotely sensed data for ecosystem analyses: combining hierarchy theory and scene models. *Environmental Management*, **31**, 429–441.

Pickett, S. T. A. & Cadenasso, M. L. (1995) Landscape ecology: spatial heterogeneity in ecological systems. *Science*, **269**, 331–334.

Pickett, S. T. A. & White, P. S. (1985) *The Ecology of Natural Disturbance and Patch Dynamics*. New York NY: Academic Press.

Pielou, E. C. (1975) *Ecological Diversity*. New York: Wiley.

Pindyck, R. S. & Rubenfeld, D. L. (1981) *Econometric Models and Econometric Forecasts*. New York, USA: McGraw Hill.

Pineda, E. & Lobo, J. M. (2009) Assessing the accuracy of species distribution models to predict amphibian species richness patterns. *Journal of Animal Ecology*, **78**, 182–190.

Platts, P. J., McClean, C. J., Lovett, J. C., & Marchant, R. (2008) Predicting tree distributions in an East African biodiversity hotspot: model selection, data bias and envelope uncertainty. *Ecological Modelling*, 218, 121–134.

Plotnick, R. E., Gardner, R. H., & O'Neill, R. V. (1993) Lacunarity indices as a measure of landscape texture. *Landscape Ecology*, **8**.

Pontius, R. G. & Schneider, L. C. (2001) Land-cover change model validation by an ROC method for the Ipswich watershed, Massachusetts, USA. *Agriculture Ecosystems & Environment*, **85**, 239–248.

Posillico, M., Alberto, M. I. B., Pagnin, E., Lovari, S., & Russo, L. (2004) A habitat model for brown bear conservation and land use planning in the central Apennines. *Biological Conservation*, **118**, 141–150.

Potts, J. M. & Elith, J. (2006) Comparing species abundance models. *Ecological Modelling*, **199**, 153–163.

Poulos, H. M., Camp, A. E., Gatewood, R. G., & Loomis, L. (2007) A hierarchical approach for scaling forest inventory and fuels data from local to landscape scales in the Davis Mountains, Texas, USA. *Forest Ecology and Management*, **244**, 1–15.

Poyry, J., Luoto, M., Heikkinen, R. K., & Saarinen, K. (2008) Species traits are associated with the quality of bioclimatic models. *Global Ecology and Biogeography*, **17**, 403–414.

Prasad, A. M., Iverson, L. R., & Liaw, A. (2006) Newer classification and regression techniques: bagging and random forests for ecological prediction. *Ecosystems*, **9**, 181–199.

Prates-Clark, C. D., Saatchi, S. S., & Agosti, D. (2008) Predicting geographical distribution models of high-value timber trees in the Amazon Basin using remotely sensed data. *Ecological Modelling*, **211**, 309–323.

Prentice, I. C., Bartlein, P. J., & Webb, T., III (1991) Vegetation and climate change in eastern North-America since the Last Glacial Maximum. *Ecology*, **72**, 2038–2056.

Prentice, I. C., Cramer, W., Harrison, S. P., Leemans, R., Monserud, R. A., & Solomon, A. M. (1992) A global biome model based on plant physiology and dominance, soil properties and climate. *Journal of Biogeography*, **19**, 117–134.

Preston, C. D., Pearman, D. A., & Dines, T. D. (Eds.) (2002) *New Atlas of the British and Irish Flora*. Oxford, UK: Oxford University Press.

Preston, F. W. (1948) The commonness, and rarity, of species. *Ecology*, **29**, 254–283.

Preston, K., Rotenberry, J. T., Redak, R. A., & Allen, M. F. (2008) Habitat shifts of endangered species under altered climate conditions: importance of biotic interactions. *Global Change Biology*, **14**, 2501–2515.

Pulliam, H. R. (1988) Sources, sinks, and population regulation. *American Naturalist*, **132**, 652–661.

Pulliam, H. R. (2000) On the relationship between niche and distribution. *Ecology Letters*, **3**, 349–361.

Pulliam, H. R., Dunning, J., & Liu, J. (1992) Population dynamics in complex landscapes: a case study. *Ecological Applications*, **2**, 165–177.

Quinlan, J. R. (1993) *C4.5: Programs for Machine Learning.* San Mateo, CA: Morgan Kaufman Publishers.

Quinn, P., Beven, K., Chevallier, P., & Planchon, O. (1991) The prediction of hillslope flow paths for distributed hydrological modeling using digital terrain models. *Hydrological Processes,* **5,** 59–79.

Quinn, G. P. & Keough, M. J. (2002) *Experimental design and data analysis for biologists,* Cambridge, UK, Cambridge University Press.

R Development Core Team (2004) R: A language and environment for statistical computing. 2.0.1 ed. Vienna, Austria http://rrr.R-project.org, R Foundation for Statistical Computing.

Rabinowitz, D. (1981) Seven forms of rarity. In Synge, H. (Ed.) *The Biological Aspects of Rare Plant Conservation.* Chichester, UK: Wiley, pp. 205–217.

Rahbek, C. & Graves, G. R. (2001) Multiscale assessment of patterns of avian species richness. *Proceedings of the National Academy of Sciences, USA,* **98,** 4534–4539.

Raleigh, R. F., Zuckerman, L. D., & Nelson, P. C. (1986) Habitat suitability index models and instream flow sustainability curves: Brown trout, revised. U.S. Fish Wildlife Service Biological Report 82(10.124).

Randin, C. F., Dirnbock, T., Dullinger, S., Zimmermann, N. E., Zappa, M., & Guisan, A. (2006) Are niche-based species distribution models transferable in space? *Journal of Biogeography,* **33,** 1689–1703.

Rangel, T. F. L. V. B., Diniz-Filho, J. A. F., & Bini, L. M. (2006) Towards an integrated computational tool for spatial analysis in macroecology and biogeography. *Global Ecology and Biogeography,* **15.**

Rasmussen, C. (2006) Distribution of soil organic and inorganic carbon pools by biome and soil taxa in Arizona. *Soil Science Society of America Journal,* **70,** 256–265.

Raso, G., Vounatsou, P., Singer, B. H., N'Goran, E. K., Tanner, M., & Utzinger, J. (2006) An integrated approach for risk profiling and spatial prediction of *Schistosoma mansoni*-hookworm coinfection. *Proceedings of the National Academy of Sciences of the United States of America,* **103,** 6934–6939.

Raul, R. C., Novillo, C. J., Millington, J. D. A., & Gomez-Jimenez, I. (2008) GIS analysis of spatial patterns of human-caused wildfire ignition risk in the SW of Madrid (Central Spain). *Landscape Ecology,* **23,** 341–354.

Ray, N. & Burgman, M. A. (2006) Subjective uncertainties in habitat suitability maps. *Ecological Modelling,* **195,** 172–186.

Ree, R. H., Moore, B. R., Webb, C. O., & Donoghue, M. J. (2005) A likelihood framework for inferring the evolution of geographic range on phylogenetic trees. *Evolution,* **59,** 2299–2311.

Reese, G. C., Wilson, K. R., Hoeting, J. A., & Flather, C. H. (2005) Factors affecting species distribution predictions: a simulation modeling experiment. *Ecological Applications,* **15,** 554–564.

Regan, H. M., Colyvan, M., & Burgman, M. A. (2002) A taxonomy and treatment of uncertainty for ecology and conservation biology. *Ecological Applications,* **12,** 618–628.

Rehfeldt, G. E., Crookston, N. L., Warwell, M. V., & Evans, J. S. (2006) Empirical analyses of plant–climate relationships for the western United States. *International Journal of Plant Sciences,* **167,** 1123–1150.

Reineking, B. & Schroder, B. (2006) Constrain to perform: Regularization of habitat models. *Ecological Modelling*, **193**, 675–690.

Rejmanek, M. & Richardson, D. M. (1996) What attributes make some plant species more invasive?. *Ecology*, **77**, 1655–1661.

Richardson, D. M. & Thuiller, W. (2007) Home away from home – objective mapping of high-risk source areas for plant introductions. *Diversity and Distributions*, **13**, 299–312.

Ridgeway, G. (1999) The state of boosting. *Computing Science and Statistics*, **31**, 172–181.

Ripley, B. D. (1981) *Spatial Statistics*. New York: John Wiley and Sons.

Ripley, B. D. (1996) *Pattern Recognition and Neural Networks*. Cambridge, UK: Cambridge University Press.

Rissler, L. J. & Apodaca, J. J. (2007) Adding more ecology into species delimitation: Ecological niche models and phylogeography help define cryptic species in the black salamander (*Aneides flavipunctatus*). *Systematic Biology*, **56**, 924–942.

Robertson, M. P., Villet, M. H., & Palmer, A. R. (2004) A fuzzy classification technique for predicting species distributions: applications using invasive alien plants and indigenous insects. *Diversity and Distributions*, **10**, 461–474.

Rocchini, D., Ricotta, C., & Chiarucci, A. (2007) Using satellite imagery to assess plant species richness: The role of multispectral systems. *Applied Vegetation Science*, **10**, 325–331.

Rodrigues, A. S. L., Andelman, S. J., Bakarr, M. I. *et al.* (2004) Effectiveness of the global protected area network in representing species diversity. *Nature*, **428**, 640–643.

Rodriguéz, J. P., Brotons, L., Bustmante, J., & Seoane, J. (2007) The application of predictive modelling of species distribution to biodiversity conservation. *Diversity and Distributions*, **13**, 243–251.

Romero-Calcerrada, R. & Luque, S. (2006) Habitat quality assessment using Weights-of-Evidence based GIS modelling: the case of *Picoides tridactylus* as species indicator of the biodiversity value of the Finnish forest. *Ecological Modelling*, **196**, 62–76.

Rondinini, C., Stuart, S. & Boitani, L. (2005) Habitat suitability models and the shortfall in conservation planning for African vertebrates. *Conservation Biology*, **19**, 1488–1497.

Root, T. L., Price, J. T., Hall, K. R., Schneider, S. H., Rosenzweig, C., & Pounds, J. A. (2003) Fingerprints of global warming on wild animals and plants. *Nature*, **421**, 57–60.

Rosenblatt, A. E., Gold, A. J., Stolt, M. H., Groffman, P. M., & Kellogg, D. Q. (2001) Identifying riparian sinks for watershed nitrate using soil surveys. *Journal of Environmental Quality*, **30**, 1596–1604.

Rosenzweig, C., Karoly, D., Vicarelli, M. *et al.* (2008) Attributing physical and biological impacts to anthropogenic climate change. *Nature*, **453**, 353.

Rotenberry, J. T., Knick, S. T., & Dunn, J. E. (2002) A minimalist approach to mapping species habitat: Pearson's planes of closest fit. In Scott, J. M., Heglund, P. J., Morrison, M. L. *et al.* (Eds.) *Predicting Species Occurrences: Issues of Accuracy and Scale*. Covelo, CA: Island Press, pp. 281–289.

Rotenberry, J. T., Preston, K. L., & Knick, S. T. (2006) GIS-based niche modeling for mapping species habitat. *Ecology*, **87**, 1458–1464.

Roura-Pascual, N., Suarez, A. V., McNyset, K. *et al.* (2006) Niche differentiation and fine-scale projections for Argentine ants based on remotely sensed data. *Ecological Applications*, **16**, 1832–1841.

Royle, J. A., Kery, M., Gautier, R., & Schmid, H. (2007) Hierarchical spatial models of abundance and occurrence from imperfect survey data. *Ecological Monographs*, **77**, 465–481.

Rubio, A. & Sanchez-Palomares, O. (2006) Physiographic and climatic potential areas for *Fagus sylvatica* L. based on habitat suitability indicator models. *Forestry*, **79**, 439–451.

Rüger, N., Schlueter, M., & Matthies, M. (2005) A fuzzy habitat suitability index for *Populus euphratica* in the Northern Amudarya delta (Uzbekistan). *Ecological Modelling*, **184**, 313–328.

Running, S. W., Justice, C. J., Salomonson, V. *et al.* (1994) Terrestrial remote sensing science and algorithms planned for EOS/MODIS. *International Journal of Remote Sensing*, **15**, 3587–3620.

Rushton, S. P., Ormerod, S. J., & Kerby, G. (2004) New paradigms for modelling species distributions? *Journal of Applied Ecology*, **41**, 193–200.

Rykiel, E. J., Jr. (1996) Testing ecological models: the meaning of validation. *Ecological Modelling*, **90**, 229–244.

Saatchi, S., Buermann, W., Ter Steege, H., Mori, S., & Smith, T. B. (2008) Modeling distribution of Amazonian tree species and diversity using remote sensing measurements. *Remote Sensing of Environment*, **112**, 2000–2017.

Saathoff, E., Olsen, A., Sharp, B., Kvalsvig, J. D., Appleton, C. C., & Kleinschmidt, I. (2005) Ecologic covariates of hookworm infection and reinfection in rural Kwazulu-Natal/South Africa: A geographic information system-based study. *American Journal of Tropical Medicine and Hygiene*, **72**, 384–391.

Sattler, T., Bontadina, F., Hirzel, A. H., & Arlettaz, R. (2007) Ecological niche modelling of two cryptic bat species calls for a reassessment of their conservation status. *Journal of Applied Ecology*, **44**, 1188–1199.

Saveraid, E. H., Debinski, D. M., Kindscher, K., & Jakubauskas, M. E. (2001) A comparison of satellite data and landscape variables in predicting bird species occurrences in the Greater Yellowstone Ecosystem, USA. *Landscape Ecology*, **16**, 71–83.

Schadt, S., Revilla, E., Wiegand, T. *et al.* (2002) Assessing the suitability of central European landscapes for the reintroduction of Eurasian lynx. *Journal of Applied Ecology*, **39**, 189–203.

Scheller, R. M., van Tuyl, S., Clark, K., Hayden, N. G., Hom, J., & Mladenoff, D. J. (2008) Simulation of forest change in the New Jersey Pine Barrens under current and pre-colonial conditions. *Forest Ecology and Management*, **255**, 1489–1500.

Schmit, C., Rousevell, M. D. A., & La Jeunesse, I. (2006) The limitations of spatial land use data in environmental analysis. *Environmental Science & Policy*, **9**, 174–188.

Schulz, T. T. & Joyce, L. A. (1992) A spatial application of a marten habitat model. *Wildlife Society Bulletin*, **20**, 74–83.

Scott, J. M., Davis, F., Csuti, B. *et al.* (1993) Gap analysis: a geographical approach to protection of biological diversity. *Wildlife Monographs*, **123**, 1–41.

Scott, J. M., Heglund, P. J., Morrison, M. L. *et al.* (Eds.) (2002) *Predicting Species Occurrences: Issues of Accuracy and Scale.* Covelo, CA: Island Press.

Scull, P., Franklin, J., & McArthur, D. (2003) Predictive soil mapping: a review. *Progress in Physical Geography,* **27,** 171–197.

Scull, P., Franklin, J., & Chadwick, O. A. (2005) The application of classification tree analysis to soil type prediction in a desert landscape. *Ecological Modelling,* **181,** 1–15.

Segurado, P. & Araújo, M. B. (2004) An evaluation of methods for modelling species distributions. *Journal of Biogeography,* **31,** 1555–1568.

Segurado, P., Araújo, M. B., & Kunin, W. E. (2006) Consequences of spatial auto-correlation for niche-based models. *Journal of Applied Ecology,* **43,** 433–444.

Sellars, J. D. & Jolls, C. L. (2007) Habitat modeling for *Amaranthus pumilus:* An application of light detection and ranging (LIDAR) data. *Journal of Coastal Research,* **23,** 1193–1202.

Sergio, C., Figueira, R., Draper, D., Menezes, R., & Sousa, A. J. (2007) Modelling bryophyte distribution based on ecological information for extent of occurrence assessment. *Biological Conservation,* **135,** 341–351.

Seto, K. C., Fleishman, E., Fay, J. P., & Betrus, C. J. (2004) Linking spatial patterns of bird and butterfly species richness with Landsat TM derived NDVI. *International Journal of Remote Sensing,* **25,** 4309–4324.

Sheppard, J. K., Lawler, I. R., & Marsh, H. (2007) Seagrass as pasture for seacows: landscape-level dugong habitat evaluation. *Estuarine Coastal and Shelf Science,* **71,** 117–132.

Shi, H., Laurent, E. J., Lebouton, J. *et al.* (2006) Local spatial modeling of white-tailed deer distribution. *Ecological Modelling,* **190,** 171–189.

Shriner, S. A., Simons, T. R., & Farnsworth, G. L. (2002) A GIS-based habitat model for Wood Thrush, *Hyocichla mustelina,* in Great Smokey Mountains National Park. In Scott, J. M., Heglund, P. J., & Morrison, M. L. (Eds.) *Predicting Species Occurrences: Issues of Accuracy and Scale.* Covelo, CA: Island Press, pp. 529–535.

Shugart, H. H. (1998) *Terrestrial Ecosystems in Changing Environments,* Cambridge, UK: Cambridge University Press.

Sitch, S., Smith, B., Prentice, I. C. *et al.* (2003) Evaluation of ecosystem dynamics, plant geography and terrestrial carbon cycling in the LPJ dynamic global vegetation model. *Global Change Biology,* **9,** 161–185.

Skidmore, A. K. (1989) An expert system classifies eucalypt forest types using thematic mapper data and a digital terrain model. *Photogrammetric Engineering and Remote Sensing,* **55,** 1449–1464.

Skidmore, A. K. (1990) Terrain position as mapped from a gridded digital elevation model. *International Journal of Geographic Information Systems,* **4,** 33–49.

Skov, F. & Svenning, J. C. (2004) Potential impact of climatic change on the distribution of forest herbs in Europe. *Ecography,* **27,** 366–380.

Smith, G. M. & Wyatt, B. K. (2007) Multi-scale survey by sample-based field methods and remote sensing: A comparison of UK experience with European environmental assessments. *Landscape and Urban Planning,* **79,** 170–176.

Smith, P. A. (1994) Autocorrelation in logistic regression modelling of species distributions. *Global Ecology and Biogeography Letters,* **4,** 47–61.

Smith, T. M., Shugart, H. H., & Woodward, F. I. (1997) *Plant Functional Types.* Cambridge, UK: Cambridge University Press.

Soberón, J. (2007) Grinnellian and Eltonian niches and geographic distributions of species. *Ecology Letters*, **10**, 1115–1123.

Soberón, J. & Peterson, A. T. (2004) Biodiversity informatics: managing and applying primary biodiversity data. *Philosophical Transactions of the Royal Society of London Series B-Biological Sciences*, **359**, 689–698.

Soberón, J. & Peterson, A. T. (2005) Interpretation of models of fundamental ecological niches and species distributional areas. *Biodiversity Informatics*, **2**, 1–10.

Speight, J. G. (1974) A parametric approach to landform regions. *Special Publications Institute of British Geographers*, **7**, 213–230.

Stanbury, K. B. & Starr, R. M. (1999) Applications of Geographic Information Systems (GIS) to habitat assessment and marine resource management. *Oceanologica Acta*, **22**, 699–703.

Stehman, S. V. & Czaplewski, R. L. (1998) Design and analysis for thematic map accuracy assessment: fundamental principles. *Remote Sensing of Environment*, **64**, 331–344.

Steiner, F. M., Schlick-Steiner, B. C., VanDerWal, J. *et al.* (2008) Combined modelling of distribution and niche in invasion biology: a case study of two invasive *Tetramorium* ant species. *Diversity and Distributions*, **14**, 538–545.

Stephenson, C. M., MacKenzie, M. L., Edwards, C., & Travis, J. M. J. (2006) Modelling establishment probabilities of an exotic plant, *Rhododendron ponticum*, invading a heterogeneous, woodland landscape using logistic regression with spatial autocorrelation. *Ecological Modelling*, **193**, 747–758.

Steyaert, L. T. (1996) Status of land data for environmental modeling and challenges for geographic information systems in land characterization. In Goodchild, M. F., Steyaert, L. T., Parks, B. O. *et al.* (Eds.) *GIS and Environmental Modeling: Progress and Research Issues.* Fort Collins, CO: GIS World Books, pp. 17–27.

Stickler, C. M. & Southworth, J. (2008) Application of multi-scale spatial and spectral analysis for predicting primate occurrence and habitat associations in Kibale National Park, Uganda. *Remote Sensing of Environment*, **112**, 2170–2186.

Stigall, A. L. & Lieberman, B. S. (2006) Quantitative palaeobiogeography: GIS, phylogenetic biogeographical analysis, and conservation insights. *Journal of Biogeography*, **33**, 2051–2060.

Stockwell, D. R. B. (1999) Genetic algorithms II. In Fielding, A. H. (Ed.) *Machine Learning Methods for Ecological Applications.* Boston: Kluwer, pp. 123–144.

Stockwell, D. R. B. (2006) Improving ecological niche models by data mining large environmental datasets for surrogate models. *Ecological Modelling*, **192**, 188–196.

Stockwell, D. R. B. & Noble, I. R. (1992) Induction of sets of rules from animal distribution data – a robust and informative method of data-analysis. *Mathematics and Computers in Simulation*, **33**, 385–390.

Stockwell, D. R. B., Arzberger, P., Fountain, T., & Helly, J. (2000) An interface between computing, ecology, and biodiversity: environmental informatics. *Korean Journal of Ecology*, **23**, 101–106.

Stockwell, D. R. B., Beach, J. H., Stewart, A., Vorontsov, G., Vieglais, D., & Pereira, R. S. (2006) The use of the GARP genetic algorithm and Internet grid computing in the Lifemapper world atlas of species biodiversity. *Ecological Modelling*, **195**, 139–145.

Stockwell, D. R. B. & Peters, D. P. (1999) The GARP modelling system: problems and solutions to automated spatial prediction. *International Journal of Geographical Information Science*, **13**, 143–158.

Stockwell, D. R. B. & Peterson, A. T. (2002) Effects of sample size on accuracy of species distribution models. *Ecological Modelling*, **148**, 1–13.

Stoms, D. & Estes, J. E. (1993) A remote sensing research agenda for mapping and monitoring biodiversity. *International Journal of Remote Sensing*, **14**, 1839–1860.

Stoms, D. M., Davis, F. W., & Cogan, C. B. (1992) Sensitivity of wildlife habitat models to uncertainties in GIS data. *Photogrammetric Engineering and Remote Sensing*, **58**, 843–850.

Store, R. & Kangas, J. (2001) Integrating spatial multi-criteria evaluation and expert knowledge for GIS-based habitat suitability modelling. *Landscape and Urban Planning*, **55**, 79–93.

Strahler, A. H. (1980) The use of prior probabilities in maximum likelihood classification of remotely sensed data. *Remote Sensing of Environment*, **10**, 135–163.

Strahler, A. H. (1981) Stratification of natural vegetation for forest and rangeland inventory using Landsat digital imagery and collateral data. *International Journal of Remote Sensing*, **2**, 15–41.

Strickland, B. K. & Demarais, S. (2006) Effectiveness of the State Soil Geographic Database (STATSGO) to predict white-tailed deer morphometrics in Mississippi. *Wildlife Society Bulletin*, **34**, 1264–1272.

Suarez-Seoane, S., Osborne, P. E., & Rosema, A. (2004) Can climate data from METEOSAT improve wildlife distribution models? *Ecography*, **27**, 629–636.

Suarez-Seoane, S., de la Morena, E. L. G., Prieto, M. B. M., Osborne, P. E., & de Juana, E. (2008) Maximum entropy niche-based modelling of seasonal changes in little bustard (*Tetrax tetrax*) distribution. *Ecological Modelling*, **219**, 17–29.

Sundermeyer, M. A., Rothschild, B. J., & Robinson, A. R. (2006) Assessment of environmental correlates with the distribution of fish stocks using a spatially explicit model. *Ecological Modelling*, **197**, 116–132.

Swanson, F. J., Kratz, T. K., Caine, N., & Woodmansee, R. G. (1988) Landform effects on ecosystem patterns and processes. *BioScience*, **38**, 92–98.

Swets, J. A. (1988) Measures of the accuracy of diagnostic systems. *Science*, **240**, 1285–1293.

Syphard, A. D. & Franklin, J. (in press-a) Species functional type affects the accuracy of species distribution models for plants in southern California. *Journal of Vegetation Science*.

Syphard, A. D. & Franklin, J. (in press-b) Differences in spatial predictions among species distribution modeling methods vary with species traits and environmental predictors. *Ecography*.

Syphard, A. D., Clarke, K. C., & Franklin, J. (2007) Simulating fire frequency and urban growth in southern California coastal shrublands, USA. *Landscape Ecology*, **22**, 431–445.

Syphard, A. D., Franklin, J., & Keeley, J. E. (2006) Simulating the effects of frequent fire on southern California coastal shrublands. *Ecological Applications*, **16**, 1744–1756.

Syphard, A. D., Radeloff, V. C., Keuler, N. S. *et al.* (2008) Predicting spatial patterns of fire on a southern California landscape. *International Journal of Wildland Fire*, **17**, 602–613.

Telesco, R. L., van Manen, F. T., Clark, J. D., & Cartwright, M. E. (2007) Identifying sites for elk restoration in Arkansas. *Journal of Wildlife Management*, **71**, 1393–1403.

ter Braak, C. J. F. (1987) The analysis of vegetation-environment relationships by canonical correspondence analysis. *Vegetatio*, **69**, 69–77.

Termansen, M., McClean, C. J., & Preston, C. D. (2006) The use of genetic algorithms and Bayesian classification to model species distributions. *Ecological Modelling*, **192**, 410–424.

Thatcher, C. A., van Manen, F. T., & Clark, J. D. (2006) Identifying suitable sites for Florida panther reintroduction. *Journal of Wildlife Management*, **70**, 752–763.

Thenkabail, P. S., Enclona, E. A., Ashton, M. S., Legg, C., & De Dieu, M. J. (2004) Hyperion, IKONOS, ALI, and ETM plus sensors in the study of African rainforests. *Remote Sensing of Environment*, **90**, 23–43.

Theobald, D. M. (2007) Using GIS to generate spatially balanced random survey designs for natural resource applications. *Environmental Management*, 134–146.

Thogmartin, W. E., Sauer, J. R., & Knutson, M. G. (2004) A hierarchical spatial model of avian abundance with application to Cerulean Warblers. *Ecological Applications*, **14**, 1766–1779.

Thomaes, A., Kervyn, T., & Maes, D. (2008) Applying species distribution modelling for the conservation of the threatened saproxylic Stag Beetle (*Lucanus cervus*). *Biological Conservation*, **141**, 1400–1410.

Thomas, C. D., Cameron, A., Green, R. E. *et al.* (2004) Extinction risk from climate change. *Nature*, **427**, 145–148.

Thomas, K., Keeler-Wolf, T., & Franklin, J. (2002) A comparison of fine- and coarse-resolution environmental variables toward predicting vegetation distribution in the Mojave Desert. In Scott, J. M., Heglund, P. J., Morrison, M., Raphael, M., Haufler, J., & Wall, B. (Eds.) *Predicting Species Occurrences: Issues of Accuracy and Scale*. Covello, CA: Island Press, pp. 133–139.

Thompson, L. M., van Manen, F. T., Schlarbaum, S. E., & Depoy, M. (2006) A spatial modeling approach to identify potential Butternut restoration sites in Mammoth Cave National Park. *Restoration Ecology*, **14**, 289–296.

Thompson, S. K. (1992) *Sampling*. New York: John Wiley & Sons.

Thomson, J., Weiblen, G., Thomson, B., Alfaro, S., & Legendre, P. (1996) Untangling multiple factors in spatial distributions: lilies, gophers and rocks. *Ecology*, **77**, 1698–1715.

Thorn, J. S., Nijman, V., Smith, D., & Nekaris, K. A. I. (2009) Ecological niche modelling as a technique for assessing threats and setting conservation priorities for Asian slow lorises (Primates: *Nycticebus*). *Diversity and Distributions*, **15**, 289–298.

Thuiller, W. (2004) Patterns and uncertainties of species range shifts under climate change. *Global Change Biology*, **10**, 2020–2027.

Thuiller, W. (2007) Biodiversity – climate change and the ecologist. *Nature*, **448**, 550–552.

Thuiller, W., Albert, C., Araújo, M. B. *et al.* (2008) Predicting global change impacts on plant species distributions: future challenges. *Perspectives in Plant Ecology Evolution and Systematics*, **9**, 137–152.

Thuiller, W., Araújo, M. B., & Lavorel, S. (2003) Generalized models vs. classification tree analysis: Predicting spatial distributions of plant species at different scales. *Journal of Vegetation Science*, **14**, 669–680.

Thuiller, W., Araújo, M. B., Pearson, R. G., Whittaker, R. J., Brotons, L., & Lavorel, S. (2004a) Biodiversity conservation – uncertainty in predictions of extinction risk. *Nature*, **430**.

Thuiller, W., Brotons, L., Araújo, M. B., & Lavorel, S. (2004b) Effects of restricting environmental range of data to project current and future species distributions. *Ecography*, **27**, 165–172.

Thuiller, W., Lavorel, S., Sykes, M. T., & Araújo, M. B. (2006) Using niche-based modelling to assess the impact of climate change on tree functional diversity in Europe. *Diversity and Distributions*, **12**, 49–60.

Tilman, D. (1982) *Resource Competition and Community Structure.* Princeton, New Jersey: Princeton University Press.

Tobalske, C. (2002) Effects of spatial scale on the predictive ability of habitat models for the Green Woodpecker in Switzerland. In Scott, J. M., Heglund, P. J., Morrison, M. L. *et al.* (Eds.) *Predicting Species Occurrences: Issues of Accuracy and Scale.* Covelo, CA: Island Press, pp. 197–204.

Tobler, W. (1979) Cellular geography. In Gale, G. O. S. (Ed.) *Philosophy in Geography.* Dordrecht: Reidel, pp. 379–389.

Tognelli, M. F. & Kelt, D. A. (2004) Analysis of determinants of mammalian species richness in South America using spatial autoregressive models. *Ecography*, **27**, 427–436.

Tomlin, C. D. (1990) *Geographic Information Systems and Cartographic Modeling.* Englewood Cliffs, NJ: Prentice-Hall Publishers.

Torres, L. G., Read, A. J., & Halpin, P. (2008) Fine-scale habitat modeling of a top marine predator: do prey data improve predictive capacity? *Ecological Applications*, **18**, 1702–1717.

Townsend, P. A. & Walsh, S. J. (2001) Remote sensing of forested wetlands: application of multitemporal and multispectral satellite imagery to determine plant community composition and structure in southeastern USA. *Plant Ecology*, **157**, 129–149.

Traill, L. W. & Bigalke, R. C. (2006) A presence-only habitat suitability model for large grazing African ungulates and its utility for wildlife management. *African Journal of Ecology*, **45**, 347–354.

Trivedi, M. R., Berry, P. M., Morecroft, M. D., & Dawson, T. P. (2008) Spatial scale affects bioclimate model projections of climate change impacts on mountain plants. *Global Change Biology*, **14**, 1089–1103.

Tsoar, A., Allouche, O., Steinitz, O., Rotem, D., & Kadmon, R. (2007) A comparative evaluation of presence-only methods for modelling species distribution. *Diversity and Distributions*, **13**, 397–405.

Tucker, C. J. (1979) Red and infrared linear combinations for monitoring vegetation. *Remote Sensing of Environment*, **8**, 127–150.

Tucker, K., Rushton, S. P., Sanderson, R. A., Martin, E. B., & Blaiklock, J. (1997) Modelling bird distributions – a combined GIS and Bayesian rule-based approach. *Landscape Ecology*, **12**, 77–93.

Tukey, J. W. (1977) *Exploratory Data Analysis*. Reading, MA: Addison-Wesley Pub. Co.

Turner, J. C., Douglas, C. L., Hallam, C. R., Krausman, P. R., & Ramey, R. R. (2004) Determination of critical habitat for the endangered Nelson's bighorn sheep in southern California. *Wildlife Society Bulletin*, **32**, 427–448.

Turner, M. G. & Gardner, R. H. (1991) *Quantitative Methods in Landscape Ecology: The Analysis and Interpretation of Landscape Heterogeneity*, New York, Springer-Verlag.

Turner, M. G., Gardner, R. H., & O'Neill, R. V. (2001) *Landscape Ecology in Theory and Practice*. New York: Springer-Verlag.

Turner, W., Spector, S., Gardiner, N., Fladeland, M., Sterling, E., & Steininger, M. (2003) Remote sensing for biodiversity science and conservation. *Trends in Ecology & Evolution*, **18**, 306–314.

Tyre, A. J., Tenhumberg, B., Field, S. A., Niejalke, D., Parris, K., & Possingham, H. P. (2003) Improving precision and reducing bias in biological surveys: estimating false-negative error rates. *Ecological Applications*, **13**, 1790–1801.

Underwood, E. C., Klinger, R., & Moore, P. E. (2004) Predicting patterns of non-native plant invasions in Yosemite National Park, California, USA. *Diversity and Distributions*, **10**, 447–459.

Unitt, P. (2004) *San Diego County Bird Atlas*: San Diego, California: San Diego Natural History Museum.

US Fish and Wildlife Service (1981) Standards for the development of habitat suitability index models. Government Printing Office, Washington, DC, US Department of the Interior, Fish and Wildlife Service, Division of Ecological Services.

Valavanis, V. D., Georgakarakos, S., Kapantagakis, A., Palialexis, A., & Katara, I. (2004) A GIS environmental modelling approach to essential fish habitat designation. *Ecological Modelling*, **178**, 417–427.

Vallecillo, S., Brotons, L., & Thuiller, W. (2009) Dangers of predicting bird species distributions in response to land cover changes. *Ecological Applications*, **19**, 538–549.

van Horne, B. (1983) Density is a misleading indicator of habitat quality. *Journal of Wildlife Management*, **47**, 893–901.

van Manen, F. T., Clark, J. D., Schlarbaum, S. E., Johnson, K., & Taylor, G. (2002) A model to predict the occurrence of surviving butternut trees in the southern Blue Ridge Mountains In Scott, J. M., Heglund, P. J., Morrison, M. L. *et al.* (Eds.) *Predicting Species Occurrences: Issues of Accuracy and Scale*. Covelo, CA: Island Press, pp. 491–497.

Van Niel, K. P. & Austin, M. P. (2007) Predictive vegetation modeling for conservation: impact of error propagation from digital elevation data. *Ecological Applications*, **17**, 266–280.

Van Niel, K. & Laffan, S. W. (2003) Gambling with randomness: the use of pseudo-random number generators in GIS. *International Journal of Geographical Information Science*, **17**, 49–68.

Van Niel, K. P., Laffan, S. W., & Lees, B. G. (2004) Effect of error in the DEM on environment variables for predictive vegetation modelling. *Journal of Vegetation Science*, **15**, 747–756.

Vaughan, I. P. & Ormerod, S. J. (2005) The continuing challenge of testing species distribution models. *Journal of Applied Ecology*, **42**, 720–730.

Vayssiéres, M. P., Plant, R. E., & Allen-Diaz, B. H. (2000) Classification trees: an alternative non-parametric approach for predicting species distributions. *Journal of Vegetation Science*, **11**, 679–694.

Vaz, S., Martin, C. S., Eastwood, P. D. *et al.* (2008) Modelling species distributions using regression quantiles. *Journal of Applied Ecology*, **45**, 204–217.

Venables, W. M. & Ripley, B. D. (1994) *Modern Applied Statistics with S-Plus.* New York: Springer-Verlag.

Venier, L. A. & Pearce, J. L. (2007) Boreal forest landbirds in relation to forest composition, structure, and landscape: implications for forest management. *Canadian Journal of Forest Research*, **37**, 1214–1226.

Ver Hoef, J. M., Cressie, N., Fisher, R. N., & Case, T. J. (2001) Uncertainty and spatial linear models for ecological data. In Hunsaker, C. T., Goodchild, M. F., Friedl, M. A., & Case, T. J. (Eds.) *Spatial Uncertainty in Ecology: Implications for Remote Sensing and GIS Applications.* New York: Springer, pp. 214–237.

Verbyla, D. L. & Litaitis, J. A. (1989) Resampling methods for evaluating classification accuracy of wildlife habitat models. *Environmental Management*, **13**, 783–787.

Verner, J., Morrison, M. L. & Ralph, C. J. (1986) *Wildlife 2000: Modeling Habitat Relationships of Terrestrial Vertebrates.* Madison, WI: University of Wisconsin Press.

Vernier, P. R., Schmiegelow, F. K. A., & Cumming, S. G. (2002) Modeling bird abundance from forest inventory data in the boreal mixed-wood forests of Canada. In Scott, J. M., Heglund, P. J., Morrison, M. L. *et al.* (Eds.) *Predicting species Occurrences: Issues of Accuracy and Scale.* Covelo, CA: Island Press, pp. 559–572.

Vincent, P. J. & Haworth, J. M. (1983) Poisson regression models of species abundance. *Journal of Biogeography*, **10**, 153–160.

Virkkala, R., Luoto, M., Heikkinen, R. K., & Leikola, N. (2005) Distribution patterns of boreal marshland birds: modelling the relationships to land cover and climate. *Journal of Biogeography*, **32**, 1957–1970.

Wade, P. R. (2000) Bayesian methods in conservation biology. *Conservation Biology*, **14**, 1308–1316.

Wagner, H. H. & Fortin, M. J. (2005) Spatial analysis of landscapes: Concepts and statistics. *Ecology*, **86**, 1975–1987.

Walker, P. A. (1990) Modelling wildlife distributions using a geographic information system: kangaroos in relation to climate. *Journal of Biogeography*, **17**, 279–289.

Walker, P. A. & Cocks, K. D. (1991) HABITAT: a procedure for modeling a disjoint environmental envelope for a plant or animal species. *Global Ecology & Biogeography Letters*, **1**, 108–118.

Walker, P. A., & Moore, D. (1988) SIMPLE: An inductive modelling and mapping tool for spatially-oriented data. *International Journal of Geographical Information Systems*, **2**, 347–363.

Walther, G. R., Post, E., Convey, P. *et al.* (2002) Ecological responses to recent climate change. *Nature*, **416**, 389–395.

Wang, Q., Tenhunen, J. D., Schmidt, M. G., Kolcun, O., Droesler, M., & Reichstein, M. (2006) Estimation of total, direct and diffuse PAR under clear skies in complex alpine terrain of the National Patk Berchtesgaden, Germany. *Ecological Modelling*, **196**, 149–162.

Ward, D. F. (2007) Modelling the potential geographic distribution of invasive ant species in New Zealand. *Biological Invasions*, **9**, 723–735.

Waring, R. H., Coops, N. C., Fan, W., & Nightingale, J. M. (2006) MODIS enhanced vegetation index predicts tree species richness across forested ecoregions in the contiguous USA. *Remote Sensing of Environment*, **103**, 218–226.

Warren, T. L., Betts, M. G., Diamond, A. W., & Forbes, G. J. (2005) The influence of local habitat and landscape composition on cavity-nesting birds in a forested mosaic. *Forest Ecology and Management*, **214**, 331–343.

Waser, L. T. & Schwarz, M. (2006) Comparison of large-area land cover products with national forest inventories and CORINE land cover in the European Alps. *International Journal of Applied Earth Observation and Geoinformation*, **8**, 196–207.

Waser, L. T., Stofer, S., Schwarz, M., Kuechler, M., Ivits, E., & Scheidegger, C. (2004) Prediction of biodiversity – regression of lichen species richness on remote sensing data. *Community Ecology*, **5**, 121–133.

Watrous, K. S., Donovan, T. M., Mickey, R. M., Darling, S. R., Hicks, A. C., & Von Oettingen, S. L. (2006) Predicting minimum habitat characteristics for the Indiana bat in the Champlain Valley. *Journal of Wildlife Management*, **70**, 1228–1237.

Webb, T., III (1992) Past changes in vegetation and climate: lessons for the future. In Peters, R. L. & Lovejoy, T. E. (Eds.) *Global Warming and Biological Diversity*. New Haven, CT: Yale University Press.

Weibel, R. & Heller, M. (1991) Digital terrain modeling. In Maguire, D. J., Goodchild, M. F., & Rhind, D. W. (Eds.) *Geographic Information Systems*. Harlow, UK: Longman, pp. 269–297.

Weins, J. A. (1989) Spatial scale in ecology. *Functional Ecology*, **3**, 311–332.

Wessels, K. J. & van Jaarsveld, A. S. (1998) An evaluation of the gradsect biological survey method. *Biodiversity and Conservation*, **7**, 1093–1121.

Whaley, S. D., Burd, J. J., & Robertson, B. A. (2007) Using estuarine landscape structure to model distribution patterns in nekton communities and in juveniles of fishery species. *Marine Ecology-Progress Series*, **330**, 83–99.

Whittaker, R. H. (1956) Vegetation of the Great Smoky Mountains. *Ecological Monographs*, **26**, 1–80.

Whittaker, R. H. (1960) Vegetation of the Siskiyou Mountains, Oregon and California. *Ecological Monographs*, **30**, 279–338.

Whittaker, R. H. (1965) Dominance and diversity in land plant communities. *Science*, **147**, 250–260.

Whittaker, R. H. (1967) Gradient analysis of vegetation. *Biological Review*, **42**, 207–264.

Whittaker, R. H. (1970) *Communities and Ecosystems*. London: The Macmillan Company.

Whittaker, R. H., Levin, S. A., & Root, R. B. (1973) Niche, habitat and ecotope. *The American Naturalist*, **107**, 321–338.

Wiegland, C. L., Richardson, A. J., & Nixon, P. R. (1986) Spectral components analysis: a bridge between spectral observations and agrometeorological crop models. *IEEE Transactions on Geoscience and Remote Sensing*, **GE-24**, 83–89.

Wiens, J. J. (2004) Speciation and ecology revisited: phylogenetic niche conservatism and the origin of species. *Evolution*, **58**, 193–197.

Wiens, J. J. & Graham, C. H. (2005) Niche conservatism: integrating evolution, ecology, and conservation biology. *Annual Review of Ecology Evolution and Systematics*, **36**, 519–539.

Wilby, R. L. & Schimel, D. S. (1999) Scales of interaction in eco-hydrological relations. In Baird, A. J. & Wilby, R. L. (Eds.) *Eco-Hydrology: Plants And Water in Terrestrial and Aquatic Environments.* London: Routledge, pp. 39–77.

Williams, N. S. G., Hahs, A. K., & Morgan, J. W. (2008) A dispersal-constrained habitat suitability model for predicting invasion of alpine vegetation. *Ecological Applications*, **18**, 347–359.

Wilson, J. & Gallant, J. (2000) *Terrain Analysis: Principles and Applications.* New York: John Wiley & Sons.

Wilson, J. P., Inskeep, W. P., Wraith, J. M., & Snyder, R. D. (1996) GIS-based solute transport modeling applications: Scale effects of soil and climate data input. *Journal of Environmental Quality*, **25**, 445–453.

Wilson, K. A., Westphal, M. I., Possingham, H. P., & Elith, J. (2005) Sensitivity of conservation planning to different approaches to using predicted species distribution data. *Biological Conservation*, **122**, 99–112.

Wintle, B. A. & Bardos, D. C. (2006) Modeling species-habitat relationships with spatially autocorrelated observation data. *Ecological Applications*, **16**, 1945–1958.

Wintle, B. A., Elith, J., & Potts, J. M. (2005) Fauna habitat modelling and mapping: a review and case study in the Lower Hunter Central Coast region of NSW. *Austral Ecology*, **30**, 719–738.

Wintle, B. A., McCarthy, M. A., Parris, K. M., & Burgman, M. A. (2004) Precision and bias of methods for estimating point survey detection probabilities. *Ecological Applications*, **14**, 703–712.

Wintle, B. A., McCarthy, M. A., Volinsky, C. T., & Kavanagh, R. P. (2003) The use of Bayesian model averaging to better represent uncertainty in ecological models. *Conservation Biology*, **17**, 1579–1590.

Wisz, M. S., Dendoncker, N., Madsen, J. et al. (2008a) Modelling pink-footed goose (*Anser brachyrhynchus*) wintering distributions for the year 2050: potential effects of land-use change in Europe. *Diversity and Distributions*, **14**, 721–731.

Wisz, M. S., Hijmans, R. J., Li, J., Peterson, A. T., Graham, C. H., & Guisan, A. (2008b) Effect of sample size on the performance of species distribution models. *Diversity and Distributions* **14**, 763–773.

Wollan, A. K., Bakkestuen, V., Kauserud, H., Gulden, G., & Halvorsen, R. (2008) Modelling and predicting fungal distribution patterns using herbarium data. *Journal of Biogeography*, **35**, 2298–2310.

Wolter, P. T., Mladenoff, D. J., Host, G. E., & Crow, T. R. (1995) Improved forest classification in the Northern Lake States using multi-temporal Landsat imagery. *Photogrammetric Engineering and Remote Sensing*, **61**, 1129–1143.

Wood, S. N. (2006) *Generalized Additive Models: An Introduction with R.* New York: CRC Press.

Woodward, F. I. (1987) *Climate and Plant Distribution*. Cambridge, UK: Cambridge University Press.

Woodward, F. I. & Williams, B. G. (1987) Climate and plant distribution at global and local scales. *Vegetatio*, **69**, 189–197.

Wright, D. J. (1999) Getting to the bottom of it: Tools, techniques, and discoveries of deep ocean geography. *Professional Geographer*, **51**, 426–439.

Wright, D. J. & Goodchild, M. F. (1997) Data from the deep: Implications for the GIS community. *International Journal of Geographical Information Science*, **11**, 523–528.

Wu, H. & Huffer, F. W. (1997) Modelling the distribution of plant species using the autologistic regression model. *Environmental and Ecological Statistics*, **4**, 49–64.

Wu, X. B. & Smeins, F. E. (2000) Multiple-scale habitat modeling approach for rare plant conservation. *Landscape and Urban Planning*, **51**, 11–28.

Wulder, M. (1998) Optical remote-sensing techniques for the assessment of forest inventory and biophysical parameters. *Progress in Physical Geography*, **22**, 449–476.

Wulder, M. A. & Franklin, S. E. (Eds.) (2003) *Remote Sensing of Forest Environments: Concepts and Case Studies*. Boston: Kluwer Academic Publishers.

Wulder, M. A., Hall, R. J., Coops, N. C., & Franklin, S. E. (2004) High spatial resolution remotely sensed data for ecosystem characterization. *Bioscience*, **54**, 511–521.

Xu, C. G., Gertner, G. Z., & Scheller, R. M. (2007) Potential effects of interaction between CO_2 and temperature on forest landscape response to global warming. *Global Change Biology*, **13**, 1469–1483.

Yang, X., Skidmore, A. K., Melick, D. R., Zhou, Z., & Xu, J. (2006) Mapping non-wood forest product (matsutake mushrooms) using logistic regression and a GIS expert system. *Ecological Modelling*, **198**, 208–218.

Yee, T. W. & MacKenzie, M. (2002) Vector generalized additive models in plant ecology. *Ecological Modelling*, **157**, 141–156.

Yee, T. W. & Mitchell, N. D. (1991) Generalized additive models in plant ecology. *Journal of Vegetation Science*, **2**, 587–602.

Yesson, C. & Culham, A. (2006) Phyloclimatic modeling: combining phylogenetics and bioclimatic modeling. *Systematic Biology*, **55**, 785–802.

Zaniewski, A. E., Lehmann, A., & Overton, J. M. (2002) Predicting species spatial distributions using presence–only data: a case study of native New Zealand ferns. *Ecological Modelling*, **157**, 261–280.

Zeilhofer, P., dos Santos, E. S., Ribeiro, A. L. M., Miyazaki, R. D., & Dos Santos, M. A. (2007) Habitat suitability mapping of *Anopheles darlingi* in the surroundings of the Manso hydropower plant reservoir, Mato Grosso, Central Brazil. *International Journal of Health Geography*, **6**, 7.

Zhang, L., Ma, Z., & Guo, L. (2008) Spatially assessing model errors of four regression techniques for three types of forest stands. *Forestry*, **81**, 209–225.

Zhao, C. Y., Nan, Z. R., Cheng, G. D., Zhang, J. H., & Feng, Z. D. (2006) GIS-assisted modelling of the spatial distribution of Qinghai spruce (*Picea crassifolia*) in the Qilian Mountains, northwestern China based on biophysical parameters. *Ecological Modelling*, **191**, 487–500.

Zielinski, W. J., Truex, R. L., Dunk, J. R., & Gaman, T. (2006) Using forest inventory data to assess fisher resting habitat suitability in California. *Ecological Applications*, **16**, 1010–1025.

Zimmermann, N. E., Edwards, T. C., Moisen, G. G., Frescino, T. S., & Blackard, J. A. (2007) Remote sensing-based predictors improve distribution models of rare, early successional and broadleaf tree species in Utah. *Journal of Applied Ecology*, **44**, 1057–1067.

Zinner, D., Pelaez, F., & Torkler, F. (2002) Distribution and habitat of grivet monkeys (*Cercopithecus aethiops aethiops*) in eastern and central Eritrea. *African Journal of Ecology*, **40**, 151–158.

Zobel, M. (1992) Plant species coexistence – the role of historical, evolutionary and ecological factors. *Oikos*, **65**, 314–320.

Index

Printed in the United States
By Bookmasters